MW00354758

Citizen Science

Citizen Science

Innovation in Open Science,
Society and Policy

Edited by Susanne Hecker, Muki Haklay,
Anne Bowser, Zen Makuch,
Johannes Vogel & Aletta Bonn

First published in 2018 by
UCL Press
University College London
Gower Street
London WC1E 6BT

Available to download free: www.ucl.ac.uk/ucl-press

Text © Authors, 2018
Images © Copyright holders named in captions, 2018

The authors have asserted their rights under the Copyright, Designs and Patents Act 1988 to be identified as authors of this work.

A CIP catalogue record for this book is available from The British Library.

This book is published under a Creative Commons 4.0 International license (CC BY 4.0). This license allows you to share, copy, distribute and transmit the work; to adapt the work and to make commercial use of the work providing attribution is made to the authors (but not in any way that suggests that they endorse you or your use of the work). Attribution should include the following information:

Hecker, S., Haklay, M., Bowser, A., Makuch, Z., Vogel, J. & Bonn, A. 2018. *Citizen Science: Innovation in Open Science, Society and Policy*. London: UCL Press. https://doi.org/10.14324/111.9781787352339

Further details about Creative Commons licenses are available at http://creativecommons.org/licenses/

ISBN: 978-1-78735-235-3 (Hbk.)
ISBN: 978-1-78735-234-6 (Pbk.)
ISBN: 978-1-78735-233-9 (PDF)
ISBN: 978-1-78735-236-0 (epub)
ISBN: 978-1-78735-237-7 (mobi)
ISBN: 978-1-78735-238-4 (html)
DOI: https://doi.org/10.14324/111.9781787352339

Foreword

The world is facing unprecedented social, environmental and economic challenges that will require policymakers, business, scientists and citizens to open up to one another and find new ways of collaborating. In our digital age, we are reinventing the way knowledge is produced, distributed and acted upon. And an approach based on citizen science will be part of this new relationship between science and society.

The current increase in citizen science shows clearly the societal desire to participate more actively in knowledge production, knowledge assessment and decision-making. At the same time, scientists, research organisations and research funders are discovering the benefits of opening research to society by actively collaborating with citizens. There has been a significant rise in public participation in research in recent times, with citizens becoming engaged in the process of knowledge co-creation. This is not just a passive role, but actively setting the agenda, crowdsourcing via web platforms, and collecting and analysing a broad spectrum of scientific data. To invent new innovative ways to tackle societal challenges we need to involve those most affected – the citizens themselves.

I very much welcome these developments. The Commission is supporting them through its Open Science Agenda as well as through actions funded under the EU's Horizon 2020 framework programme. The recent 'Lab – Fab – App' report on maximising the impact of EU research and innovation recommended greater mobilisation and involvement of citizens in future EU research and innovation programmes through stimulating co-design and co-creation.

This book brings together some of the key insights into citizen science, highlighting what is already happening and exploring its potential to create new forms of knowledge generation, transfer and use and to foster the civic engagement of science. As a part of the open science agenda, citizen science contributes to the idea of a more innovative, inclusive, future-oriented and democratic Europe.

Carlos Moedas
European Commissioner for Research, Science and Innovation

Preface

Citizen science is becoming a global movement. Although there is a long history of co-operation between members of the general public and professionals, only now are its social benefits and transformative power the subjects of political and scientific debates. Citizen science is growing as a network of different players and is undergoing a self-identification process, making itself known in discussions about quality criteria, the role of the humanities, and its relationship to the concepts of Responsible Research and Innovation (RRI) and open science.

Our normative idea of citizen science encompasses democratic governance in Europe and the rest of the world relying on the informed decisions of its citizens and the shared understanding of science as one if its foundation – a contribution especially valuable given struggles to identify shared meanings for Europe and global citizenship.

One part of the evolution of citizen science, like other movements, involves the desire to gain legitimacy, and one way to achieve this is through institutionalisation. The European Citizen Science Association (ECSA), Citizen Science Association (CSA) based in United States, and Australian Citizen Science Association (ACSA), as well as national networks in countries including Germany, New Zealand, China and Austria, are transitioning from lose networks to legal entities. The self-reflection taking place within the citizen science community is also increasing, as shown by the increasing amount of research conducted on citizen science and its impacts. Citizen science is not only institutionalising, but professionalising. Practitioners exchange experiences, tips and tricks, but also consider societal and political impacts: What do the participants – citizens and scientists – learn? Do their attitudes and routines change? How does citizen science impact policy? Does citizen science impact the innovative potential of a society, and how can this be measured?

Another characteristic of a movement is the process of developing a self-identity and joint understanding – in this case, of citizen science. One example of this identity forming is demonstrated by the Ten Principles of Citizen Science developed by the ECSA principles group, chaired by Lucy Robinson (Robinson et al. in this volume), and translated into more than 20 different languages. As citizen science associations, we aim to promote sustainability through citizen science, build competence centres for citizen science and develop participatory methods for co-operation, empowerment and impact.

Our vision is for citizen science to advance its integrative power, to develop tools and find resources to approach and integrate marginalised groups, and for the concerns and findings of citizens to be taken seriously by different scientific communities and in the political arena.

Katrin Vohland, Claudia Göbel, ECSA, Jennifer Shirk, CSA &
Jessie Oliver, ACSA

European Citizen Science Association
(ECSA https://ecsa.citizen-science.net/)

Citizen Science Association based in United States
(CSA http://citizenscience.org/)

Australian Citizen Science Association
(ACSA https://csna.gaiaresources.com.au/)

Acknowledgements

The process of scoping, writing and compiling diverse perspectives for this volume has been a rewarding and inspiring journey for all involved. This volume brings together the diverse perspectives of 121 authors including researchers from the natural, social and computational sciences, educators in formal and informal contexts, policy experts and policymakers as well as non-governmental organisations (NGOs). These authors span 16 countries and represent 82 organisations. We sincerely thank all authors for joining this stimulating journey of the production of this volume and for the many discussions throughout the process. We are also indebted to the many hundreds of volunteers as well as project managers and policy advisers who have engaged in the citizen science projects presented in this volume. The case studies, we believe, help to ground the more theoretical perspectives and offer concrete examples of research in action. We hope that this book will inspire continued dialogue on the intersections of citizen science and policy.

We also thank our reviewers, who lent their scientific expertise and practitioner experience to enhance all chapters in this book. We are grateful to Chris Penfold and to the excellent support staff at UCL Press for helpful advice and encouraging guidance. A very special thanks goes to Madeleine Hatfield as our editor for fantastic improvements to the flow of the text and to Olaf Herling for the professional graphic design and for re-drawing some of the figures. Without the efforts of a wide range of collaborators, this synthesis would not have been possible.

This volume developed out of a productive international conference by the European Citizen Science Association (ECSA) and the German 'Citizen create Knowledge – Knowledge creates Citizens' (GEWISS) capacity-building programme in May 2016 under the title 'Citizen Science – Innovation in Open Science, Policy and Society', organised by some of the editors and authors of this volume in Berlin. The ECSA conference which attracted over 360 participants from more than 240 organisations and 30 countries allowed for rich and fruitful discussions on key topics from science and policy with keynotes and panel sessions, interactive workshops and a marketplace with posters. In addition, a ThinkCamp and a range of live demonstrations that included robots and hacking and presentations of local Berlin grassroots science fostered discussions. The European Citizen Science Association (ECSA) fosters a vibrant citizen science community across Europe and works closely with its partner organisations Citizen Science Association (CSA) and Australian Citizen Science Association (ACSA). Members from all organisations have actively contributed to this volume.

Funding for the conference was provided by the German Research Foundation (DFG grant agreement No BO 1919/2-1) and over 24 international partner organisations from science and policy:

Helmholtz Centre for Environmental Research (UFZ); German Centre for Integrative Biodiversity Research (iDiv) Halle-Jena-Leipzig; Friedrich-Schiller-Universität Jena; Museum für Naturkunde Berlin; Berlin-Brandenburg Institute of Advanced Biodiversity Research (BBIB); Wissenschaft im Dialog; Center for Ecology & Hydrology (CEH); Muséum National d'Histoire Naturelle Paris; Museum for Natural History London; Museo di Storia Naturale della Marremma; Fondazione Grosseta Cultura; University College London (UCL); Austrian Federal Ministry of Education, Science and Research (BMBWF); Leibniz-Institute of Freshwater Ecology and Inland Fisheries (IGB); Leibniz-Institute for Zoo and Wildlife (IZW); Leibniz Centre for Agricultural Landscape Research (ZALF); EarthWatch Institute; Zoological Research Museum Alexander Koenig; Haus der Zukunft/Futurium Berlin; Helmholtz-Gemeinschaft; Irstea; British Ecological Society (BES); Science Hack Day Berlin. The ThinkCamp was sponsored by Writelatex Limited.

The German DFG grant also supported the development of this book. The GEWISS project was funded by the German ministry of education and research (BMBF grant agreement No 01508444). Further support for the development of this volume was provided through funding from the European Union's Horizon 2020 research and innovation programme under grant agreement No 709443 - Doing It Together Science (DITOS) and grant agreement No 6417 - ECOPOTENTIAL. Anne Bowser's contributions were supported by the Alfred P. Sloan Foundation.

The editors have used their best endeavours to ensure URLs provided for external websites are correct and active at the time of going to press. The publisher has no responsibility for websites and cannot guarantee that contents will remain live or appropriate.

Susanne Hecker
Helmholtz Centre for Environmental Research – UFZ
German Centre for Integrative Biodiversity Research (iDiv)
Halle-Jena-Leipzig, Germany

Mordechay (Muki) Haklay
UCL, UK

Anne Bowser
Woodrow Wilson International Center for Scholars
Washington, DC, US

Zen Makuch
Imperial College London, UK

Johannes Vogel
Museum für Naturkunde Berlin
Humboldt University, Berlin, Germany

Aletta Bonn
Helmholtz Centre for Environmental Research – UFZ
Friedrich Schiller University Jena
German Centre for Integrative Biodiversity Research (iDiv)
Halle-Jena-Leipzig, Germany

Contents

PART II Innovation in science with and for society

PART IIA Case studies

PART III Innovation at the science-policy interface

Conclusions

List of figures

List of tables

List of contributors

Miriam Aczel is a President's PhD Scholar at Imperial College London's Centre for Environmental Policy. Her research interests include international energy science and policy, environmental and health impacts of shale gas extraction, and citizen science.

Janice Ansine is a senior project manager for citizen science, STEM Faculty, Open University, UK, who manages innovative, accessible web-based initiatives facilitating engagement, research and learning around biodiversity, and supports informal–formal education and outreach STEM activities.

Arnoud Apituley works at the Royal Netherlands Meteorological Institute (KNMI). His main interest is in the study of air quality and climate change, combining LIDAR and satellite data. He was involved in the iSPEX experiment.

Robert Arlinghaus is a professor of integrative fisheries management at Humboldt-Universität zu Berlin in Germany. He studies recreational fisheries using inter- and transdisciplinary research approaches.

Isabelle Arpin is a senior research scientist at the French National Research Institute of Science and Technology for Environment and Agriculture (Irstea). She studies contemporary changes in ways of managing and investigating nature.

Heidi L. Ballard is an associate professor of environmental science education and faculty director of the Center for Community and Citizen Science (CCS) at University of California, Davis, USA. She studies learning and conservation through CCS.

Frederic Bartumeus is a research professor on theoretical and computational ecology at the Catalan Institute for Research Studies who also works at the Centre for Advanced Studies of Blanes (CEAB-CSIC) and at Consorci Centre de Recerca Ecològica i Aplicacions Forestals (CREAF).

Mathias Becker is a senior researcher at the European Institute for Participatory Media with a focus on Social Innovation and Social Impact.

Arne J. Berre is a chief scientist at SINTEF and an associated professor at the University of Oslo. His research interest is in systems interoperability and standards with a focus on big data, IoT and machine learning/AI – with applications including geographic information, bio technologies and citizen science.

Aletta Bonn is a professor of ecosystem services at the Helmholtz Centre for Environmental Research – UFZ, Leipzig, and the Friedrich Schiller University Jena within the German Centre of Integrative Biodiversity Research (iDiv). A co-founding board member of ECSA, her research focuses on biodiversity-society interactions, global change and citizen science.

Rick Bonney is the director of the Public Engagement in Science Program at the Cornell Lab of Ornithology, a founder of the Citizen Science Association, and editor of the journal *Citizen Science: Theory and Practice*.

Anne Bowser is the director of innovation at the Woodrow Wilson International Center for Scholars in Washington, DC. She is interested in expanding public participation in and ownership of science and technology, like through crowdsourcing and citizen science.

Miriam Brandt is a scientific coordinator at the IZW in Berlin, Germany. She currently leads several research projects developing novel instruments for knowledge transfer and investigating the impact of citizen science.

Peter Brenton leads the field data collection and citizen science support functions of the Atlas of Living Australia, which provides national infrastructure for biodiversity data in Australia. Brenton has a strong interest in data standards and in making data accessible, reusable and available for research, policy and management.

Martin Brocklehurst is an independent environmental consultant and ecologist, with international experience on regulation and its application in the public and private sectors. She is a founding member of the European Citizen Science Association and the Citizen Science Global Partnership.

Neil D. Burgess is with the Center for Macroecology, Evolution, and Climate at the University of Copenhagen and also with United Nations Environment Programme-World Conservation Monitoring Centre, in Cambridge, UK.

Linda Carton is an assistant professor of Spatial Planning at Radboud University Nijmegen, Institute for Management Research (IMR). Carton's research is focused on sustainable governance arrangements, citizen science, community planning, participatory sensing and innovative mapping practices.

Jade Cawthray is a PhD student at University of Dundee, Scotland. Her PhD research explores how co-created citizen science practices can support communities in taking action on the issues and challenges that matter to them.

Luigi Ceccaroni is a manager in Engagement & Science at Earthwatch in Oxford, UK. Since 2014, he has been a member of the Board of Directors and the Steering Committee of the European Citizen Science Association.

Irene Celino is a research manager at Cefriel, where she leads the Knowledge Technologies research group, working on Human Computation for the active involvement of human contributors and Semantic Interoperability of data and services.

Colin Chapman is a senior data manager within the Department for Natural Resources at the Welsh Government. His expertise includes working at the policy/citizen science interface, as well as with multiple stakeholder groups in the citizen science arena.

Oscar Corcho is a full professor at Universidad Politécnica de Madrid. He co-leads the Ontology Engineering Group, working on the areas of Ontological Engineering, Semantic Web, Open Data, Open Science. He coordinates the EU project STARS4ALL.

Indiana Coronado is a member of the department of biology, faculty of sciences and technology, Herbarium of the National Autonomous University of Nicaragua, León, Nicaragua. She is also a research fellow at the Missouri Botanical Garden in Saint Louis, Missouri.

Finn Danielsen is an ecologist with NORDECO. He collaborates with natural and social science colleagues from many countries. He pilots locally based natural resource monitoring programmes, and he examines their reliability and usefulness for decision-making.

Jeroen Devilee is social scientist and advisor at the Dutch National Institute for Public Health and the Environment (RIVM). His interests include citizen science, stakeholder engagement in knowledge production, risk governance and identification of new risks.

Daniel Dörler is a researcher at the University of Natural Resources and Life Sciences in Vienna, Austria. He is founder and coordinator of the Citizen Science Network Austria and its associated platform *Österreich forscht*. Dörler is a pioneer in Open Innovation in Science in Austria, a member of various advisory boards and an organiser of the Austrian Citizen Science Conference.

Richard Edwards is an emeritus professor of education at the University of Stirling, UK. He has researched and published extensively on adult learning outside formal educational institutions, including learning through citizen science.

Martin Enghoff is sociologist with NORDECO in Denmark. He works with community involvement in natural resource management and rural development processes.

Lisa Garbe is a researcher at the University of St. Gallen, Switzerland. She works on telecommunication and Internet ownership and control, conducting large-N analyses covering Africa, and occasionally spends time in Africa for field research.

Claudia Göbel is project manager at the European Citizen Science Association and guest researcher at Museum für Naturkunde Berlin. She works on the organisation of citizen science, links to open science and policy engagement.

Margaret Gold is a project officer at the European Citizen Science Association for the WeObserve project. She specialises in designing mobile and web-based platforms for citizen science, community-building and engagement, and running creative collaboration events.

François Grey is the director of digital strategy at University of Geneva and heads Citizen Cyberlab, a partnership on rethinking public participation in research, with CERN and the UN Institute for Training and Research.

Michel Grothe works at Geonovum in the Netherlands and is a programme manager of INSPIRE. He is involved in the Smart Emission citizen science project in the city of Nijmegen, specialising in open data.

Mordechai (Muki) Haklay is a professor of geographic information science at UCL. His research interests include public access to environmental information, citizen science, participatory GIS and mapping, and Human-Computer Interaction (HCI) for geospatial technologies.

John Harlin, a former mountain guide and outdoor writer, directs the Alpine Institute at the Leysin American School in Switzerland and is co-chair of ECSA's working group on citizen science in education.

Susanne Hecker is a researcher at the Helmholtz Centre for Environmental Research – UFZ and the German Centre for Integrative Biodiversity Research (iDiv) in Halle-Jena-Leipzig, Germany. Her research focus is on science communication in citizen science at the science-society-policy interface and she organised the first ECSA Conference.

Florian Heigl is a postdoc at the University of Natural Resources and Life Sciences in Vienna, Austria. He is also founder of the project Roadkill and initiator and coordinator of the Citizen Science Network Austria and its associated platform *Österreich forscht*. He is a member of various advisory boards and an active networker and organiser of the annual Austrian Citizen Science Conference.

Thomas Hillman is an associate professor of IT and learning at the University of Gothenburg. His research investigates the transformation of technologies and epistemic practices with a focus on learning processes in large-scale online communities.

Franz Hölker is a senior scientist at IGB Berlin and an associate professor at FU Berlin. He is researching the biological impacts of light pollution on a wide range of biological processes, partly together with citizen scientists.

Sune Holt is a climate change adaptation consultant at the Inter-American Development Bank. He is interested in global development, eco-sustainable solutions and biodiversity conservation.

Alan Irwin is a professor in the Department of Organization at Copenhagen Business School. His research field is Science, Technology and Innovation Studies with special focus on scientific governance and research policy.

Per Moestrup Jensen is an associate professor at the University of Copenhagen, Denmark. He works with wildlife biology and associated zoonotic diseases. He teaches basic ecology and zoology in a number of courses.

Christoph Keller is a full professor of experimental astrophysics at the Leiden Observatory in the Netherlands. His research centres on polarised light. He is one of the developers of the citizen science instrument iSPEX.

Barbara Kieslinger is a senior researcher at the Centre for Social Innovation – ZSI, Vienna, Austria. She studies online social networking practice, digital social innovation and public participation processes.

Heidy Kikillus is a research fellow at Victoria University of Wellington in New Zealand. Her interests include urban ecology and biosecurity, specifically pets that may become pests.

Sarah Kirn is the education programs strategist for the Gulf of Maine Research Institute, where she designed and developed the Vital Signs citizen science programme. She serves on the board of the Citizen Science Association.

Laure Kloetzer is an assistant professor in Sociocultural Psychology at the University of Neuchâtel, Switzerland, where she studies how research contributes to social innovation and institutional transformation. She leads the ECSA and EU COST Action working groups on Learning & Education.

Frank Kresin is a managing director of the DesignLab at the University of Twente, fellow of Waag Society, and board member of V2_, Tetem and The Mobile City. He is interested in citizen science and empowerment.

Christopher Kyba is a physicist at the GFZ German Research Centre for Geosciences in Potsdam, Germany. His work focuses on quantification of artificial light in outdoor environments.

Poppy Lakeman-Fraser is the senior coordinator of a UK-wide public engagement network, OPAL. She is a citizen science practitioner, has a background in global change ecology and was the first coordinator of ECSA.

Anne Land-Zandstra is an assistant professor of science communication at Leiden University, the Netherlands. Her research interest includes the motivations, expectations and learning impact of citizen scientists.

Marie-Elise Lecoq is the lead architect, analyst and developer with GBIF France and is largely responsible for the development of the GBIF France biodiversity data portal. Lecoq is a strong advocate for open data, data sharing and community-driven infrastructure

Didier Guy Leibovici is a member of the Nottingham GeoSpatial Institute, University of Nottingham, who has interests in geo-computational data analytics, including scientific workflows, uncertainty and spatial analysis. His research, for a range of environmental domains, looked into data quality of crowdsourcing.

Chris Leonhard has an interest in citizen science and enquiry-based learning. He is the former head of science at Leysin American School and the co-founder of the LETS Study Leysin project.

Chunming Li is a researcher at IUE, CAS in Xiamen, China. His research interests are environmental sensing network, participatory sensing and environmental management.

Grégoire Loïs works at the National History Museum in Paris, France. In the ecology lab, he works on citizen science initiatives dedicated to feed macro-ecology research activities and policy tools such as large-scale indicators.

Monique Luckas is the head of press and public relations at Futurium in Berlin, Germany. She studied communication science and is fascinated by communication structures in citizen science projects.

Dana Mahr is an historian and sociologist of science and medicine at the University of Geneva. Her research focuses on participatory phenomena in various areas and historical periods. In addition, she works on experiential knowledge and risk.

Karen Makuch is a lecturer in Environmental Law at Imperial College London, Centre for Environmental Policy, where she teaches international environmental law. Her research areas include inter alia, environmental roles and rights of children, and citizen science and fracking regulation.

Zen Makuch is an international barrister who directs the Sustainable Transitions and Food Systems research initiatives at Imperial College London. A co-founding board member of ECSA, he designed and applied the EU environmental protection/biodiversity conservation implementation strategies.

Ilona Marenbach is a certified sociologist who worked as editor and editor-in-chief of Radio Multikulti, a multicultural-multilingual broadcast programme in Germany. She was deputy Editor-in-Chief of radioeins and is now head of the multimedia science editorial office of Rundfunk Berlin-Brandenburg (rbb).

Andreas Matheus is the managing director of Secure Dimensions GmbH (Germany), focusing on Single-Sign-On security solutions, based upon open standards for distributed systems, such as spatial data infrastructure for smart cities and citizen science.

Katherine Mathieson is chief executive of the British Science Association, an independent charity that is seeking to make science a fundamental part of culture and society.

Suvodeep Mazumdar is a lecturer at Sheffield Hallam University. He conducts inter-disciplinary research on highly engaging, interactive and visual mechanisms in conjunction with complex querying techniques for understanding large complex datasets.

Diarmuid McDonnell is a research fellow in social policy, University of Birmingham. His research explores the determinants of nonprofit misconduct and vulnerability, and utilises linked administrative data derived from charity regulators internationally.

Rosy Mondardini is the managing director of the Citizen Science Center Zurich, a joint initiative of ETH Zurich and the University of Zurich to engage scientists and the public in next-generation citizen science projects.

Susana Nascimento is a researcher and policy analyst at the European Commission, Joint Research Centre, EU Policy Lab. She works on future oriented technology, blockchain and DLTs, artificial intelligence, transdisciplinarity, and open science and technology.

Erik Noordijk is an air-quality researcher at RIVM, where he is involved in the MAN (ammonia measurement) network in which nature managers and volunteers of 85 nature areas co-operate with RIVM on a voluntary basis.

Jasminko Novak is a professor of information systems at the University of Applied Sciences Stralsund and leads the European Institute for Participatory Media. Collective intelligence and participatory systems are among the core areas of his research.

Erinma Ochu is a transdisciplinary researcher and social entrepreneur in the Ecosystems and Environment Research Centre at the University of Salford. She co-designs immersive experiences to offer new perspectives on global challenges.

Alessandro Oggioni is a researcher at Consiglio Nazionale delle Ricerche (CNR, Italy) - Institute for Electromagnetic Sensing of the Environment (IREA). Oggioni specialises in data management, spatial data infrastructure, Sensor Web Enablement and other web-based services.

Jessie Oliver researches how to design interactive citizen science technologies to find species in environmental acoustic recordings, and she advocates for citizen science nationally and globally through her work with the Australian Citizen Science Association.

Roger Owen is a principal scientist at the Scottish Environment Protection Agency specialising in international research, natural capital and citizen science. Owen was the first co-chair of the EPA Network Interest Group on Citizen Science.

John Palmer is a tenure-track professor at Pompeu Fabra University in Barcelona. He uses citizen science in his work on human mobility and migration, social segregation and disease ecology.

Alison Parker is an ecologist and citizen science enthusiast, currently supporting scientists and exploring citizen science for environmental policy as an ORISE postdoctoral fellow hosted by the US Environmental Protection Agency in Washington DC, USA.

Dan Patton has been teaching science for 10 years. He is passionate about getting kids learning science through authentic, hands-on activities with real-world applications, such as citizen science, STEM and entrepreneurship.

Eleonore Pauwels is the director of the Anticipatory Intelligence (AI) Lab with the Science and Technology Innovation Program at the Woodrow Wilson International Center for Scholars. She is a writer and international science policy expert, who specialises in the governance and democratisation of converging technologies.

Taru Peltola is a senior research scientist at the Finnish Environment Institute. She studies knowledge practices in environmental policy, transformative outcomes of citizen science and citizen science in contentious societal issues.

Lisa Pettibone is a former postdoc at the Natural History Museum in Berlin, Germany. She coordinated the capacity-building programme 'GEWISS' and among other things published a guide on how to conduct citizen science. Pettibone's research interests are sustainability sciences.

Tina Phillips is a social science researcher based at the Cornell Lab of Ornithology in Ithaca, New York, USA. Her work focuses on building evaluation capacity for and researching socio-ecological outcomes from citizen science.

Jaume Piera is a researcher at the ICM-CSIC and CREAF, both in Barcelona, Spain. His research is focused on environmental monitoring systems, mostly based on citizen science. He is member of the ECSA steering committee.

Michael Køie Poulsen is a Danish socio-ecologist with NORDECO. He supports local communities in the Arctic and developing regions in designing and implementing monitoring programmes with focus on biodiversity and living resources.

Anett Richter is a postdoctoral scholar at the Helmholtz Centre for Environmental Research – UFZ and the German Centre for Integrative Biodiversity Research (iDiv) in Halle-Jena-Leipzig, Germany. She led the development of Citizen Science Strategy 2020 for Germany and coordinated the capacity-building programme 'GEWISS'. Richter's research is about citizen science as a concept and understanding the people involved in citizen science.

Jeroen Rietjens is an instrument scientist working at SRON Netherlands Institute for Space Research as performance lead on the development of airborne and space-based highly accurate multi-angle polarimeters for climate and air-quality research.

Lucy Robinson is citizen science manager at the Natural History Museum London. As a citizen science practitioner for 10 years, she champions best practices and is PI on the LEARN CitSci research programme.

Jose Miguel Rubio Iglesias is a geospatial engineer working at the European Environment Agency. He is a co-chair of the EPA Network Interest Group on Citizen Science and works on the implementation of the INSPIRE Directive.

Ricardo M. Rueda is with the Department of Biology, Faculty of Sciences and Technology, National Autonomous University of Nicaragua, León, Nicaragua. He works with Flora of Nicaragua and studied in Saint Louis, Missouri.

Sven Schade is a scientific project officer at the European Commission's Joint Research Centre (JRC). He provides scientific advice to policy on topics such as citizen science, the use of new data sources, and digital transformation.

Teresa Schäfer is a researcher at the Centre for Social Innovation - ZSI, Vienna, Austria. Her work focuses on participation processes in digital social innovations and the assessment of its impact.

Bernard Schiele is a researcher and a professor at UQAM (Montreal). He works on the socio-dissemination of S&T. He is also a founding member and current member of the Scientific Committee of the PCST network.

Sibylle Schroer is scientific coordinator for research on light pollution at the IGB. She connects scientists and supports transdisciplinary communication. She is creating and expanding networks of citizen scientists and all other stakeholders.

Linda See is a research scholar at the International Institute for Applied Systems Analysis in Austria whose research interests intersect the fields of citizen science. She has volunteered in geographic information and Earth Observation.

Fermin Serrano Sanz is the commissioner for knowledge economy and innovation at the Government of Aragon. He was a member of the team that developed the White Paper for the European Citizen Science Strategy and actively promotes and supports citizen science in Europe.

Andrea Sforzi is the director of the Maremma Natural History Museum in Grosseto, Italy. He actively promotes environmental citizen science initiatives, as well as national and international CS networking. He is a founding member of ECSA.

Lea Shanley served as an Obama White House Presidential Innovation Fellow, as founder and co-chair of the Federal Crowdsourcing and Citizen Science Community of Practice, and as founding director of the Wilson Center's Commons Lab.

Jennifer Shirk works at the Cornell Lab of Ornithology and as Director of the Citizen Science Association. She explores practices and connects people across disciplines to support innovations and integrity in both research and engagement.

Andrea Sieber is a scientist at the University of Klagenfurt, Austria. Her current research topics are cultural sustainability, landscape, participatory research and education.

Melanie Smallman is lecturer in Science and Technology Studies at UCL and co-director of the Responsible Research and Innovation Hub. Her research looks at public attitudes toward technologies and how these views influence science and policy.

Frans Snik works at Leiden University on optical technology for astronomy and Earth observation. He led the iSPEX project that involved several thousand participants all over Europe to measure air pollution with a smartphone add-on.

Leysin American School high school students travel from their homes around the world to be educated in the mountains of Switzerland.

Erik Tielemans is the head of department Research & Development Environmental Quality, National Institute for Public Health and the Environment (RIVM).

John Tweddle is the head of the Angela Marmont Centre for UK Biodiversity at the Natural History Museum London. His interests span citizen science theory and practice, wildlife identification training and urban nature conservation research.

Elizabeth Tyson is a research fellow at Conservation X Labs in Washington, DC. She builds capacity and conducts research and outreach on behalf of open innovation, citizen science and crowdsourcing as methods for driving innovative solutions to our environmental problems.

Paul van Genuchten is a software engineer at GeoCat BV (the Netherlands). He works with various organisations and communities on the GeoNetwork opensource project, an open standards–based catalogue for datasets with a spatio-temporal focus.

Roy van Grunsven is an ecologist and entomologist from IGB and Dutch Butterfly Conservation. He studies the effect of artificial light on moths and other organisms and frequently works with citizen scientists in monitoring programmes.

Edith van Putten is an environmental researcher at RIVM. She is an expert in building sensors and is involved in several citizen science projects concerning air quality and radiation.

Arnold van Vliet is a biologist at Wageningen University in the Netherlands who focuses on monitoring, analysing, forecasting and communicating climate impact on biodiversity and society and on how society can adapt.

Ella Vogel is the projects development officer for the National Biodiversity Network in the UK. Her primary focus is on improving open access to biodiversity information across the UK via the NBN Atlas portal.

Johannes Vogel is director general of the Museum für Naturkunde Berlin and professor for Biodiversity and Public Science at the Humboldt University. He is chair of the European Citizen Science Association and the EU commissions Open Science Policy Platform.

Katrin Vohland heads the research programme 'public engagement with science' at the Museum für Naturkunde Berlin. She conducts research on the interface between biodiversity policy, science and society, with a strong emphasis on citizen science.

Hester Volten works at the National Institute of Public Health and the Environment (the Netherlands) as an air-quality scientist who specialises in citizen science. Among others, she was involved in the iSPEX project.

Stephanie von Gavel was the business development manager of the Atlas of Living Australia (ALA) for several years. She was instrumental as a founding board member in the establishment of the Australian Citizen Science Association (ACSA) and continues to contribute to driving its strategic, institutional and policy directions.

Jan Vonk works at RIVM on measurements of particulate matter within the Dutch national air-quality monitoring network, and is always on the lookout for new applications and approaches, including the use of sensors.

Marita Voogt coordinates the RIVM programme 'Innovation of Environmental Monitoring'. She is the contact person for citizen communities in the Netherlands as well as jointly responsible for the content of the knowledge and data portal.

Wolfgang Wägele is the director of the Leibniz-Institute for Animal Biodiversity, also known in the German Rhine region as Museum Koenig. He promotes the co-operation between expert taxonomists among citizen scientists, ecologists and governmental agencies to improve biodiversity monitoring in Germany.

Ulrich Walz is a professor for landscape ecology and GIS at the University of Applied Sciences (HTW) in Dresden. He works on the theme of landscape changes and their impacts on landscape functions and biodiversity.

Wolfgang Wende is a scientist at IOER Dresden and professor at Technische Universität Dresden, Germany. He works on landscape change and management and he integrates the landscape focus with citizen science.

Joost Wesseling is senior scientist in Environmental Quality at the National Institute for Public Health and the Environment (RIVM). More recently, he has been involved in studies about the application of sensors for environmental monitoring.

Sarah West is a senior researcher at the Stockholm Environment Institute at the University of York. She is part of SEI's Practice and Research in Citizen Science (SPARCS) group: designing, conducting and researching citizen science projects across the world.

Jamie Williams is a principal consultant for Environment Systems (UK), an environmental data company, where he works on the interface between research and development, and industry, with a particular interest in innovation and applied science.

Daniel Wyler is a retired professor of theoretical physics at the University of Zurich, Switzerland. He has a strong interest in participatory research and has established a center of citizen science in Zurich at the university and the ETH.

1

Innovation in open science, society and policy – setting the agenda for citizen science

Susanne Hecker[1,2], Muki Haklay[3], Anne Bowser[4],
Zen Makuch[5], Johannes Vogel[6] and Aletta Bonn[1,2,7]

[1] Helmholtz Centre for Environmental Research – UFZ, Leipzig, Germany
[2] German Centre for Integrative Biodiversity Research (iDiv) Halle-Jena-Leipzig, Germany
[3] University College London, UK
[4] Woodrow Wilson International Center for Scholars, Washington DC, US
[5] Imperial College London, UK
[6] Museum für Naturkunde Berlin, Germany
[7] Friedrich Schiller University Jena, Germany

corresponding author email: susanne.hecker@idiv.de

In: Hecker, S., Haklay, M., Bowser, A., Makuch, Z., Vogel, J. & Bonn, A. 2018. *Citizen Science: Innovation in Open Science, Society and Policy.* UCL Press, London. https://doi.org/10.14324/111.9781787352339

> *Simply generating and communicating scientific knowledge is not sufficient [to combat biodiversity loss] . . . Knowledge of traditional and 'ordinary' citizens [brings] possibilities for innovation.*
>
> (Turnhout et al. 2012, 454)

> *This is renaissance, your dentist now an authority on butterflies and you (in retrospect this happened so pleasantly, watching clouds one afternoon) connected by Twitter to the National Weather Service. This is revolution, breaking down barriers between expert and amateur, with new collaborations across class and education. Pygmy hunters and gatherers use smartphones to document deforestation in the Congo Basin. High school students identify fossils in soils from ancient seas in upstate New York. Do-it-yourself biologists make centrifuges at home.*
>
> *This is falling in love with the world, and this is science, and at the risk of sounding too much an idealist, I have come to believe they are the same thing.*
>
> (Russell 2014, p. 11)

Background

Citizen science is a rapidly growing field with expanding legitimacy. Often seen as a cluster of activities under a larger umbrella of concepts, including 'open science' and 'open innovation', citizen science expands public participation in science and supports alternative models of knowledge production. This includes strengthening scientific research by engaging with a variety of topics and information sources, and fostering cross- or trans-disciplinary knowledge production. Citizen science can expand stakeholder participation and introduce new perspectives and information as well as new partnerships. Many projects are opening up cutting-edge areas of science, such as gene editing and synthetic biology, to new audiences, enabling a wider discussion about their societal implications. In these ways, citizen science projects are often initiated to address an immediate problem or research question, while also building capacity for communities to participate in science and shape policy decision-making and implementation in the longer term (see box 1.1).

Research in citizen science takes a diverse approach where the balance between scientific, educational, societal and policy goals varies across projects (see Kieslinger et al. in this volume). A common, shared goal is to collect and analyse information that is scientifically valuable. This distinguishes citizen science from areas such as experiential learning or environmental education for sustainability, although learning and other educational goals and outcomes are additional valuable aims and contributions of some citizen science projects.

It is easy to think of citizen science as a new phenomenon. However, it actually has historic roots (see Mahr et al. in this volume), which have been recently invigorated by evolving digital technologies such as networked mobile devices that connect people easily and effectively with the scientific community and with their peers. The growth of citizen science has also been driven by the public's desire to be actively involved in scientific processes. This may be a result of recent societal trends, including the rise in tertiary education (see Haklay in this volume) and the increasing value placed in science, as well as the wish to actively participate in providing evidence to help manage urgent societal problems.

The rich history of citizen science extends across a range of areas, notably astronomy, biology and biodiversity monitoring, environmental monitoring and public health (see Mahr et al. in this volume). Recent projects have also explored opportunities for engagement in transportation,

Box 1.1. Citizen science: Definitions

The concept of citizen science is often attributed to two distinct sources. In 1995, Alan Irwin used the term to refer to a paradigm where research goals were collaboratively determined by professional scientists and the public in the UK (Irwin 1995). Around the same time, Rick Bonney began to use the same term to refer to numerous projects at the Cornell Lab of Ornithology in the United States, which involved members of the public in avian research (Bonney 1996). Many more recent definitions have been offered since then, with various degrees of alignment to these early roots (Eitzel et al. 2017). Especially notable for this volume are the definitions advanced by governments and other policy-making bodies.

In the United States, for example, citizen science was defined at the national level first by John Holdren, Director of the White House Office of Science and Technology Policy (OSTP) under the Obama Administration (Holdren 2015). Holdren's memo defined citizen science broadly, as a process where 'the public participates voluntarily in the scientific process, addressing real-world problems in ways that may include formulating research questions, conducting scientific experiments, collecting and analyzing data, interpreting results, making new discoveries, developing technologies and applications, and solving complex problems'. This definition was later picked up in the US Citizen Science and Crowdsourcing Act, signed into law in 2017.

The European Commission has used the definition from the Oxford English Dictionary, defining citizen science as 'scientific work undertaken by members of the general public, often in collaboration with or under the direction of professional scientists and scientific institutions'. It also notes that 'Citizen Science is often linked with outreach activities, science education or various forms of public engagement with science' (European Commission 2016c, p. 54).

To complement the definition of citizen science, the European Citizen Science Association has also developed *Ten Principles of Citizen Science* (Robinson et al. in this volume).

irrigation and agriculture and energy production (including bioenergy) among other topics (Lisjak, Schade & Kotsev 2017). Citizen science projects related to public policy matters are now touching on many agendas, from environmental protection, to health and education, to research and innovation. Those intended to drive innovation have particularly led to collaboration across the spectrum of science, medicine and engineering disciplines, as well as the social sciences. Citizen science is also encouraging interaction between practitioners and key societal stakeholders and public policymakers, although this remains limited (see Nascimento et al. in this volume).

In parallel, citizen science is becoming more widely discussed and accepted within the scientific community as an appropriate research approach to answer specific research questions and meet scientific demands. Thousands of scientific projects involve millions of citizens who are investing extensive time, energy and resources in research supported by new technologies (Bonney et al. 2014). While many volunteers wish to contribute to scientific research and be included in policy responses and decision-making, many are also motivated by an interest in research and the integration of science and society (Rotman et al. 2012). Researchers and policymakers further express the need for improved evidence, participation and knowledge as the legitimate bases for decision-making, which supports the demand for citizen science (see for example Smallman in this volume).

The growth in citizen science projects has seen them operating at multiple spatial scales, from neighbourhood and village concerns over environmental issues, to continental-scale monitoring of trends. This also means that the policy implications of citizen science projects can range across jurisdictions, from the international (e.g., the United Nations Framework Convention on Climate Change) to the regional (e.g., the European Union), national (e.g., the US federal climate change policy) and sub-national (e.g., California's low-carbon transport policies), to community groups and ad-hoc organisations established by concerned citizens (e.g., water or air quality monitoring). Citizen science can also cover the entire temporal range of scientific enquiry, from short-term initiatives to address current issues (such as mapping accident-prone traffic spots in a city, or litter or invasive species) to long-term monitoring (e.g., weather or animal populations). These diverse backdrops require theoretical and practical understanding of how citizen science operates both in general and specific contexts.

Citizen science contributions to policy

Citizen science is unusual as a developing research field in that there is acute awareness among its practitioners about the importance of its wider current and potential societal and – increasingly – political impacts. Unlike most other academic fields, the conversations taking place inside the community are outward-looking. A good example of this is the way guidance documents are being developed through bottom-up approaches, which reach out to the practitioner and research community and to policymakers. Examples include the Socientize *White Paper on Citizen Science for Europe*, intended to reach out to a wider group of researchers and policymakers at the European level (Serrano Sanz et al. 2014), and the European Citizen Science Association's *Ten Principles of Citizen Science,* which both speak to practitioners in the field and provide orientation for science-based policy by establishing universal principles for citizen science projects (Robinson et al. in this volume). The *Greenpaper Citizen Science Strategy 2020 for Germany* (Bonn et al. 2016; Richter et al. in this volume) is another strategic-political document developed in collaboration with more than a hundred scientific organisations and universities, non-governmental organisations (NGOs), learned societies, science shops, the media, individual researchers and members of the public (see box 2.1).

In the political sphere, the value of citizen science is starting to be recognised at the European Union policy level (Hyder et al. 2015; Smallman in this volume) and by European Member States (box 1.2), as well as by national governments in the United States (see box 1.1) and Australia. Global NGOs, including the United Nations Environment Programme (UNEP), have also voiced their support. Citizen science therefore enables both traditional as well as modern avenues of engagement between science and society (Mazumdar et al. in this volume). In fact, in 2013 an internal European Commission document stated that the 'development of communication technologies through the internet creates highly valuable opportunities for citizen science and crowdsourcing, offering enhanced levels of participation in assessing (and determining) the success of EU environment policies' (European Commission 2013, p.4).

Governments and policymakers need evidence and scientifically reliable, up-to-date information to identify, formulate, implement and evaluate policies. They are obliged to fulfil regulations, such as those on environmental monitoring and assessments under EU directives like the Habitats Directive or the Birds Directive, or international conventions like the Convention on Biological Diversity. Citizen science provides the

Box 1.2. National contexts: Spotlight on Germany

We want to involve citizens and stakeholders from civil society consistently in the discussion about future projects and the design of research agendas. We want to develop new forms of citizen participation and the communication of science and merge them into an overall concept. (2013)

   We want to intensify the dialogue of economy, policy, science and society, trial new formats of participation of civil society and strengthen science communication. (2018)

Extracts from coalition contracts of the
German government

In Germany, a two-year citizen science capacity-building programme was implemented in 2014–2016 by the Ministry of Research and Education (BMBF) to assess the potential and challenges of citizen science. Researchers from all fields – citizens, civil society organisations and scientific institutions – contributed their ideas and experiences to the enhancement of citizen science in a programme that built on dialogue and participation. The resulting *Greenpaper Citizen Science Strategy 2020 for Germany* (Bonn et al. 2016; Richter et al. in this volume) received significant attention from policy and international citizen science networks. A subsequent outcome of the capacity-building programme and the strategy was a federal BMBF funding scheme to support citizen science projects in 2017.

opportunity for policymakers to create programmes with scientific researchers to support these obligations, or to draw upon existing initiatives. In the United States, for example, the annual State of the Birds report – produced by the North American Bird Conservation Initiative (NABCI), a consortium of federal agencies and NGOs – utilises contributions from the citizen science project eBird to assess the status and health of key species, and promotes birds as indicators of overall environmental health and human well-being. This report is used to help the US government offer progress reports on international commitments, including the Migratory Bird Treaty, and to evaluate and refine domestic policy, for example, policy related to land use.

At the same time, policymakers want to ensure the societal relevance of their actions, and this calls for stronger engagement with society (European Commission 2013; see also Nascimento et al. and Parker & Owen, both in this volume). For example, the legislation of the EU's Environmental *Acquis Communautaire* includes periodic mandatory monitoring and reporting requirements, much of which can be standardised for citizen science participation in such diverse activities as species/habitat monitoring, and air or water quality monitoring. Outside of Europe, other political bodies outline similar values. In the United States, for example, citizen science is broadly portrayed as accelerating research and addressing politically relevant social needs by drawing on previously untapped resources, namely the public.

Citizen science and societal relevance

Historically, innovation opportunities have been available to a minority of the population, particularly privileged staff in the research and development sections of major firms or public sector institutions. In contrast, citizen science offers innovative potential at the science-society interface by drawing in many millions of participants worldwide. Research is literally 'opened' up to members of society and they often become part of the whole process, thus making science more inclusive. This allows members of the public to learn about, understand and discuss scientific methods, standards and values, developing their overall scientific literacy. This can increase public awareness of the value of scientific research in addressing problems faced in everyday life as well as global challenges. Citizen science can therefore positively influence society by providing opportunities for learning, empowerment, enjoyment of nature, social engagement or enhanced scientific capital (see Edwards et al. in this volume). Ideally, therefore, citizen science can contribute to good citizenship and, in turn, progressive societies.

Collaboration with members of society also offers the opportunity for scientists to make their research more relevant and to extend its impact. Citizen science practitioners are given the chance to become ambassadors for science (Druschke & Seltzer 2012) as they interact directly with members of society while at the same time benefiting from participants' expertise, knowledge and engagement. This brings the need for new forms of science communication in citizen science to better integrate collective science, society and policy aims and ambitions whilst the level of public collaboration and direct interaction varies considerably across citizen science projects (see Haklay; Novak et al.; Gold & Ochu, all in this volume) and needs to be considered accordingly.

One way in which practitioners and researchers are furthering the influence of citizen science is through the creation of dedicated associations, such as those in the United States, Europe, and Australia (Storksdieck et al. 2016). Similar needs can also be seen among practitioners at research universities (Wyler & Haklay in this volume). These new organisations and structures act as catalysts that allow communication with policymakers, as well as a forum for discussion among practitioners.

Open science and citizen science

'Open science' and 'open-access' approaches are at the forefront of new frameworks for research and innovation (see European Commission 2016c). Citizen science is recognised as an important element in the conceptualisation of open science, which has gained importance as part of the rethinking of how science relates to wider societal goals. Open science is a framework for how scientists interact with one another and how the public engages with, and is engaged in, science (European Commission 2016c). Open science engages with issues such as accessible data and publications, open evaluation and policies as well as developing its own tools. This includes open access, which is driven by the understanding that publicly funded research should be accessible to all members of society. The open science imperative of sharing information and results from publicly funded research has led to the promotion of the open access publication model (where scientific publications are freely available rather than subject to expensive subscription rates) as well as open-data repositories (where datasets are made freely available to other potential users).

Citizen science has a role to play in both the open science and open-access movements, and is in turn driven by them. The call for open-access research publication resonates with citizen science. Without it, members of the public who participate in research may well be deprived of the fruits of their participation, and without access to (other) research literature, citizens may not be equipped to conduct or analyse their own research.

However, citizen science means going beyond publishing data and results. Arguably, it is changing the way that science is done by opening up research throughout the process; from idea generation and planning to conducting the research and disseminating outputs (see Haklay in this volume). Citizen science also tends to call for open science communication in which multiple forms of media are used throughout the research process. Collaboration with the mass media in particular further demands novel forms of partnerships and could lead to a different approach to news-making (see Hecker et al. 'Stories' in this volume).

Citizen science is therefore both an aim and an enabler of open science (see Smallman in this volume). It contributes to open science by involving citizens in research, opening up the process of creating new knowledge through participation. In turn, this produces greater understanding of science through open information and communication. Engagement in citizen science can also stimulate active participation in policy-making.

Scope of this volume

This volume discusses the current and potential future contribution of citizen science and scientific innovation to a more productive and open science-society-policy interface. The chapters identify experience-based solutions that could be applied in different contexts. The emphasis is on identifying solutions to promote a vibrant citizen science community by bringing together major stakeholders and individuals to improve research, understanding and engagement in society, policy, education, innovation and academia.

Previous edited volumes have considered the value of citizen science in environmental research (Dickinson & Bonney 2012), the potential for citizen science to bridge the science-society gap (Cavalier & Kennedy 2016), and how citizen science can advance research through knowledge acquisition and transfer (Ceccaroni & Piera 2017). However, this volume adds to the discussion in that it focuses on the value of citizen science for informing policy whilst also contributing to education, scientific knowledge and societal organisation. The collective imperative is to understand and shape a world characterised by accelerated change and multiple grand challenges across the policy landscape.

Structure and content of this volume

The volume is structured in five main sections dedicated to the innovative potential of citizen science for science, policy and society. It also includes a section dedicated to case studies illustrating best practice examples.

Section I: Innovation in citizen science – setting the scene
Section II: Innovation in science with and for society
Section IIa: Case studies
Section III: Innovation at the science-policy interface
Section IV: Innovation in technology and environmental monitoring
Section V: Innovation in science communication and education

The chapters provide up-to-date scientific background information and show the variety of citizen science research from the natural to the social sciences, covering its practical application and technology design. Some chapters provide applications of citizen science in various contexts illustrated with case studies; others reflect on citizen science theory and concepts and their application.

This book is mainly written from a European perspective as the idea for the volume originated at the First European Citizen Science Conference held in Berlin, Germany, in 2016 (see Hecker et al. 'Innovation in Citizen Science' 2018). At the same time, the book includes international perspectives (see figure 1.1 for the global network of contributing authors) and authors were encouraged to include international case studies and, where appropriate, enlarge their insights and conclusions to a wider view. Throughout, the chapters offer critical reflection, guidance and best

Fig. 1.1 Global network of contributing authors to this volume (lines indicate connections between co-authors)

practice examples of citizen science that can be applied to international contexts.

Section I: Innovation in citizen science – setting the scene

The first section sets the scene by introducing key elements for innovation in citizen science, including standards to ensure high-quality research approaches; integration with, and contribution to, science; the nature of participation; supporting technology and infrastructures; and evaluation.

As baseline for citizen science projects, *Robinson et al.* identify and explain the Ten Principles of Citizen Science as the product of international collaboration within the citizen science community. The ten principles present a framework of standards to foster excellence in all aspects of citizen science. Focusing on the innovative potential of citizen science, *Shirk and Bonney* highlight the strengths of citizen science for data collection, processing capacity, public engagement and policy. They then apply their expertise to highlight scientific innovations emerging in different research contexts.

Haklay casts light on the nature and relevance of participation in science-linked decision-making, with useful lessons for policy-making processes and their participants. He advocates for a differentiated understanding of engagement in citizen science and identifies different types of participation according to participants' knowledge and engagement levels. *Brenton et al.* explain how the important revolution in information technology infrastructures enables and supports citizen science. They provide guidance on how to select and use appropriate digital tools, so that data is fit for purpose and can be used to meet existing demands, for example, also by government agencies. *Kieslinger et al.* provide an open framework for evaluating citizen science projects. The evaluation criteria they identify apply both to the process and outcomes of citizen science projects.

Section II: Innovation in science with and for society

Focussing on relevant and up-to-date topics in innovation in science, with a special focus on society, contributions in section II address how citizen science is embedded in science. It addresses questions such as: How can citizen science lead to empowerment and enhance scientific literacy to benefit individuals, communities and society? What is the potential for inclusive participation across society, especially when citizen science involves individuals and communities typically left out of science and policy-making? What are the innovation opportunities and challenges where support is needed, both technically and socially? And how can citizen science best be integrated at the science-society interface of the higher education system?

The chapter on citizen science studies by *Mahr et al.* reflects upon the many heterogeneous projects, methodologies and communities aiming to co-produce reflexiveness and dialogue between citizen science practitioners and researchers. In a highly topical work, *Danielsen et al.* discuss the inclusion of indigenous and local knowledge (ILK) in citizen science for science-based land management and its mutual benefits for participants and science. The chapter also explains the relevant conditions for knowledge exchange with indigenous communities in management and decision-making in the research process. *Novak et al.* discuss different forms of citizen engagement in participatory digital social innovation related to do-it-yourself (DIY) science and participatory citizen science, and illustrate common challenges and experiences. The chapter also situates new knowledge in participation models framed by democratic and economic discussions. The following chapter by *Gold* showcases creative collaboration in citizen science and the evolution of ThinkCamp events. It draws useful conclusions for further similar activities. Closing the section, *Wyler and Haklay* discuss the potential of citizen science to be integrated into university research and the related opportunities and challenges. Their chapter points to a template for achieving such an integration in the service of civil society.

Section IIa: Case studies

To illustrate citizen science and its various formats and capacity-building initiatives, this section presents four case studies in more depth to be read alongside case studies presented in the chapters themselves (see table 1.1). The case studies in this section aim to highlight citizen science projects from a broader geographic range. Two cover different geographical areas and the other two address different topics of global importance.

The case study by *Li* provides several examples to illustrate activities within the spectrum of citizen science on the Chinese mainland. Public participation in China is becoming a growing movement supported by IT technologies and greater interest in citizen science. The multilingual landscape of citizen science in Europe is the focus of the snapshot by *Hecker et al.*, presenting results of the first European explorative survey on citizen science projects as a baseline for the European Open Science monitor. *Piera and Ceccaroni* present a case study of stakeholder engagement around water quality through the Citclops project, offering

Table 1.1 Overview on selected case studies presented in this volume

Case study name	Subject area	Location/ geographical scope	Relation to policy	Website	Page number in this volume
The Atlas of Living Australia	Biodiversity data infrastructure	National, Global Australia	Providing data for domestic and international research, policy and planning	http://www.ala.org.au/	pp. 73–9
GBIF France	Biodiversity data infrastructure	National France	Providing data for domestic and international research, policy and planning	http://portail.gbif.fr	pp. 75–7
UK National Biodiversity Network	Biodiversity data infrastructure	National UK	Providing data for domestic and international research, policy and planning	https://nbnatlas.org/	pp. 77–9
Exploring participation in volunteer computing	Social science	Local Geneva, Switzerland	Understanding participants and their characteristics		p. 105
Addressing democratisation/economisation tension in citizen science	Social science	International Europe	Identifying the political goals of citizen science		p. 107
Indigenous knowledge in biodiversity management	Conservation	Local Bosawás Biosphere Reserve, Nicaragua	Including of traditional ecological knowledge (TEK) in environmental management		pp. 113–9

(continued)

Table 1.1 (continued)

Case study name	Subject area	Location/ geographical scope	Relation to policy	Website	Page number in this volume
Hybrid LetterBox	Design	—	Demonstrating the co-design of a public solution by researchers and citizens	http://www.design-research-lab.org/projects/hybrid-letter-box/	pp. 127–8
Magenta Traffic Flow	Design	Local Florence, Italy	Addressing traffic issues	http://www.magentalab.it/	p. 140
Crowd2Map Tanzania	Mapping	National Tanzania	Collecting data to address a social problem	https://crowd2map.wordpress.com/	p. 144
ECSA 2016 ThinkCamp	Design and planning	Global	Developing ideas and shared visions	https://sites.google.com/a/gold-mobileinnovation.co.uk/ecsa2016--citsci-thinkcamp/About-the-Think-Camp/home	pp. 158–62
FLOAT Beijing	Environmental monitoring and management	Local Beijing, China	Recording evidence for policy implementation	http://f-l-o-a-t.com/	p. 186
Xiangjiang Watcher	Environmental monitoring and management	Local Hunan, China	Monitoring water quality		p. 187
Computing for Clean Water	Research	Local Beijing, China	Using volunteer computing to raise awareness and engagement	https://www.worldcommunitygrid.org/research/c4cw/overview.do	pp. 187, 189

Name	Field	Scope	Description	URL	Pages
Citclops/EyeonWater	Environmental monitoring and management (water quality)	Global	Using stakeholder mapping to connect public to the policy process	https://eyeonwater.wordpress.com/	pp. 204–8
Global Mosquito Alert Consortium	Invasive species; Public health	Global	Tackling disease-vector mosquitoes worldwide; Working with UNEP	https://ecsa.citizen-science.net/global-mosquito-alert	pp. 213–5
MYGEOSS	Research	European Union	Developing smart internet applications for the local environment	https://www.ecologyandsociety.org/vol15/iss1/art12/	pp. 222–3
Making Sense: Advances and Experiments in Participatory Sensing	ICT for the local environment	European Union International	Developing software and hardware to address local environmental issues	https://www.rri-tools.eu/-/making-sense-advances-and-experiments-in-participatory-sensing	p. 237
Challenge-Driven Innovation, Vinnova (Sweden)	Innovation research	National Sweden	Sharing citizen-developed principles for innovation project research aimed at citizen-selected social challenges	https://www.vinnova.se/en/	pp. 249–50
Xplore Health	Education	International European Union	Empowering secondary school students to participate in Responsible Research and Innovation (RRI) processes and in RRI decision-making	https://www.xplorehealth.eu	p. 251

(continued)

Table 1.1 (continued)

Case study name	Subject area	Location/ geographical scope	Relation to policy	Website	Page number in this volume
Living Knowledge Exchange Network and Partnerships	Community-based innovation research	International	Integrating community/civil society organisations into research for innovation	http://www.livingknowledge.org/	pp. 275–7
See it? Say it!	Environmental monitoring and management (pollution)	National Ireland	Developing a smart phone application from the Irish EPA for people to identify and report pollution	http://www.epa.ie/ enforcement/report/seeit/	pp. 294–6
The Clean Air Coalition of Western New York Volatile Organic Compound (VOC) study	Environmental monitoring and management (VOC contamination)	Local New York, USA	Regulating point source pollution beyond EPA-accepted standards	http://www.cacwny.org/ campaigns/tonawanda/	p. 296
Enviroza	Environmental monitoring and management (pollution)	National Slovakia	Developing Slovak Environment Agency-sponsored game for students to identify contaminated sites	http://www.enviroza.sk/	pp. 297–8
Collaborative research on sustainable fish stocking in Germany	Fisheries manage-ment	National Germany	Experimentally evaluating new stocking practices	http://www.besatz-fisch.de/	p. 316
Participatory technology development through Fab Labs and TechShop	Open source technology	Global, National United States	Demonstrating value of open source approaches to government entities such as the US-based NASA	http://www.fabfoundation.org/ https://www.techshop.info/	p. 317

Project	Theme	Scope	Purpose	URL	Pages
Measuring Ammonia in Nature (MAN) network	Environmental monitoring and management	National The Netherlands	Leveraging volunteer contributions as an input for policy	https://man.rivm.nl/	pp. 340–1
iSPEX	Environmental monitoring and management	National The Netherlands	Demonstrating the value of citizen science in government	http://ispex-eu.org	pp. 342–4
Waag Society Amsterdam Smart Citizens Lab	Environmental monitoring and management; open source technology	National The Netherlands	Helping policymakers measure pollutants that are not yet regulated	https://waag.org/en/project/amsterdam-smart-citizens-lab	pp. 344–5
Nijmegen Smart Emission project	Environmental monitoring and management; Spatial data	Local Nijmegen, The Netherlands	Demonstrating how openness, transparency and feedback in co-created projects help bring the public and policymakers closer together	https://smartemission.ruhosting.nl/	pp. 345–6
Ik heb last (I suffer now)	Health; Environmental monitoring and management (air quality)	National The Netherlands	Linking human health issues to environmental policy priorities	http://ikheblastapp.nl/	pp. 346–7
Measuring ALAN and skyglow from the ground	Environmental monitoring and management (light pollution)	Global	Informing regulatory planning instruments/ Ensuring compliance/ Promoting advocacy	http://www.verlustdernacht.de/astronomyskyglow.html	pp. 355–8

(continued)

Table 1.1 (continued)

Case study name	Subject area	Location/ geographical scope	Relation to policy	Website	Page number in this volume
Propage programme	Urban biodiversity management	Local Grenoble, France	Promoting inclusiveness of research process and social learning		pp. 372–9
National Sampling of Small Plastic Debris, Supported by Children	Ocean management; Marine litter	National Chile	Developing environmental stewardship through education		pp. 397–9
LETS (Local Elevation Transect Survey)	Environmental Monitoring and management; Climate change	Local Leysin, Switzerland	Developing Environmental stewardship through Formal education		pp. 417–24
OPAL Bugs Count Survey	Environmental monitoring and management	National United Kingdom	Providing environmental data on invertebrates to support management	https://www.opalexplorenature.org/bugscount	pp. 434–5
German Barcode of Life Project (GBOL)	Environmental monitoring	National Germany	Building an inventory of all species in Germany to support management	www.bolgermany.de	pp. 435–6
Orchid Observers project	Biological Recording Digitalisation of historical specimen data	National United Kingdom	Making historical data available for climate change research and policy	https://www.orchidobservers.org/	pp. 439–40

Project	Topic	Scale / Location	Purpose	Website	Page
Foxes in Berlin	Urban Wildlife Ecology	Local Berlin, Germany	Media relation/ Public communication/ Raising awareness	https://www.rbb-online.de/fuechse/	pp. 449–50
Cat Tracker	Cat management and behaviour	Global, National New Zealand	Monitoring environmental and social data leading to new legislation	www.cattracker.nz	pp. 451–2
De Natuurkalender	Science communication	National The Netherlands	Communicating to public and specific target groups about science policy issues	www.naturetoday.com	p. 453
BreadTime	Cultural heritage; Oral history; Intergenerational dialogue	Local Lesachtal, Austria	Saving local intangible cultural heritage and supporting intergenerational dialogue	www.lesachtalerbrot.wordpress.com	pp. 454–5
Landschaft im Wandel	Landscape change; Biodiversity; Public perception	Regional Germany	Reconnecting people with landscapes, surroundings and history to promote understanding of landscape change and policy	www.landschaft-im-wandel.de	pp. 456–7
Landscape and You-th – Tracing Flax	Cultural heritage; Oral history; Intergenerational dialogue	Local Lesachtal, Austria	Meeting the aims of the Framework Convention on the Value of Cultural Heritage for Society	www.lesachtalerflachs.wordpress.com	p. 458

a framework for understanding and engaging with a variety of stakeholder needs that will benefit a wide range of citizen science projects. The fourth case study, by *Palmer et al.*, is of the Global Mosquito Alert Consortium. It discusses how local projects can engage in mosquito-vector monitoring with a range of partners, and also share interoperable data to advance continental or even global research in epidemiology, biodiversity and other domains.

Section III: Innovation at the science-policy interface

This section addresses citizen science at the science-policy interface. The idea that citizen science can lead to better policy formulation, implementation and assessment is critical to this volume. Contributions here relate to questions highly relevant to policy: What are the opportunities for citizen science to feed into better decision-making? What are the synergies and opportunities brought by the science policy priorities of Responsible Research and Innovation (RRI), open science and citizen science? What are the benefits and challenges of citizen science for nature conservation? And in terms of the bigger picture, how can these benefits and challenges be addressed by strategic developments and feed environmental endeavours?

Nascimento et al. offer a high-level exploration of the value of citizen science for empowering citizens while leading to better and more transparent governments. They also assess citizen science's risks. Accordingly, citizen science is shown as a key means of advancing open science. This chapter also references citizen science in action in key areas such as biodiversity monitoring and the identification and monitoring of alien species. Promoting empowerment and behaviour change are identified as key benefits of citizen science. *Smallman* introduces the concept of RRI as a cross-cutting theme of the EU's Horizon 2020 programme that can be used to align the priorities of scientists, policymakers and the public at large. This chapter is important in the context of the current post-factual discourse, which underplays science and the related need to balance scientific objectivity with the competing challenges of market-friendly or politically derived claims in the name of science. Historically, citizen science has been particularly valuable in shaping conservation policy and monitoring outcomes, as demonstrated by *Ballard et al.* This chapter discusses how and why citizen science can contribute to beneficial social and environmental outcomes. *Richter et al.* discuss the challenges and benefits of capacity-building programmes that have developed in Europe with citizen science in mind. They highlight the need to build citizen science projects in line with the five main steps of capacity-building: identifying and engaging different actors; assessing capacities and needs for citizen science in context; developing visions, missions and action plans; building resources such as websites and guidance, and considering implementation and evaluation. *Parker and Owen* show the increasing importance of citizen science for environmental monitoring and its use for Environmental Protection Agencies (EPAs) in Europe and the United States. They discuss the accompanying needs and challenges, illustrated through four case studies. Finally, they point to citizen science as having the potential to transform environmental protection through work with government agencies to generate knowledge and find solutions.

Section IV: Innovation in technology and environmental monitoring

Chapters in section IV discuss technologies for citizen science and environmental monitoring. In doing so, they address the following questions: How can digital technologies be harnessed to enhance citizen science participation and delivery? What policy and technical solutions can mobile sensor technology offer citizen science? How is data quality ensured, and how can different protocols for ensuring data quality be developed and applied to support fitness for purpose in research and policy deliverables? How can this contribute to advances in environmental monitoring in accordance with existing and emerging regulations? How can these technologies be implemented in monitoring citizen science projects, and what are the challenges of doing so?

Based on assertions that technology will revolutionise the practice of citizen science, the first chapter in this section, by *Mazumdar et al.*, reviews a wide range of technologies available for data collection, data analysis and improving the citizen science experience through new opportunities for interactive participation. One particularly important area of consideration pertains to data standards. *Williams et al.* discuss the role of standards and open data for promoting interoperability and therefore the reuse of data, especially when information is contextualised with metadata that is accurate and up-to-date. Low-cost tools and data standards are also important to the EPAs discussed by *Volten et al.*, who note that agencies increasingly utilise small-sensor networks for environmental monitoring. As explained by *Schroer et al.*, technologies such as mobile applications and co-ordinated, interoperable activities are critical to monitoring artificial light pollution and understanding the impacts on human health and the biosphere.

Section V: Innovation in science communication and education

Section V focuses on citizen science innovation in science communication and education. Relevant questions to this topic are: How should primary, secondary and tertiary education and further training opportunities be addressed to optimise citizen science knowledge and participation? Can citizen science benefit learning across different ages and stages in individual learning curricula? How can communication in citizen science improve policy impact? What opportunities and challenges does citizen science provide for scientists collaborating with the media?

Reflecting on learning, education and empowerment through citizen science, *Peltola and Arpin* discuss the need to apply effective techniques in citizen science to ensure inclusiveness for less experienced or privileged participants. *Edwards et al.* offer a societal perspective discussing the potential of citizen science to create science capital as part of the relationship between people's dispositions towards science, participation in science-related activities and science-related outcomes, including learning. *Makuch and Aczel* provide insights into the effects of engaging children in science on their learning processes and understanding of science, as well as on safeguarding the environment. Following this, *Harlin et al.* discuss the opportunities and challenges of using citizen science in schools, both through theoretical reflection and practical case study. *Sforzi et al.* focus on the role of natural history museums as another form of societal institution and a traditional space for two-way learning and education. These organisations offer new opportunities and formats for participant engagement in the contemporary development of citizen science as a network and research field. In the closing chapter, *Hecker et al.* assess the innovative potential, limits and opportunities for science communication in citizen science through case and best practice studies from Europe and New Zealand. They highlight opportunities to engage the public via techniques including storytelling and visualisation.

Outlook

This volume demonstrates that citizen science is growing, both in terms of the number of projects and the volume of peer-reviewed research generated from these activities. This trend will continue along with the increasing influence of citizen science on policy agenda-setting, formation, implementation and assessment. Citizen science communities of practice are at the same time becoming increasingly formalised through associations, including established organisations in the United States (the Citizen Science Association – CSA), Europe (the European Citizen Science Association – ECSA) and Australia (the Australian Citizen Science Association – ACSA). Nascent networks are also emerging in other regions, including in Africa and Asia, with the intent of creating associations to convene researchers and practitioners in these areas.

Much of citizen science has been driven by immediate interest and curiosity as well as the practical need to develop science and provide evidence for the policy arena. Greater understanding of the impacts of citizen science in the field of open science and policy is now starting to emerge and is underpinned by sound evidence. In this way, we can build on a stronger understanding of the drivers of citizen science for success.

The conclusion to this volume draws on findings by all chapters and synthesises them to offer recommendations for citizen science practitioners, researchers, educators and policymakers to develop the field of citizen science and advance innovation in open science, society and policy.

At a time in history in which society faces unprecedented grand challenges which require informed, inclusive policy responses across our nations, this book aims to provide a further catalyst for discussions and collaboration among organisations, scientists, practitioners and other stakeholders that are interested in and will gain from citizen science.

Part I
Innovation in citizen science – setting the scene

2
Ten principles of citizen science

Lucy Danielle Robinson[1], Jade Lauren Cawthray[2], Sarah Elizabeth West[3], Aletta Bonn[4,5,6] and Janice Ansine[7]

[1]*Angela Marmont Centre for UK Biodiversity, The Natural History Museum, London, UK*
[2] *Duncan of Jordanstone College of Art & Design, University of Dundee, UK*
[3] *Stockholm Environment Institute, University of York, UK*
[4] *Helmholtz Centre for Environmental Research – UFZ, Leipzig, Germany*
[5] *Friedrich-Schiller-University Jena, Germany*
[6] *German Centre for Integrative Biodiversity Research (iDiv) Halle-Jena-Leipzig, Germany*
[7] *The Open University, UK*

corresponding author email: l.robinson@nhm.ac.uk

In: Hecker, S., Haklay, M., Bowser, A., Makuch, Z., Vogel, J. & Bonn, A. 2018. *Citizen Science: Innovation in Open Science, Society and Policy.* UCL Press, London. https://doi.org/10.14324/111.9781787352339

Highlights

- The Ten Principles of Citizen Science were developed by an international community of citizen science practitioners and researchers to set out their shared view of the characteristics that underpin high-quality citizen science. They are currently available in 26 languages.
- The Ten Principles provide a framework against which to assess new and existing citizen science initiatives with the aim of fostering excellence in all aspects of citizen science.
- At a time when citizen science is rapidly expanding but not yet mainstreamed within traditional research or policy processes, the Ten Principles provide governments, decision-makers, researchers and project leaders with a common set of core principles to consider when funding, developing or assessing citizen science projects.

Introduction

Citizen science is a flexible concept that has been adapted and applied within diverse situations and disciplines. The rapid expansion of citizen science programmes globally presents researchers and citizen science practitioners with incredible opportunities as well as a challenge: creating cohesion and identifying a common purpose globally, whilst also supporting and enhancing the further expansion, independence, creativity and bottom-up nature of citizen science. Networks such as the global Citizen Science Association (CSA), the European Citizen Science Association (ECSA) and the Australian Citizen Science Association (ACSA) provide forums for the exchange of knowledge and ideas, identification of shared goals, networking and developing best practice. In 2015, the ECSA working group on 'Sharing best practice and building capacity for citizen science' developed a document outlining Ten Principles of Citizen Science. Drawing from the collective experiences of many ECSA members, this series of statements set out the key principles which ECSA believes underlies good practice in citizen science, regardless of the academic discipline or cultural context in which it is applied. Used internationally and currently available in 26 languages, the Ten Principles of Citizen Science provide an important starting point for discussion and debate. This chapter introduces the Ten Principles and their development. It gives examples of good practice and explores how the Principles may challenge current working practices to drive excellence in citizen science, maximising the benefits for science, citizen scientists and wider society. Finally, the chapter considers the policy and innovation potential of the Ten Principles in a rapidly expanding and diversifying field.

Developing the Ten Principles of Citizen Science

The ECSA working group on 'Sharing best practice and building capacity for citizen science' is chaired by the Natural History Museum London and its members come from universities, natural history museums and not-for-profit organisations, representing researchers, citizen science practitioners and networking or co-ordination bodies (see also Sforzi et al. in this volume about the role of museums in citizen science). The working group aims to facilitate the exchange of knowledge, experience, innovation and lessons learned in the field of citizen science, both within and beyond the ECSA membership. The group's first task was to develop a series of principles or characteristics that underpin responsible and impactful citizen science projects, with the aims of supporting those new to citizen science to deliver high-quality projects and providing a benchmark against which to examine existing citizen science programmes. These became the Ten Principles of Citizen Science and were designed to be applicable across a broad spectrum of citizen science activities.

Box 2.1. The Ten Principles of Citizen Science

(for other languages see https://ecsa.citizen-science.net/engage-us
/10-principles-citizen-science)

1. Citizen science projects actively involve citizens in scientific endeavour that generates new knowledge or understanding.
    Citizens may act as contributors, collaborators or as project leaders and have a meaningful role in the project.
2. Citizen science projects have a genuine science outcome.
    For example, answering a research question or informing conservation action, management decisions or environmental policy.
3. Both the professional scientists and the citizen scientists benefit from taking part.
    Benefits may include the publication of research outputs, learning opportunities, personal enjoyment, social benefits, satisfaction through contributing to scientific evidence, for example, to address local, national and international issues, and through that, the potential to influence policy.
4. Citizen scientists may, if they wish, participate in multiple stages of the scientific process.
    This may include developing the research question, designing the method, gathering and analysing data, and communicating the results.
5. Citizen scientists receive feedback from the project.
    For example, how their data are being used and what the research, policy or societal outcomes are.
6. Citizen science is considered a research approach like any other, with limitations and biases that should be considered and controlled for.
    However unlike traditional research approaches, citizen science provides opportunity for greater public engagement and democratisation of science.
7. Citizen science project data and metadata are made publicly available and where possible, results are published in an open-access format.
    Data sharing may occur during or after the project, unless there are security or privacy concerns that prevent this.

(continued)

8. Citizen scientists are acknowledged in project results and publications.
9. Citizen science programmes are evaluated for their scientific output, data quality, participant experience and wider societal or policy impact.
10. The leaders of citizen science projects take into consideration legal and ethical issues surrounding copyright, intellectual property, data-sharing agreements, confidentiality, attribution and the environmental impact of any activities.

Developed between 2013 and 2015, the scope and structure of the Ten Principles were initially informed by reference to existing sets of principles from related disciplines (European Commission 2008; Wing 2014). A longlist of potential principles was generated by working group members before being rationalised and distilled to the 10 most universally applicable. These were presented for consultation with ECSA members and the wider citizen science community multiple times over two years at ECSA General Assemblies, via the ECSA website, e-newsletter and a popular blog written by an ECSA Steering Committee member, with iterative feedback and edits throughout this time. This extensive feedback process led to the Principles becoming more universal (relevant to a diversity of disciplines, projects and audiences), actionable (rather than theoretical), inclusive of individual, societal and policy outcomes, and targeted towards citizen science practitioners (rather than citizen scientists or policymakers). The length of each core Principle was shortened but clarification statements were added to each.

The Ten Principles of Citizen Science were published on the ECSA website in September 2015 (see box 2.1). At the time of writing, the Ten Principles of Citizen Science have been translated by ECSA members into 26 languages to make them accessible to non-English speakers, and this continues to expand.

Global impact of the Ten Principles of Citizen Science

No systematic review has yet been conducted to measure the extent of use and impact of the Principles, but ECSA headquarters and the working group are recording known uses to create a bank of case studies. To date,

Fig. 2.1 The Museo di Storia Naturale della Maremma (Natural History Museum of Maremma, Italy) displays the Ten Principles of Citizen Science in their 'Citizen Science Corner' gallery to inspire visitors to participate in local projects. (Source: © Andrea Sforzi)

Box 2.2. Case study: How the Ten Principles of Citizen Science informed a US policy brief

Dr Lea Shanley

The US Federal Community of Practice for Crowdsourcing and Citizen Science (CCS) is a self-organised grassroots group of more than 350 federal employees representing 60 federal organisations. It seeks to expand and improve the US government's use of crowdsourcing, citizen science and public participation techniques to enhance agency missions and to improve scientific and societal outcomes.

In 2015, the CCS leadership worked closely with the White House Office of Science and Technology Policy to help shape a policy memo that would guide and encourage the use of these open science and innovation approaches across the federal government. Drawing from the Ten Principles of Citizen Science, the CCS leadership incorporated three core principles into the text of the memo. The memo (Office of Science and Technology Policy 2015) was released on 30 September 2015 as part of the White House's Forum on Citizen Science (Gustetic, Honey & Shanley 2015), co-organised by the CCS.

The principles detailed in the memo emphasised openness, accessibility, meaningful participation and recognition for contributions to ensure that the use of citizen science and crowdsourcing 'is appropriate and leads to [the] greatest value and impact' (Office of Science and Technology Policy 2015). The White House memo directs agencies to adhere to three principles, summarised as:

- *Data quality:* Data collected are credible, usable and fit for purpose;
- *Openness:* Datasets, code, applications and technologies used are transparent, open and available to the public, consistent with applicable intellectual property, security and privacy protections; and
- *Public participation:* Participation should be fully voluntary, volunteers should be acknowledged for their contributions and should know how their contributions are meaningful to the project and how they, as volunteers, will benefit from participating.

the Ten Principles have been used in a wide variety of settings, including to inform further development of best practice guidelines for citizen science (including League of European Research Universities 2016; see also Wyler & Haklay in this volume), on Wikipedia to set out ethical considerations in citizen science (Wikipedia 2017), in public-facing museum displays about citizen science (figure 2.1) and to inform government policy, as in the case study of a US White House policy memo described in box 2.2.

Implementing the Ten Principles of Citizen Science: Successes and challenges

The Ten Principles of Citizen Science are intended to both support and challenge the citizen science practitioner community. Whilst some Principles are implemented within every citizen science project, others are more challenging to incorporate and require a greater investment of time and resources to fulfil. This section examines each Principle in turn, assessing the extent to which the citizen science community is currently meeting it and identifying where there are opportunities to improve practice. The chapters in this volume explore many of these themes in greater depth.

1. Citizen science projects actively involve citizens in scientific endeavour that creates new knowledge or understanding.

At the heart of all citizen science projects is the involvement of citizens in real scientific endeavour. Whilst this Principle refers to scientific endeavour in particular, there are many 'citizen science' projects focusing on other disciplines including the arts, geography and social history (see www.zooniverse.org/projects for a range of examples; and see also Mahr et al. in this volume). With many thousands of projects active globally (SciStarter [2017] lists over 1,500 projects) this represents millions of citizen scientists (Roy et al. 2012; Theobald et al. 2015). These impressive levels of participation notwithstanding, citizen science initiatives tend to be less successful at engaging communities that are historically underrepresented in science, including (but not limited to) certain minority ethnic groups and people from lower socioeconomic backgrounds (Pandya 2012; West, Pateman & Dyke 2016; West & Pateman 2016; see also Peltola & Arpin; Haklay; both in this volume). Significant opportunities remain to collaborate with a greater diversity of participants that are truly reflective of wider society and that also bring new and different knowledge (Danielsen et al. in this volume). Guidance on how project leaders may approach this is emerging (Pandya 2012; Ruzic et al. 2016), and new formats can be found to engage in person (Gold & Ochu in this volume) or through digital technologies (e.g., Novak et al. in this volume). The widening participation agenda is not unique to citizen science and is likely to require a range of long-term changes to be successful, including (but not limited to) greater flexibility in the range of opportunities available, for example, time commitment and prior skills required (see Haklay in this volume), new approaches to publicity and recruitment of participants, language translation of project materials and more participatory project development to ensure project activities and community priorities are better aligned (West & Pateman 2016).

2. Citizen science projects have a genuine science outcome.

This is what distinguishes citizen science from pure education and outreach programmes. Citizen science projects – while also serving learning goals (see e.g., Edwards et al.; Harlin et al.; Makuch & Aczel, all in this volume) – are increasingly resulting in research publications in a wide range of discipline-specific journals, with the number of peer-reviewed publications growing rapidly year on year (Follett & Strezov 2015). Science outcomes delivered by citizen science may also include the development of scientific specimen collections, for example for natural history museums (Sforzi et al. in this volume), tracking progress towards global biodiversity targets (Chandler et al. 2017), implementing changes to science policy and achieving conservation outcomes (see Ballard et al. 2017 for examples). However, there are still some projects that do not use the data collected for scientific purposes, thereby failing to realise the scientific benefits of the project. For example, biological records collected at 15 per cent of the BioBlitz events surveyed in the UK were not passed on to recommended data repositories (Postles & Bartlett 2014). This may be due to lack of staff or financial resources to publish the findings and attain other scientific outputs, uncertainty over the quality of the data, or poor study design resulting in data unsuited to the scientific need. A strong motivation to harness the public engagement benefits of citizen science can also lead to scientific rigour being compromised (see Lakeman-Fraser et al. 2016 for a discussion of this trade-off). However, achieving and maximising science outcomes from citizen science projects is a cornerstone of this field and an essential element in maintaining trust with the citizens that participate.

3. Both the professional scientists and the citizen scientists benefit from taking part.

To be sustainable, citizen science must be mutually beneficial for all parties involved. Benefits may be wide ranging, including scientific outcomes (Shirk & Bonney in this volume), social interaction, improved well-being, career development, learning and empowerment (e.g., Bela et al. 2016; Haklay in this volume; Edwards et al. in this volume). Whilst a limited number of resources exist to support the measurement and identification of these benefits (Phillips et al. 2014; Blaney et al. 2016), a broad evidence base of the benefits of participating in citizen science for all parties is lacking. Literature examining the impacts of citizen science has focused attention on the scientific or educational impacts (see Silva et al. 2016). In order for all parties to benefit, parity or overlap in their expectations and motivations for participating is required. West and Pateman (2016) provide a review and guidance on identifying and meeting citizen scientists' motivations, and Geoghegan et al. (2016) examine the motivations of participants and other stakeholders (see also Richter et al. in this volume). These reviews indicate that the numerous motivations for participating should be considered throughout the project lifecycle; ultimately, long-term project success depends on all stakeholders reaping the benefits. Researchers from other disciplines, including those from the social sciences (see also Mahr et al. in this volume), are encouraged to collaborate with citizen science programme leaders to gather more evidence on the benefits a citizen science approach offers for all involved.

4. Citizen scientists may, if they wish, participate in multiple stages of the scientific process.

The dominant method for engaging the public in scientific research is the 'contributory' method, where the public solely collect and submit data to research projects. However, the citizen science community recognises that a multitude of benefits is likely if the public is more deeply involved in scientific research, through 'collaborative' and 'co-created' methods (for an example of the latter, see Collins 2016; see also Novak et al. in this volume). Involving participants in more stages of the research process can foster a greater sense of ownership for the participants, and benefit the research by incorporating local knowledge and expertise (Corburn 2007). However, little is published on the practice and impacts of collaborative and co-created citizen science, and additional research and sharing of evaluations in this area would be welcome. Some pressing questions include: What do the different citizen science approaches (contributory, collaborative and co-created) achieve for science and for citizens? How can collaborative or co-created projects be run at a large scale whilst maintaining a close personal connection between the scientists and participants? And how can citizens be actively supported to participate in aspects of the scientific process beyond data collection and processing?

5. Citizen scientists receive feedback from the project.

There are many ways of giving feedback to volunteers, for example via social media, websites, maps, e-newsletters, celebratory events, blogs and meet-ups. Good feedback brings many benefits. It shares the outcomes of the science, justifies why people spent their time on the project, encourages repeat participation (Segal et al. 2015), explains the science research in more detail, and creates a personal connection between the citizen scientists and the project/research team (Rotman et al. 2012). It is also a way of showing participants that their contribution is recognised; an important feature for many (Rotman et al. 2012). There is evidence that feedback is a motivator for more participation (Singh et al. 2014), and there is great potential for project leaders to both speed up and improve the quality of their feedback, for example, by making it more personalised. Tools such as Natural Language Generation are being developed to automate the process of giving instant, personalised feedback (see, for example, Wal et al. 2016), helping project leaders to better manage large-scale communication with participants.

6. Citizen science is considered a research approach like any other, with limitations and biases that should be considered and controlled for.

Citizen-collected data are still sometimes criticised for being of lower accuracy, biased or of uncertain quality, which limits their use for many scientific purposes (see Williams et al. in this volume). However, in many cases, citizens gather data that are of equal quality to professionally collected data (Lewandowski & Specht 2015; Kosmala et al. 2016) and all data, including those collected by professional scientists, have an error rate or some degree of variation between observers. Citizen science project leaders have a responsibility to control, measure and report data quality and quality assurance procedures, to demonstrate the validity and reliability of the data (for discussion, see Williams et al. in this volume). Innovations in technology can support data validation and verification in environmental monitoring, for example Mazumdar et al., Volten et al., Schroer et al., all in this volume. A citizen science approach, however, will not be appropriate for all research questions and the 'Choosing and Using Citizen Science' guide supports researchers in making this assessment (Pocock et al. 2014b).

7. Citizen science project data and metadata are made publicly available and where possible, results are published in an open-access format.

Citizen science is an example of open science – a movement within the academia to make science research, data and outputs accessible to all. Whilst the principles of open science are welcomed within the citizen science community (both CSA and ECSA have working groups on open data; see Smallman et al. in this volume), in practice there is still a long way to go. This situation is not unique to the field of citizen science but is found across the sciences where time, resources, infrastructure and incentives are not always available to support open-data sharing (Tenopir et al. 2011). There have been many successes in the global sharing of citizen science data (for example Chandler et al. 2017) but still too few citizen science projects give participants direct access to the resulting dataset, and few project websites clearly describe if/how data will be shared with national and international databases. Cleaning, formatting and archiving data requires resources and infrastructure, and this vital step must be planned into project timescales and funding at the outset. The time lag between data collection and the publishing of results in academic journals remains a challenge for citizen science projects where participants may have to wait several years to see the 'final results' of the project. Researchers may also have to navigate data embargoes, a lack of institutional repositories for datasets and open-access publishing fees (Tenopir et al. 2011). However, new technologies and increased availability of repositories for data and publications are making this process ever easier, and the opportunities afforded by opening up citizen science data are significant. There may also be a role for citizen science, and citizen scientists, in the wider sharing of project outputs and findings within and beyond the research community using non-traditional approaches. This could include non-science outlets such as local newspapers, NGO/association newsletters, special interest journals (e.g., gardening/angling magazines) or online communication and visualisation through story telling (Hecker et al. 'Stories' in this volume).

8. Citizen scientists are acknowledged in project results and publications.

The contributions of citizen scientists are usually recognised throughout the lifetime of a project via project communications, the awarding of badges or certificates, events and many other routes. However, this does not always carry through to more academic project outputs. Acknowledging citizen scientists in project publications and other academic outputs is relatively easy to achieve but often overlooked. The volunteer hours donated to any given project are significant and should be celebrated! Appropriate levels of acknowledgement will vary by project and participant role, but – as a minimum – a generic thank you statement covering all volunteers should be included in publications and presentations wherever possible. Acknowledging large numbers of participants individually has been known, for example Lee et al. (2014) included 37,000 co-authors in their published paper on the EteRNA project, and whilst this is a rather extreme example, acknowledging individual participants may be appropriate where they have given significant input to a project (although data protection and ethical issues should be considered when disseminating personal information of participants). Data papers listing all contributors can also be published in data journals (e.g., http://www.forschungsdaten.org/index.php/Data _Journals), which can be cited in subsequent analyses and publications.

9. Citizen science programmes are evaluated for their scientific output, data quality, participant experience and wider societal or policy impact.

Project evaluation is typically under-resourced, and as a result, some outcomes of citizen science projects are not fully identified, measured or reported (Ballard et al. 2017), despite potentially significant scientific, societal, policy, community and individual outcomes. Time constraints, a lack of established evaluation criteria (but see Kieslinger et al. in this volume) and a lack of understanding and confidence in how to conduct evaluation may prevent practitioners from collecting evidence of their successes and failures (for an example of this within environmental education, see West 2014). Training in evaluation methods and prioritisation of evaluation as part of the project delivery process would assist in collecting this evidence, as would greater interdisciplinary collaborations with academics in the social sciences and education fields to study the wider impacts and outcomes of participation in citizen science (see Mahr et al. in this volume). Research focused on the learning outcomes of citizen science is growing and some supporting resources for project leaders already exist, including practitioner guides (e.g., Phillips et al. 2014) and academic literature, in particular the new journal *Citizen Science: Theory and Practice* (Bonney, Cooper & Ballard 2016), which provides a route for project leaders to share tools and strategies for evaluation and learning research. Societal and policy impacts are equally as important as research and education outcomes, as citizen science projects can provide substantial input to policy formulation and implementation (Nascimento et al.; Owen & Parker, both this volume). Evaluation needs to consider this adequately even though such indirect impacts may at times be hard to assess. The citizen science community should therefore be encouraged to prioritise evaluation, including sharing details of less successful ventures, because the field cannot advance rapidly and effectively without self-reflection.

10. The leaders of citizen science projects take into consideration legal and ethical issues surrounding copyright, intellectual property, data-sharing agreements, confidentiality, attribution and the environmental impact of any activities.

Involving volunteers in any activity requires careful consideration for their health and well-being, their rights as individuals and an awareness of the power balance between volunteers and other parties involved in any given project. Resnik, Elliott and Miller (2015) provide a useful framework for addressing ethical issues in citizen science, and the CSA supports a working group on ethics. Many citizen science projects involve online activity, in which participants register for an online account, submit personal details about themselves, upload and share images and other content to which they hold the intellectual property, and collaborate with others. The gathering, processing and sharing of these types of data must be approached sensitively and with an understanding of the legal and ethical implications (see also Williams et al. in this volume). This may be a particularly sensitive issue in projects that deal with medical data (see Hoffman 2014 for an analysis of the benefits and risks). Scassa and Chung (2015b) provide a useful guide for considering intellectual property rights in citizen science projects and Bowser and Wiggins (2015) address privacy issues.

Conclusion

At a time when citizen science is rapidly expanding but not yet mainstreamed within traditional research or policy processes, the Ten Principles provide governments, decision-makers, researchers and project leaders with a common set of core principles to consider when funding, developing, implementing or assessing citizen science projects/programmes. Imposition of a top-down set of standards for citizen science would be incongruent with its naturally bottom-up, flexible nature, but the Ten Principles may nonetheless serve the same aim of promoting excellence in science research, environmental protection, and public engagement and active involvement in the scientific and policy processes. Strategic national and international developments (see box 2.2 and Richter et al. in this volume) may provide examples and lead to action plans of how policymakers could make practical use of the Principles to drive widespread support for this approach.

Reviewing the Ten Principles of Citizen Science has highlighted the enormous amount of excellent work currently underway in this sector. The appetite for sharing good practice and learning lessons from others to maximise the benefits for science, policy, society and the individuals involved is inspiring. Widening participation, maximising and reporting data quality, and ensuring data and publications are made available in open-access formats remains challenging for this field. Innovative, non-traditional approaches will be required to move beyond the current state of the art. Later chapters of this book share some of these innovations and it is hoped that the reader finds these, together with the Ten Principles, inspiring and instructive.

In a rapidly moving field, best practice, too, will evolve and develop, and in time an 11th or 12th principle may be added to this current suite. In particular, developments in the fields of ethics, technologies and open data will strongly influence views of 'best' practice in coming years. Such innovations and advances in the field of citizen science, and the new challenges and opportunities they present, are to be welcomed.

Acknowledgements

We would like to thank ECSA members and the wider community for their invaluable constructive feedback during the development of the Ten Principles, in particular the members of the ECSA working group for 'Sharing best practice and building capacity for citizen science'. We would also like to extend our gratitude to those members of the ECSA community who have translated the Ten Principles of Citizen Science.

3
Scientific impacts and innovations of citizen science

Jennifer L. Shirk[1] and Rick Bonney[1]

[1] *Cornell Lab of Ornithology, US*

corresponding author email: jls223@cornell.edu

In: Hecker, S., Haklay, M., Bowser, A., Makuch, Z., Vogel, J. & Bonn, A. 2018. *Citizen Science: Innovation in Open Science, Society and Policy*. UCL Press, London. https://doi.org/10.14324 /111.9781787352339

Highlights

- Citizen science makes distinct, novel and innovative contributions to scientific knowledge and can connect scientific research with public engagement to inform policy.
- Different scientific disciplines are advancing distinct research techniques, such as computational modelling, to draw useful insights from opportunistic datasets and technologies that support new approaches to engagement.
- New scientific knowledge can be gained when citizen science puts research in the hands of people who have insights and concerns previously not addressed by academia, NGOs or government agencies.
- Citizen science may be an optimal strategy to address policy priorities, including indicators and outcomes set by high-profile treaties such as the Convention on Biological Diversity.
- Cross-disciplinary networking can advance innovations and practices around concerns shared by all disciplines employing citizen science approaches.

Introduction

From the Ten Principles of Citizen Science (Robinson et al., this volume), we can see that pursuing scientific outcomes is an integral element of citizen science. Citizen science can make distinct, novel and innovative contributions to scientific understandings. In doing so, citizen science opens both new opportunities and new appreciations for the ways that science can engage public insight and conduct policy-relevant research. This chapter focuses on the scientific impacts and innovations across the diverse field of citizen science. It highlights the general strengths of citizen science for data collection and processing capacity, public engagement and policy, then looks to scientific innovations emerging (or in some cases being rediscovered) from different disciplinary domains.

Although the history of citizen science often focuses on environmental sciences, a rich tradition of similar research approaches is found in disciplines as varied as astronomy, meteorology and public health. Citizen science is also rapidly expanding across research domains both within and beyond the sciences, as a collaborative approach to knowledge building (see also Mahr et al. in this volume). As the field of citizen science grows, its use continues to advance discovery, foster innovation and expand the boundaries of knowledge, which can in turn reveal new ways to connect research and public engagement for policy relevance, especially when taking the opportunity to explore and connect advancements across different disciplines.

Citizen science as a distinct means of research

Citizen science depends upon the thoughtful and meaningful engagement of the public in scientific investigations. At its core, citizen science draws upon the strengths of scientific traditions, employing systematic observations and/or enquiries to produce information that can be confirmed by others. What sets citizen science apart from other research approaches is that it rejects the notion that only credentialed and/or paid scientists can take part in, lead or shape how questions are asked, data are collected, results are interpreted or findings are used (see also Haklay in this volume; Novak in this volume for more on participatory approaches). In doing so, citizen science opens up research to public input and insights, and through the combination of engagement and rigorous research, it can broaden opportunities to inform and influence policy (Vann-Sander, Clifton & Harvey 2016).

Scientific significance: Public engagement has enabled the expansion of data collection and data processing capacities (see Wyler & Haklay in this volume). In a 2016 article on the game-changing nature of internet-enabled citizen science, Watson and Floridi describe how citizen science projects can be designed to enhance the 'reliability, scalability, and connectivity' of information. By engaging tens, thousands, and even millions of participants, citizen science can offer both human and statistical power. With observers available around the clock and around the globe, citizen science can yield observations at unprecedented temporal and geographic scales and can produce data of sufficient quality for research (Kosmala et al. 2016; and see also Williams et al. in this volume) and for evidence-based decision-making (McKinley et al. 2017). Paired with powerful and novel computational and modelling techniques, this research approach can generate useful insights even when a dataset has known limitations, such as gaps in reporting times or species that are challenging to detect (Kelling et al. 2015).

While Watson and Floridi point to the role that technology plays in these enhancements, citizen science can be a powerful strategy for distributed collaboration without technology and also at much smaller scales (see also Peltola & Arpin in this volume; Danielsen et al. in this volume). Mobilising a committed corps of 20 volunteers in a watershed, for example, can vastly enhance the capacity for local monitoring to capture and document events of concern or to have confidence in the stability of a system. What is critical in research at every scale is not to have the most data, nor even the most precise data, but to have data of known quality and data that are fit to purpose (Ellett & Mayio 1990, 23; Vaughan et al. 2003).

The practice of citizen science has also brought new technologies, new data analysis techniques and new questions. Citizen science can make historic data available for analysis (e.g., Miller-Rushing & Primack 2008; Ellwood et al. 2016) and can lead to combined datasets accessible for wider use (Schmucki et al. 2016; see also Williams et al. in this volume). Perhaps most importantly, citizen science puts science in the hands of people who have insights and concerns previously not addressed by academia or agencies (Ottinger 2016). Citizen science thus provides avenues for interrogating topics that have both scientific and social relevance – a prime nexus for informing policy (McKinley et al. 2017), whether for the environment, health, public safety or any of an increasing number of topics.

Public engagement: Scientific advancements through citizen science have only been possible because of a willingness to think differently about who is involved in the research process, how those participants engage and what they bring to the research endeavour. Beyond engaging the public in the process of data collection, citizen science opens doors to broader knowledge exchange about the research in question (McKinley, Briggs & Bartuska 2013). Listening to participants' experiences can increase scientists' and policymakers' awareness of social concerns and influencing factors. This can be particularly important in complex settings such as conservation and medicine, where findings and implementations may be context-specific and where generalised, 'objective' knowledge may be less useful than scientific traditions generally assume. Research in all areas of exploration indicates that the more deeply participants are involved in the process of investigation – from shaping the research question to interpreting and acting on the results – the more profound the outcomes are for participant learning and for policy action (Danielsen, Burgess & Balmford 2005; Shirk et al. 2012; Stepenuck & Green 2015; and see Nascimento et al. and Smallman, both in this volume). Regardless of the depth of engagement, a significant motivator for many who choose to participate is an understanding that they are making a contribution, whether to broadening scientific understandings or to making a change in the world (Raddick et al. 2013; Alender 2015; Tsueng et al. 2016).

Policy: In an ideal world, policy decisions would be informed by evidence, but actionable evidence may not always be available, especially in cases calling for rapid or anticipatory responses (e.g., disasters, emerging diseases) or in complex systems (e.g., climate impacts, fisheries) (see, for example, Bower et al. 2017). Policy decisions are thus often made without evidence or with data not fit for purpose, and therefore against a background of uncertainty. Citizen science mobilises multiple observers and therefore has the potential to fill data gaps (Chandler et al. 2017) and to procure data in a timely manner (Vaughan et al. 2003). Careful design is required to ensure that the data collected are of appropriate and known quality for the purpose at hand (Shirk et al. 2012; Danielsen et al., 'A Multicountry Assessment', 2014; Kosmala et al. 2016). Citizen science research can also be targeted towards questions informed by policy needs or stakeholder concerns to yield the most relevant data (McKinley, Briggs & Bartuska 2013). Participants in such research, where stakes are high, have every incentive to ensure their data are defensible (Ottinger 2016). With all of these factors in mind, Danielsen et al., 'Linking Public Participation', (2014) suggest that citizen science may be an optimal strategy to address policy priorities, including indicators and outcomes set by high-profile treaties such as the Convention on Biological Diversity.

Citizen science innovations across disciplines

It is possible to see – and learn from – advancements in research impacts and innovations emerging in the different scientific disciplines where citizen science is employed. This section briefly looks at three different research domains – geophysical, biomedical and social science – to explore the scientific contributions of citizen science and the innovations that have enabled those outcomes. In doing so, it points to advances in public engagement and policy that can also be seen in these areas. It does not aspire to provide a comprehensive review, but rather to offer a glimpse into the practices and impacts in different disciplines, which may help expand thinking in the larger field.

Geophysical/Geospatial: Earth systems and earth observation research are yielding scientific advances through citizen science at both global and local scales, and advancing this work in part through innovative uses of remote sensing, social media and distributed sensors. An entire special issue of the journal *Remote Sensing* (Fritz & Fonte 2016) is devoted to sharing outcomes of citizen science including research into land cover (Laso Bayas et al. 2016), forest biomass (Molinier et al. 2016), water clarity (Busch et al. 2016) and the timing of lifecycle events (e.g., Elmore, Stylinski & Pradhan 2016) among other topics. Seismologists have refined methods to harvest streams of Tweets to improve real-time research into earthquake intensity (D'Auria & Convertito 2016) and range of perceptibility (Earle, Bowden & Guy 2012). Hydrologists have turned to social media as well, capturing photographs of flood events to estimate flow rate and depth (Le Coz et al. 2016). Geophysical scientists are also working in person with concerned communities to assess and monitor pollutants in soils and garden vegetables (Ramirez-Andreotta et al. 2013), air pollutants near gas drilling sites (Macey et al. 2014) and changes in water quantity and quality (Stepenuck & Green 2015).

These approaches to research can facilitate both rapid and collaborative policy responses to environmental change (Minson et al. 2015; Stepenuck & Green 2015). To this end, work in this domain is confronting and advancing procedures and measures that relate to issues of public engagement, such as around risk (Ramirez-Andreotta et al. 2013) and power and participation (Ramirez-Andreotta et al. 2015; Stepenuck & Green 2015).

Biomedical: In their systematic review of crowdsourced research in medical fields, Ranard et al. (2014) found papers in hematology, radiology, genomics, molecular biology and more, which describe citizen science strategies including problem-solving and the distributed surveillance of symptoms or treatment options. Innovations in online platforms for problem-solving, such as FoldIt and Zooniverse, have engaged communities of gamers-turned-analysts to advance cancer research, protein mapping, DNA sequencing and neurobiology (Kawrykow et al. 2012; Peplow 2016). What Ranard et al. label 'surveillance systems' include strategies designed to elicit patient-contributed datasets, whether through project-specific portals or social media channels, which are sufficient to explore trends in such areas as disease outbreak (Smolinski et al. 2015), drug reactions (Salathé 2016) and risk factors for disease transmission (Garcia-Martí et al. 2016). Technologies developed for the Mark2Cure project, for example, engage volunteers in mining peer-reviewed journals to identify, annotate and curate relevant papers out of a broad literature with overlapping acronyms (Tsueng et al. 2016). Innovations are not all technological – community-based participatory research (CBPR), although far from new, continues to demonstrate the significance of collaborative learning where patients, patient advocates, health workers or at-risk communities help define research goals and processes (Wallerstein & Duran 2006). Innovations in CBPR include exploring opportunities for collaborative research to organise and mobilise concerned communities to take action around their health concerns (Cohen et al. 2016), and opening up avenues for qualitative methodologies in collaborative health research (Clark & Ventres 2016).

Policy implications in this domain may most easily be seen in CBPR work, where partnerships can help confront inequities in biomedical research and services (Israel et al. 2001) and are at times even specifically driven and directed by policy concerns (Themba & Minkler 2003). The fine line between researcher and subject in citizen science in the biomedical sphere has led to an extensive conversation around research ethics (del Savio, Buyx & Prainsack 2015; Kolman 2016; Vayena & Tasioulas 2016; Woolley et al. 2016). Work in this discipline has also helped confront and advance thinking on issues including privacy (Del Savio, Buyx & Prainsack 2015); patients' rights (Woolley et al. 2016); and even the concept of patient/public 'right to science' (Vayena & Tasioulas 2016).

Social science/humanities: Although it may be less common to think of social science and humanities research in relationship to citizen science, many of the same techniques are being employed and advanced to understand archaeology, literature, history and social dynamics. In a review of crowdsourced digital humanities research, Terras (2016) describes ways in which text and image analysis, transcription and annotation are helping to research, archive and make publicly available aspects of cultural heritage that might otherwise remain locked in museum basements or lost to time (as in the case of events and ephemeral art). Archaeologist Parcak (2015) highlights the opportunities for technology – specifically remote sensing – to document geopolitical events and conduct social and behavioural research via public access to satellite images and open mapping platforms. She is pioneering the use of aerial imagery to engage the public in identifying promising sites for archaeological exploration (Gewin 2016). Satellite observations can also facilitate monitoring and research of social conflict, human rights violations and the extent and impact of environmental disasters (Zastrow 2014; Notley & Webb-Gannon 2016). Innovative technology use is also enabling human-centred research, including studies of geographic trends of sexual behaviour (Davis et al. 2016) and correlating patterns of physical exercise with barriers to accessing outdoor spaces (Rosas et al. 2016).

Some projects and platforms in this domain are designed to have clear short- or long-term policy implications, such as to facilitate dialogue and transparency (Terras 2016) or direct action and advocacy (Rosas et al. 2016). Technology can improve understanding and management of issues of privacy (Davis et al. 2016), and can also raise concerns about equity in social research (Notley & Webb-Gannon 2016). Work with direct social implications reminds us that 'the crowd' (which includes scientists, per Parcak 2015) has interests and a stake in outcomes, and therefore scholars in this domain are working to deepen understandings of how politics and objectivity are approached in relation to research and public engagement (Notley & Webb-Gannon 2016).

Transferring innovations to advance work across disciplines

The development of citizen science in diverse disciplinary contexts has implications for the larger field. While some insights and innovations are disciplinary-, context- or project-specific, many may be transferrable to other settings. Opportunities are plentiful for advancing work by transferring innovations, and examples can be seen in terms of technology, computational strategies, engagement approaches and the practice of research itself (for the practical implications, see also Williams et al. in this volume).

Technology transfer: The rapid diversification of projects on the Zooniverse platform is a primary example of technology transfer. This platform for digital image classification, designed for public processing of astronomical images, is now employed for marine science, climatology, cancer research and more (see, for example, Tinati et al. 2015). Terras (2015) points to Zooniverse as a model platform for technologies developed to enable cultural heritage research. Hardware technologies are also transferable. Sensors in smartphones, smart watches and elsewhere allow data to be captured and shared in almost any setting (for example, a phone camera can document both species sightings and cosmic ray strikes) – a 2016 *Nature* article by Cartwright offers cross-cutting advice for scientists in any discipline who are looking to leverage these tools. Where hardware tools are not available or accessible, participants have built them – tools developed by do-it-yourself community scientists to enable community-based monitoring are now being adopted by professional researchers because of their quality and affordability (Dosemagen 2017, personal communication; and see also Volten et al. in this volume). The US Forest Service also notes that public engagement in research helps with technology transfer to private landowners/resource managers otherwise left behind as the industry rapidly advances (McKinley, Briggs & Bartuska 2013).

New computational approaches: The complexity of many citizen science datasets has led to innovative applications of data analysis techniques that have utility far beyond the discipline in which they were developed. Hochachka et al. (2012) describe the early application of sophisticated 'big data' statistical analysis and modelling techniques to citizen science in ornithology, and outline the development of new, 'semi-parametric' techniques that have particular utility for any citizen science analyses where limited assumptions can be made about individual data points. Algorithms developed for analysis of data from the Zooniverse platform provide avenues for reaching consensus on image classification within large datasets, based on the consistency of a user's contributions – where choices are binary, consistent annotations are useful regardless of whether they are consistently right or consistently wrong (Shamir, Diamond & Wallin 2016). Other transferable citizen science techniques include advances in machine learning that help identify and remove data 'noise' caused by glitches (Zevin et al. 2017), improvements in pattern recognition to automate photo identification of species (Andrzejaczek et al. 2016), and new developments in protocols that enable the repurposing of volunteered geographic information if it has been collected as vector data (Mooney et al. 2016). Bridging data analysis and policy, decision support tools have also been developed to help make sense of complex data in direct relationship to policy needs and priorities (Sullivan et al. 2009).

Opening engagement: Innovative projects continue to engage the public in new ways and in new aspects of research, which can create or enhance engagement opportunities in other disciplines. Research by Tinati et al. (2015) across the Zooniverse platform suggests that the same basic engagement strategies are applicable across the platform, regardless of research discipline; in addition, they suggest that their most valuable insights and advances came from recognising and enabling the work of volunteers as peers in conducting investigations. The FoldIt project revealed the value of inviting non-scientists to assist with scientific problem-solving. In one of the first major publications to document the success of this platform for collaboratively intuiting the structure of protein molecules, the authors (including both project leaders and solvers; Khatib et al. 2011) suggest that similar online game strategies can engage people in solving other complex problems (see also Novak et al. in this volume). Non-scientists are assisting with literature searches (Tsueng et al. 2016), developing scientific tools and instrumentation (e.g., Ottinger 2016) and participating in statistical analyses (Alliance for Aquatic Resource Monitoring 2010), most of which represent new frontiers for engagement which could be relevant to any research area. More directly connected to policy prospects, Tucker et al. (2016) present a method of 'speed dating' to match academic researchers and community leaders according to common interests and to develop collaborative research proposals. Whether the research topic is earthquakes or human rights violations, projects are also advancing response times and refining mechanisms, not just to collect data, but also to provide data tailored to inform decisions (e.g., Notley & Webb-Gannon 2016).

Implications for the practice of science itself: The collaborative nature of citizen science invites new considerations about how science is accomplished and what kinds of practices make science effective (Wyler & Haklay in this volume). Some originally disruptive aspects of citizen science have begun to shape the broader scientific landscape. For example, Franzoni and Sauermann (2014) suggest that the unconventional willingness of what they call 'crowd science' initiatives to publish intermediate results may speed innovations, in contrast to traditional research where findings are published only as a culmination of research efforts. Citizen science projects have also helped to bolster movements in open data and open-access publishing. In any domain, citizen science is helping to advance how to define, facilitate and document quality across science done by any-one, reminding all researchers of the responsibility to not take data quality for granted (Newman, Roetman & Vogel 2015). Citizen science can also offer a means for pursuing integrated research such as investigations of coupled human/natural systems (Crain, Cooper & Dickinson 2014), for example cases where livelihoods and natural resources are interdependent. Finally, where policy outcomes are an impetus for public engagement, citi-zen science can help focus research efforts towards garnering knowledge that provides a basis for specific actions (McKinley et al. 2017), such as whether or not to implement a treatment for the problem at hand.

Conclusion: Implications for citizen science as a field of practice

Looking at advances within distinct research disciplines, and their trans-ferability to other contexts, shows how opportunities for cross-disciplinary networking can enhance the practice and appreciation of citizen science more broadly. Citizen science is necessarily disruptive, and is already changing how science takes shape both within scientific institutions and in communities. An inclusive community of practice, spanning diverse disciplines and definitions, can facilitate both a more rapid uptake and adaptation of relevant technologies, and bring research approaches to new purposes. Cross-disciplinary networking can also help advance prac-tice regarding concerns shared across all disciplines, such as issues of ethics, democratisation, participation and policy (e.g., Silka 2013). The Ten Principles of Citizen Science call out these and other ideas that are broadly applicable, no matter the citizen science setting (Robertson et al. in this volume).

Cross-disciplinary work can also aid citizen science by demonstrat-ing the broad *social* and *scientific* significance and relevance of public engagement. Citizen science research within any disciplinary domain is well-served when it can leverage past successes to have the greatest impact, in ways that elevate the robustness of the research, the opportu-nities for meaningful public engagement and the relevance for policy. It is also critical for the field of citizen science as a whole to reveal and promote exemplar cases from all disciplines. This will help all stakeholders (including scientific peers who do not themselves use citizen science) understand and appreciate the value of citizen science as well as the investments necessary – in science, engagement and policy applications – to ensure its success. Connecting across disciplines offers the opportunity to draw strength from others' successes as well as lessons from their innovations, and from how they creatively advance science in relation to public interests and policy concerns.

Parcak (2015) points out that scientists increasingly want to see their research make a change in the world. The utility of citizen science for policy-making may depend, according to Vann-Sander, Clifton and Harvey (2016), on moving beyond a 'science-centric' view of citizen science. This recommendation must not be mistaken as being about moving *away* from the science in citizen science, as this may risk abandoning the rigour of scientific practices and outcomes (whether those practices involve monitoring, analysis, tool-building or cataloguing) that inform policy and even the motivation driving and serving public participation. Rather, Vann-Sander, Clifton and Harvey allude to an opening up and broadening of science to include attentiveness to the multiple interests and relationships that converge through citizen science, and which are necessary to engage in effecting policy change.

Just as with the practice of science more generally, citizen science has a unique character in each different discipline, but in all disciplines, citizen science initiatives demonstrate a shared, fundamental appreciation for the process of observation and inquiry in pursuit of verifiable knowledge gains. Citizen science helps expand the pool of collaborators and knowledge contributors who engage in this process, and in doing so, can engage broader public insights and concerns, inform the policy process with more complete and relevant datasets, and bring the process of knowledge generation more closely into conversation with the policy process and issues relevant to that process. At its best, science, policy considerations and public engagement are mutually reinforcing in policy-relevant citizen science. Divorcing the science from citizen science would be a disservice to the commitments and expectations of contributors, diminish its significance for policy and limit the ways that science and citizens intersect to inform new approaches to research. It is possible to imagine a future that moves *beyond* the 'science-centric' view of citizen science in ways that maintain the integrity and utility of the science at the heart of citizen science, in service to policy and an engaged public.

4
Participatory citizen science

Muki Haklay[1]

[1] *University College London, UK*

corresponding author email: m.haklay@ucl.ac.uk

In: Hecker, S., Haklay, M., Bowser, A., Makuch, Z., Vogel, J. & Bonn, A. 2018. *Citizen Science: Innovation in Open Science, Society and Policy*. UCL Press, London. https://doi.org/10.14324 /111.9781787352339

Highlights

- Common conceptualisations of participation assume high-level participation is good and low-level participation is bad. However, examining participation in terms of high and low levels of knowledge and engagement reveals different types of value in each case.
- The spectrum of citizen science activities means some are suitable for people who have education and knowledge equivalent to PhD level, while some are aimed at non-literate participants. There are also activities suitable for micro-engagement, and others requiring deep engagement over time.
- Issues of power, exploitation and commitment to engagement need to be explored for each citizen science project, as called for by the ECSA Ten Principles of Citizen Science, in response to the need for a more nuanced view that allows different activities to emerge.

Introduction

Participation is a potent term in citizen science. In fact, it was suggested that the field should instead be called 'Public Participation in Scientific Research' (Bonney et al., 'Public Participation', 2009) with 'participation' as the differentiating element between what is now called citizen science and public engagement with science. In another example, Cooper and Lewenstein (2016) discuss two meanings of citizen science: at one end of the spectrum, they discuss 'democratic' citizen science, which originates from a book by Alan Irwin (1995) and emphasises the responsibility of science to society; while at the other end of the spectrum they describe 'participatory' citizen science as practice in which people mostly contribute observations or efforts to scientific enterprise, which originated with the work of Rick Bonney (1996) at the Cornell Lab of Ornithology. The current author's previous contribution on this topic, which focuses on a typology of participation (Haklay 2013), also attempts to bridge the 'participatory' and 'democratic' meanings of citizen science.

The term 'participation', however, remains open to multiple interpretations and, arguably, to abuse. A good example of this is the area of participatory sensing, which originated from an attempt to bring together the two meanings Cooper and Lewenstein identified. As Burke et al. (2006, 4) noted,

> Participatory Sensing begins and ends with people, both as individuals and members of communities. The type of information collected, how it is organized, and how it is ultimately used, may be determined in a traditional manner by a centrally organized body, or in a deliberative manner by the collection of participants themselves.

Despite this definition, participatory sensing is now often used to describe activities in which the people who carry out the activity have little, if any, control over the process and activity, taking it more towards the idea of people as simple bipedal sensing platforms (Nold & Francis 2017). This is not to say that participatory sensing cannot be redeemed, only that much more attention must be paid to what participation means (Haklay 2016a).

In the fields of geography, environmental studies, urban studies, development studies and public policy among others, Sherry R. Arnstein's (1969) 'A Ladder of Citizen Participation' has to many defined the meaning of participation in political and technical processes. 'Arnstein's ladder', as it became known, uses value-laden terms to describe the potential of participation. Arnstein starts her analysis with levels of 'nonparticipation', including manipulation and therapy, then moves to 'degrees of tokenism' with informing, consultation and placation, and finally reaches 'degrees of citizen power' with partnership, delegated power and citizen control. Even without going into the meaning of these levels, it is clear that Arnstein offers a strong value judgement, in which non-participation should be frowned upon, while full citizen power is the goal. Knowingly simplified, the ladder focuses on political power relationships, and it might be this simple presentation and structure that explains its lasting influence. While it has been challenged over the years (see, for example, Chilvers &

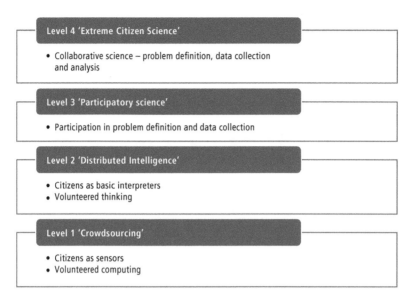

Fig. 4.1 Levels of participation in citizen science (Haklay 2013)

Kearnes 2016), Arnstein's ladder led to the development of other typologies (e.g., Wiedemann & Femers 1993). The current author is also responsible for what might, at first sight, seem to be a ladder of participation in citizen science (see figure 4.1).

From the current author's perspective, however, 'unlike Arnstein's ladder, there shouldn't be a strong value judgement on the position that a specific project takes. At the same time, there are likely benefits in terms of participants' engagement and involvement in the project to try to move to the highest level that is suitable for the specific project. Thus, we should see this framework as a typology that focuses on the level of participation' (Haklay 2013, 116). Yet, the rest of the discussion in the same paper cannot be absolved from presenting the typology as a ladder: In terms of understanding participation in scientific research as involvement in all stages of scientific enquiry, level four is the most comprehensive, while level one is the most basic. It is, therefore, easy to confuse participation in the sense of taking part in different stages of a process, and the meaning of this act of participation for the participant and project owner. Instead, it necessary to understand participation more fully, and to consider what participation means in citizen science.

This chapter therefore highlights the complex nature of participation in citizen science activities, and the need for a nuanced, detailed analysis of who participates, and how. This is the first step towards a multifaceted consideration of the role of citizen science practices in society and in science. To demonstrate the complexity of participation in citizen science, the chapter looks at two characteristics: the education levels of participants, and the way participation inequality (also known as the 90-9-1 rule) shapes the time and effort participants invest in citizen science activities. Examining these aspects reveals new insights about participation and the chapter closes with further direction on inquiry into participation.

Citizen science participation and levels of education

Among the technological and societal trends that have enabled the growth of citizen science in the past decade, increasing levels of education should be considered one of the most significant (Haklay 2013). According to Eurostat (2016), across the EU28 countries nearly 33 per cent of the population aged between 25 and 55 has tertiary (university) education, and 20 per cent of those above this age group also have tertiary education. This headline figure masks a wide variability based on the cultural and economic context of each country. For the 25 to 55 age group, in the UK 43.8 per cent hold tertiary education, Spain 38.4 per cent, France 38 per cent, Poland 32.7 per cent, Germany 28.3 per cent, with Italy the lowest with 19.1 per cent. This is part of a global trend, with UNESCO statistics recording about 200 million students currently studying in tertiary education across the world, of which about 2.5 million (about 1.25 per cent) are studying to doctoral level (UNESCO 2016). UNESCO statistics show that while participation in tertiary education in developed countries increased from 35.9 million people in 1999 to 46.8 million in 2014, participation at the doctoral level increased more moderately, from about 985,000 to about 1,343,000 people over the same period, remaining steady at about 2.8 per cent of students. The reason for paying attention to developed countries is that they are the locations where most citizen science happens.

Based on these statistics, if participation in citizen science was spread evenly across the population, about one third of participants would be expected to have tertiary education, and about 1–2 per cent to have a doctoral degree. Yet, the evidence points to a different picture. In Galaxy Zoo, a project in which participants classify galaxies and help astronomers to understand the structure of the universe, 65 per cent of participants had tertiary education and 10 per cent had doctoral-level degrees (Raddick et al. 2013; results confirmed by Curtis 2015). Curtis (2015) also found that in FoldIt (https://fold.it/), a project solving puzzles about the structure of molecules, 70 per cent of participants had tertiary education; while in Folding@home (https://folding.stanford.edu), in which people share their computing resources with scientists seeking to understand the structure of molecules, 56 per cent had tertiary education. In OpenStreetMap (https://www.openstreetmap.org/), which aims to create a free, editable digital map of the world, 78 per cent of participants hold tertiary education, with 8 per cent holding doctoral-level degrees (Budhathoki & Haythornthwaite 2013). Finally, Transcribe Bentham (https://www.ucl.ac.uk/transcribe-bentham), a digital humanities project in which volunteers transcribe the writing of nineteenth-century English philosopher Jeremy Bentham, 97 per cent of participants have tertiary education and 24 per cent hold doctoral-level degrees (Causer & Wallace 2012). While these findings are expected, it is clear that as the task complexity increases, the participation of people with higher levels of education increases – for example, Transcribe Bentham requires familiarity with a challenging transcription interface, and knowledge and interest in nineteenth-century philosophy. Across projects, the participation of people with tertiary education is at least twice the level in the general population, and the participation of people with doctoral-level education is at least three times higher.

This evidence can be interpreted in both a positive and negative light. Positively, the population with higher education has received more societal resources due to their longer period in education and deferring the period in which they are contributing to the economy and society through full-time employment. Those with doctoral-level education arguably benefited from this even more due to their longer and more specialised studies. Therefore, the opportunity to contribute to shared knowledge by volunteering to citizen science projects should be seen as a way to harness the knowledge, skills and abilities of those with higher education for a socially beneficial outcome. On the other hand, the numbers tell us that citizen science projects, even those that are based on micro-tasks and allow for a lighter level of engagement, are not reaching the wider population, and especially not enough of those without tertiary education. They are therefore not engaging across all sectors of society.

Citizen science and participation inequality

This section turns to the second characteristic of citizen science: the issue of participation inequality (see also Haklay 2016a). Participation inequality was first recognised by Hill and his team (1992) while analysing the contribution of different people to the development of digital documents. It manifests in online forums such as mailing lists, discussion forums and games (e.g., Hill et al. 1992; Mooney & Corcoran 2012; Lund, Coulton & Wilson 2011; van Mierlo 2014). It is also common in citizen science projects, such as in the iSpot (https://www.ispotnature.org/) community in which participants share images, locations and details of observed species (Silvertown et al. 2015), where the participation inequality is evident both among those who collect and share data, as well as those who classify and identify them. Across these projects, the proportion of registered people who do not contribute can reach 90 per cent or even more of the total number of participants, especially if we look at those who use the information without contributing to it. Of the remaining participants, 9 per cent or more contribute infrequently or fairly little. Finally, the last 1 per cent contribute most of the information. The phenomenon has therefore been framed as the 90-9-1 rule (Nielsen 2006). However, participation can be very skewed. As Nielsen point out, in Wikipedia, 0.003 per cent of users contribute two-thirds of the content, with a further 0.2 per cent contributing infrequently, making the relationship 99.8-0.2-0.003 per cent. For OpenStreetMap, Budhathoki (2010) suggests that the proportions are 70-29.9-0.01 per cent. Figure 4.2 is an update for Budhathoki's analysis by Wood (2014), showing that when participants in the OpenStreetMap project are sorted by their contribution, a small group of 1,000 participants dwarfs the effort of all other contributors, and only about 300,000 participants contribute more than 10 points of data although at the time there were two million registered users.

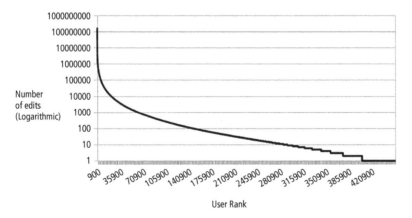

Fig. 4.2 OpenStreetMap contributions (Wood 2014)

Again, this pattern of participation has positive and negative aspects. On the positive side, some participants are highly committed and not only are they contributing to the project, but they are becoming experts in the scientific area of the project and developing many other skills (Jennett et al. 2016). As examples from biological observations show, the level of taxonomic expertise offered by dedicated volunteers can match that of credentialed experts. Notice the contrast between the educational attainment aspect noted in the previous section and the expertise gained through participation. As noted above, many projects have disproportional participation by people with credentials, such as PhDs. Yet, there are also high levels of participation by those with expertise and knowledge equivalent to credentialed experts, although not officially secured through an educational establishment. On the other hand, participation inequality demonstrates that most volunteers' depth of engagement is limited and even in the most successful projects that report high numbers of registered participants, the actual number of those who are highly engaged is small.

Knowledge and engagement in citizen science

Participation in citizen science can also be examined along many other axes: gender, ethnicity, socio-economic status or location, to name but a few. However, even for the two characteristics considered above, educational attainment and participation inequality, a complex picture of participation emerge.

For the sake of the analysis, there are citizen science projects that require a high level of knowledge from participants, so that they can understand the goals and terminology, and significant engagement, so that they can be trained to an appropriate level to ensure their effort is not wasted. In Transcribe Bentham, for example, a participant is expected to learn how to read Bentham's handwriting as well as how to use the transcription system (which does not have a user-friendly interface). Then, there are projects that require a fairly high level of knowledge but not necessarily demanding participation. Many Zooniverse projects (which share the same technological platform and principles as Galaxy Zoo mentioned above – see https://www.zooniverse.org) fall into this bracket, as they are aimed at people with some interest in astronomy but the tasks are fairly structured and short. Next, there are projects aimed at people with little knowledge but requiring high engagement. The Extreme Citizen Science group (www.ucl.ac.uk/excites) or Cybertracker foundation

Table 4.1 Engagement and skills in citizen science projects

	High engagement	Low engagement
High level of knowledge	• Highly valuable effort: research assistants • Significant time investment • Opportunities for deeper engagement (analyses, writing papers)	• Skills might contribute to data quality • Possible use of disciplinary jargon • Opportunities for lighter or deeper engagement to match time/effort constraints
Low level of knowledge	• Opportunities for education, awareness raising, other skills • Support and facilitation are necessary	• Opportunities for active engagement with science with limited effort • Potential for family/ cross-generational engagement • Outreach to marginalised groups • Potential for large temporal and spatial coverage and contribution to science

(https://www.cybertracker.org/) both work with non-literate participants recording local resources and are examples of this. Finally, there are projects aimed at participants with little knowledge and requiring low engagement. This includes science outreach projects, such as Open Air Laboratories (OPAL https://www.opalexplorenature.org/), which are structured around episodes of public engagement. The classification of projects is presented in table 4.1.

There are benefits and challenges to participation in each and every block of this classification, as described below.

High level of knowledge/high engagement: As noted above, these projects provide a way to harness highly valuable knowledge with participants acting as volunteer research assistants. There is a significant time investment by the volunteers. While in some projects high participation is linked to higher levels of education (such as doctoral level), this is not the case more generally, and the high time investment leads to the development of expertise by these dedicated volunteers. Their deep familiarity with the material brings important insights, which project managers would be wise to capture through opportunities for deeper engagement in the scientific process, such as assisting in the analysis or co-authoring papers.

High level of knowledge/low engagement: A key benefit here is the impact of well-educated participants on the outcomes of the project. For example, understanding of the principles of the scientific process can contribute to data quality, since participants can understand what the project owner is trying to achieve and the importance of rigour in carrying out the task. It can also allow the use of disciplinary jargon in the explanations and instructions to participants. The opportunities here are for lighter or deeper engagement to match time/effort constraints. This can be valuable for people who have many demands on their time but want to contribute to scientific efforts without high investment in training and learning.

Low level of knowledge /high engagement: These activities need to address the lack of representation of participants with low educational attainment. A focused effort by project organisers can provide an opportunity to educate participants, raising awareness of environmental and scientific issues, increasing knowledge of the scientific process as well as other skills such as experience with Information and Communication Technologies (ICT) or in community organisation, which can empower participants in other aspects of their lives. These projects require ongoing and targeted support for participants but can demonstrate the high potential for inclusivity in citizen science.

Low level of knowledge/low engagement: These projects bring opportunities for active engagement in science to a larger audience, which is usually not engaged in citizen science, with limited effort by the participants. Lack of awareness means these projects might still require effort from project organisers to encourage people to join the activities. This type of activity holds the potential for family and cross-generational activities that can engage parents, grandparents and children in a joint learning and exploration experience, as well as outreach to marginalised groups. Overall, this type of project can also provide a stepping stone for more in-depth participation, if desired by participants and with active encouragement by facilitators. When these types of citizen science projects are well structured and designed, they can also lead to scientific advances, especially when large spatial or temporal datasets are needed (see Palmer et al. in this volume on the Mosquito atlas, https://www.mueckenatlas.de), or image recognition in the project Chimp&See (https://www.chimpandsee.org).

In summary, this classification shows that there are important societal benefits for participation in each type of project. Scientific impacts depend on the project design and can also be attained in various forms depending on the different project types (Ballard, Phillips & Robinson in this volume). Simplistic assumptions that only full inclusion at a deep level is appropriate for citizen science projects should be avoided. Instead, they should consider how people at all levels of education and engagement gain from, and contribute to, citizen science activities.

Conclusion

This analysis argues that Arnstein's ladder should not be taken uncritically as a model for participation in citizen science, but some of its dimensions should still be considered carefully. In contrast to Arnstein's ladder, participation should be valued at many levels – from occasional contribution to deep engagement in shaping research projects and carrying them out from start to finish. Different people, with different life histories, interests and responsibilities, need the opportunity to engage in different levels of participation in citizen science. Projects should also facilitate the opportunity to move between different levels of engagement at different stages in participants' lives.

Yet, Arnstein's ladder highlights an important consideration. The core of her argument was about control and power, and citizen science can open up situations in which the effort of participants is exploited, or in which projects are conceived without allowing participants to develop deeper engagement even if they wish to do so. We should guard against such issues, and the Ten Principles of Citizen Science (Robinson in this volume) address these challenges.

From the policy perspective, the analysis of participation reveals different opportunities in each type of citizen science activity – from harnessing the knowledge of highly educated members of society, to opening new avenues for science and technology to people with limited education and lots of demands on their time. It is therefore important to support a wider range of activities within citizen science, to ensure the inclusion of people across society.

Participation in citizen science is therefore a complex and multifaceted issue that requires attention, research and theorisation. Understanding participation in citizen science also develops better understanding of the ways in which open science should operate, the importance of open-access publications to allow participants to develop their knowledge, and the need to support participants through their scientific journeys.

Acknowledgement

This chapter is based on a keynote for the first European Citizen Science Association conference, with the same title as this chapter, 'Participatory Citizen Science'. The research that provides the basis for this chapter was funded through the Engineering and Physical Science Research Council 'Extreme' Citizen Science project (EP/I025278/1), Street mobility (EP/K037323/1), European Commission FP7 Citizen Cyberlab (317705), European Commission H2020 Doing It Together Science (709443) and European Research Council ECSAnVis grant (694767).

5
Technology infrastructure for citizen science

Peter Brenton[1], Stephanie von Gavel[1],
Ella Vogel[2] and Marie-Elise Lecoq[3]

[1] *Atlas of Living Australia, CSIRO, Canberra, Australia*
[2] *National Biodiversity Network Trust (NBN), Nottingham, UK*
[3] *GBIF France, MNHN Géologie, Paris, France*

corresponding author email: Peter.Brenton@csiro.au

In: Hecker, S., Haklay, M., Bowser, A., Makuch, Z., Vogel, J. & Bonn, A. 2018. *Citizen Science: Innovation in Open Science, Society and Policy*. UCL Press, London. https://doi.org/10.14324/111.9781787352339

Highlights

- Information technology (IT) infrastructure is a vital enabler of successful citizen science projects.
- There are numerous IT tools available to citizen science projects and navigating them can be confusing. When choosing tools, it is important to consider their compliance with applicable process and data standards, their ability to connect with the information supply chain and their fitness for the required use.
- The information and data generated by citizen science projects is likely to be their most enduring and impactful legacy if they are made publicly accessible in a timely manner and in a form which is suitable for multiple downstream uses. To do this, they need to conform as much as possible to existing data and process standards.

Introduction

The chapter considers what infrastructure means in a citizen science context and characterises the types of technology-based infrastructure being used by the global citizen science community, with a focus on the environmental domain. Some issues emerging around the application of different infrastructure solutions in current use are also raised and, using some examples and case studies, existing infrastructure solutions are discussed in an 'information supply chain' framework. An information supply chain refers to the process flow or movement of a piece of information (data) from being acquired or collected, to being used in one or more transformative actions such as policy settings, physical management and/or educational or behavioural change campaigns. Invariably, this will also involve intervening processes on the data, potentially by parties other than the collectors, such as data curation, management, aggregation and analysis.

This chapter draws on the authors' experience and expertise in citizen science infrastructure in Australia and primarily in the environmental domain.

The notion of 'best practice' in the context of citizen science infrastructure is also considered, concluding that 'best practice' is relative to available solutions and practices at a given time and that it will inevitably change over time.

What is citizen science infrastructure?

The online version of the Merriam-Webster Dictionary defines 'infrastructure' as: 'the underlying foundation or basic framework (as of a system or organisation); . . . and: the resources (as personnel, buildings, or equipment) required for an activity'. Thus infrastructures are the physical structures, equipment and tools, processes, services, human capital and social networks which enable systems and enterprises to function effectively. In a citizen science context, this includes:

a. Physical kit – buildings, vehicles, telescopes, microscopes and binoculars, measuring instruments, cameras, scanners, sensors, drones and various other equipment;
b. Social assets – the organisers of projects, events and collaboration services, sponsors and funding bodies, the public participants in projects and events, and the social networks of connected individuals; and
c. Technology assets – the information technology–based platforms/tools and services used to collect, store, manage and process, share, visualise and analyse information (data and metadata) which is produced by citizen science endeavours, as well as those used to organise and manage citizen science projects and events.

This chapter deals only with the information technology–based infrastructures which support data produced by citizen science endeavours, not those used for stakeholder and event management (see also Wyler & Haklay in this volume on the infrastructure provided by universities).

Historically, data generated by citizen science projects – and indeed many non-citizen science projects, too – was often only used within the context of the project for which it was collected. However, aggregated data from multiple sources is becoming increasingly important as features of research work and as inputs to policy and management actions. It is therefore also useful to consider IT infrastructures which support citizen science in a broader context; that is, the role they play in the information supply chain. This helps us to understand the relevance and role of individual projects in contributing to new knowledge and improved management outcomes, and hence the significant role of public participation in this larger context.

Figure 5.1 shows a conceptual information supply chain model in which citizen science projects are involved in data acquisition and analysis processes. People use all sorts of tools and infrastructure to collect raw data which gets stored somewhere, usually in local databases or cloud services. However, raw data by itself has little intrinsic value or usefulness – raw data only has value and meaning when it is interpreted in conjunction with the context in, and by which, it was collected (see also Williams et al. in this volume).

The reasons for collecting raw data are many and varied, and include:

- Answering specific research questions or modelling and understanding real-world processes;
- Support social, political, environmental or economic objectives;
- Gaining personal satisfaction and fulfilment;
- Enhancing social opportunities;
- Connecting with nature; and much more.

Data aggregators procure and combine raw data from those who collect and produce it, and provide repositories in which data producers can proactively lodge their data. Aggregators typically transform inbound data to fit into a standardised data structure and add value to the raw data by providing a range of products and services to data producers and consumers.

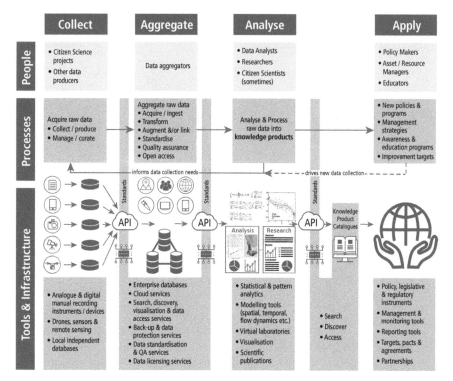

Fig. 5.1 A conceptual model for a digital information supply chain. (Source: Icon made from http://www.onlinewebfonts.com/icon fonts is licensed under CC BY 3.0)

Aggregated data is then accessed by data analysts and researchers who use tools and expertise to gain meaning and knowledge. This results in knowledge products, which can then be used to inform policy, planning and management decisions, facilitate assessment against national and international benchmarks and target measures, and many other applications. Information technology infrastructure is also used to make knowledge products more discoverable and accessible.

Policies and management actions invariably have impacts which require measurement and monitoring, which in turn drives further raw data collection. Outputs from analysis can also identify gaps in information and stimulate further focused raw data collection.

Data and procedural standards provide a common language which allows similar information from disparate sources to be efficiently aggregated and exchanged, thus giving raw data potential value, utility and impact beyond the purpose for which it was originally collected. Application interfaces (APIs) provide a simple mechanism for exchanging data between different electronic systems, facilitated by growing access to high-speed internet technologies. These are becoming increasingly important enablers by supporting 'linked data' and 'big data' approaches to understanding the complexity of the world and informing policy and management responses to complex global challenges (Ceccaroni, Bowser & Brenton 2017; Ottinger 2010).

Information technology infrastructure plays a significant and important role in the information supply chain by supporting human interactions with data, as well as enforcing standards, automating processes and performing computational functions (Wiggins et al. 2011; Newman et al. 2012), such as:

- Connecting and linking system components;
- Standardising data definitions so that there is a shared data language;
- Mobilising data from analogue (non-digital) and siloed digital systems into standardised digital formats which can be transported through and used by all tiers of the information supply chain;
- Supporting data curation and data quality improvement;
- Improving data flow and processing efficiency; and
- Much more.

Citizen science and IT infrastructure – a natural partnership

Internet and wireless technologies are enabling unprecedented access to scientific materials and facilitating mass public participation in science (Couvet et al. 2008; Hochachka et al. 2012; Newman et al. 2010).

Information technology platforms can codify and enforce rules and processes which help to improve the quality and hence the reliability, reusability and scientific trustworthiness of information generated through non-traditional scientific channels. Technology infrastructures are therefore an important enabler of citizen science and are arguably the single biggest factor driving the recent rise of citizen science and the democratisation of science generally (Nov, Arazy & Anderson 2011a).

However, it is difficult to keep pace with the constant and rapid changes in technology. Such changes generally bring improvements in usability, functionality, performance, reliability, accessibility, accuracy and precision, as well as new beneficial features. At the same time, they introduce more and potentially confusing options, creating a potentially bewildering technology landscape for citizen science project co-ordinators. The cost of hardware and sometimes software can also impede a project's uptake and benefits. Later, this chapter looks at some things to consider when choosing an IT solution for a project.

As evidenced elsewhere in this book and in many other published works, citizen science is a significant public good endeavour which provides numerous social, environmental and economic benefits in addition to enhancing science engagement and literacy. The important role of IT infrastructure in supporting citizen science makes it reasonable to consider issues such as:

- The role governments, non-governmental organisations (NGOs) and philanthropic organisations should play in facilitating access to, and reducing the cost of, technology infrastructures for citizen science;
- How citizen scientists can access and make the most effective use of technology to improve the efficiency, effectiveness, accuracy and impacts of their contributions to scientific endeavour; and,
- How technology can be used to demonstrate the impact of citizen science contributions on social, policy and management outcomes, and thus empower and enhance the engagement of the public in these areas, as well as to improve recognition of citizen science contributions in traditional science and policy circles.

Such questions are being addressed in numerous studies around the world such as Bonter & Cooper (2012); Couvet et al. (2008); Nov, Arazy & Anderson (2011a); Sequeira et al. (2014); Kaartinen et al. (2013); and many others.

The recent worldwide explosion in the number and scope of citizen science projects has seen a growing need to develop effective mechanisms to assist the public in finding, discovering and connecting with citizen science projects; and for project owners to promote and connect their projects with citizen scientists. This has resulted in the emergence of several independently developed 'Project Finders' – searchable project catalogues. Some of the open public facilities have become channels for citizen science projects to promote themselves, but with worldwide access and broadly similar functionality, they are sometimes perceived as competing with each other, which has led to some community confusion as to where they should register their projects. In an ideal world, any citizen science project registered in any catalogue system should be discoverable and accessible via any project finder – thus giving people the most comprehensive and current information possible about projects at their location, and allowing them to directly connect to projects of interest. To achieve this however, information needs to be shared between systems using common standards and protocols as described by Ceccaroni, Bowser & Brenton (2017) (and see also Williams et al. in this volume). To this end some key public catalogue managers are collaborating to develop a standard core set of data attributes for citizen science projects, as well as a standard data schema and data exchange protocols, known as the PPSR-Core project.

Information technology platforms, both desktop and mobile, facilitate vast networks of human observers, stationary autonomous sensors (e.g., camera traps, weather and environmental sampling stations, etc.), and mobile remote platform sensors (e.g., drones and satellites) to collect reasonably consistent quality spatial and temporal data. This enables large-scale spatial and temporal analyses of patterns and distributions which would otherwise be impossible using traditional scientific data collection methods (Sullivan et al. 2009). Some successful early examples of these in citizen science, such as eBird (http://ebird.org/) and Galaxy Zoo (https://www.galaxyzoo.org/), have become benchmarks for large-scale global citizen science programmes.

There are many tools currently available (box 5.1) and many more are likely to emerge in the future. This chapter does not endorse particular tools, but instead aims to illustrate the complex array of tools available. All tools have strengths and weaknesses and differ in their suitability for different projects and situations. In addition, significant gaps remain where infrastructure is not yet fully servicing the scope of requirements for technology support in the citizen science domain – for example, species identification in the biodiversity domain and portals focused on communities of interest more generally.

When choosing a tool, there are important factors to consider:

i. Is there an existing tool available at an acceptable cost? Why build a new tool when something suitable already exists or can be adapted to fit?

ii. Is the tool already connected or designed to connect and share with open data infrastructures? Most tools do not do this, but it is critical for data sharing.

iii. Are the data capture and storage structures compliant with domain-relevant standards? Most are not, and this is also critical for data sharing.

Box 5.1. Citizen science infrastructure tools

The citizen science sector has produced an impressive array of tools operating at varying spatial scales, as well as with different temporal scopes and topics of interest. These can be broadly categorised as follows:

1. **Project catalogues/finders** provide a central point of discovery and connection to citizen science projects. Examples include: CitSci.org (www.citsci.org); SciStarter (www.scistarter.com); Federal Crowdsourcing and Citizen Science Catalog (https://ccsinventory.wilsoncenter.org/); Zooniverse (https://www.zooniverse.org/projects?status=live); EU BON (http://biodiversity.eubon.eu/zh/web/citizen-science/view-all); and BioCollect (https://biocollect.ala.org.au/acsa). These facilities also support community engagement and, in some cases, data collection services. There are also commercial providers serving the citizen science community with data recording capabilities and small project catalogues.

   In addition, organisations which fund/sponsor projects often monitor their progress and have their own project catalogues, examples include: the European Commission's 'CORDIS' system (http://cordis.europa.eu/project/rcn/51266_en.html); the Alfred P. Sloane Foundation (https://sloan.org/search?q=citizen+science); the Myer Foundation (http://myerfoundation.org.au/grants/grant-finder/); the National Geographic Society (https://www.nationalgeographic.org/idea/citizen-science-projects/); and many other government, NGO and philanthropic organisations.

2. **Generic domain-agnostic tools** provide general data collection/capture capabilities for any type of science project – for example CitSci.org (http://citsci.org/cwis438/websites/citsci/home.php?WebSiteID=7); Zooniverse (https://www.zooniverse.org/); CyberTracker (http://www.cybertracker.org/); Fulcrum (http://www.fulcrumapp.com/?gclid=CMzT5IDyidICFQybvAodlWMCuQ); BioCollect (http://www.ala.org.au/biocollect/); and others.

3. **Generic domain-specific tools** provide general data collection/capture capabilities for projects within a specific area of science. There are many variations available (and a great

deal of non-compliance with standards), but typically these systems are based on a core domain-relevant data standard and/or schema such as Darwin Core (http://rs.tdwg.org /dwc/) in the biodiversity domain for species observational and collections data. Examples include: iNaturalist (http:// www.inaturalist.org/); iSpot (http://www.ispotnature.org /communities/global); Indicia (https://nbn.org.uk/news /instant-indicia/); Natusfera (http://natusfera.gbif.es/); and NatureMapr (http://naturemapr.com/). Some of these have also established large communities of users and include a range of different community-based mechanisms for verifying the accuracy and identifications of contributed records.

4. **Bespoke project-specific tools** are developed specifically for a particular project as either desktop or mobile apps, or a combination of both. Examples include: CrowdMag (https://www.ngdc.noaa.gov/geomag/crowdmag.shtml); Project Noah (http://www.projectnoah.org/); QuestaGame (https://questagame.com/); OPAL Water Survey (https:// www.opalexplorenature.org/WaterSurvey); and hundreds of others.

5. **Data transcription tools** are open platforms which facilitate crowd-sourced data transcription, enabling large amounts of data locked in analogue records to be mobilised as digital information and used in previously impossible ways. Such tools include DigiVol (https://volunteer.ala.org.au/); Trove (http://trove.nla.gov.au/); Notes from Nature (http://www .notesfromnature.org/); Ancient Lives (http://ancientlives .org/); Old Weather (http://www.oldweather.org/); the Smithsonian Transcription Centre (https://transcription.si .edu/); and others.

6. **Education, engagement and support tools** provide mainly look-up and read-only support information for specific domain areas. Examples include field guides and identification support apps such as versions of Australian museum–sponsored field guides to Australian fauna apps; various thematic versions of the Gaia Guide apps; the Waterbug App; various thematic Lucid key apps); etc. All of these are available in the Google Play and Apple iTunes app stores.

iv. How will the tool support the project and the community using it? Does it have all of the functionality and features required for the project? Can the project live with any deficiencies? Is it already used by similar communities elsewhere?

v. Is customisation required and how customisable is it?

vi. Does the tool have:

    a. A long-term future – is it sustained/maintained by an active community or vendor;

    b. A technology upgrade pathway; and

    c. User and/or technical support?

Best practice solutions

The Business Dictionary defines 'best practice' as: 'A method or technique that has consistently shown results superior to those achieved with other means, and that is used as a benchmark'. This assumes a static, or at least slow-moving environment, but technology is changing at a dizzying rate – therefore, this concept needs to be considered in the context of continuous improvement when it is applied to information technology.

Technology, like most things, does not stand still, it will always have innovators leading and pushing the boundaries of what is possible in both hardware and software, as well as early-adopter consumers with needs to be met that current solutions do not satisfy. It is both an enabler and supporter of current needs as well as a driver of new needs, because as new technologies fulfil current needs it is possible to see opportunities and applications for even newer innovations and technologies. In a nutshell: Innovators envision needs beyond the horizon and push the boundaries of the present; early-adopters consume innovations and through demand, fuel even more innovation; while old innovations become the new normal for the masses and old norms are displaced. This is how progress is made.

A multiplicity of different solutions is currently being independently developed to meet similar needs at different times and places, and the whole scene is constantly evolving. Therefore, the concept of 'best practice' solutions are only ever relative to a given point in time, essentially reflecting the solution available at a given time which best meets the requirements and needs of a demographic of consumers/users at that time.

There is unfortunately a long way to go to realise the goal of a fully connected and functioning information supply chain, but progress is being made by many dedicated people around the world. There is also a growing enthusiasm and commitment amongst many of the major global infrastructure providers to collaborate more effectively to deliver more unified (interoperating) and integrated technology platforms, as well as to build a global community of practice to maintain and enhance the platforms in the most cost-efficient and impactful ways possible (see also Williams et al. in this volume). For example, Australia's national biodiversity data aggregator, the Atlas of Living Australia, with the support of the Australian government, has developed a suite of current best-practice tools and made them freely available worldwide under open source licences.

Case studies below highlight how this 'Living Atlas' software platform (box 5.2) is now being adopted by other countries (boxes 5.3 and 5.4) and is facilitating major improvements in data quality; data mobilisation and processing efficiency; and data accessibility and reuse; as well

---

Box 5.2. Case study – The Atlas of Living Australia (ALA – www.ala .org.au)

Stephanie von Gavel, ALA Business Development

The Atlas of Living Australia was established by Australia's premier research body the Commonwealth Scientific and Industrial Research Organisation (CSIRO) and many partner organisations including museums, biological collections, research organisations and

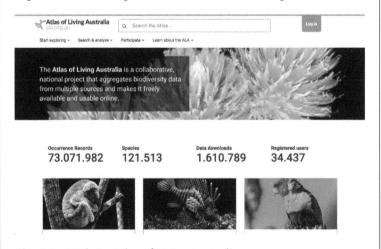

Fig. 5.2 Website Atlas of Living Australia

(continued)

government (state and federal) to provide a consistent comprehensive single point of access for Australia's biodiversity data and species information. It is funded by the Australian government via its National Collaborative Research Infrastructure Strategy (NCRIS) and is hosted by the CSIRO.

This web-based infrastructure comprises a modular suite of inter-connected databases, web applications (tools), APIs, and mobile apps. Data, which is not owned by the ALA, is also part of the infrastructure. The tools support the capture, aggregation, management, discovery, visualisation and analysis of all classes of biodiversity information. They are used for a wide range of purposes, including research, biodiversity discovery and documentation, environmental monitoring and reporting, conservation planning, biosecurity activities, education and citizen science. In addition, external enterprises and organisations are using the open infrastructure to create and enhance their own products and services. For more information on the Living Atlas platform, see http://living-atlases.gbif.org/.

Prior to the ALA, a major barrier to Australia's biodiversity research and management efforts was fragmentation and inaccessibility of data. Information was generated and siloed, housed in museums, herbaria and other collections; universities; research organisations; and government agencies, as well as with individual citizen scientists and researchers.

The ALA brings together biodiversity data and associated information from a wide variety of sources, processing and linking it together, and making it accessible from a single place in a standard format via a set of purpose-built tools and services. Accessing biodiversity data is now free and more efficient than ever before, as the ALA has already addressed a wide range of data access issues for all consumers which would otherwise have to be negotiated individually by each data consumer.

The ALA is the Australian node of the international open-data infrastructure the Global Biodiversity Information Facility (GBIF). The ALA has also 'open sourced' its software as the Living Atlas Platform to encourage the development of a collaborative community of practice around the infrastructure, and to facilitate interoperability and cost savings to the global biodiversity community. Accessible and affordable technology platforms empower and enable people to participate more actively in biodiversity knowledge-building activities. This democratises biodiversity science and develops fascination and enquiry among the next generation of scientists.

As an exemplar for open infrastructure, open data and data reuse, the Living Atlas platform is being adopted and used by an ever-increasing number of organisations both domestically and internationally. The data available via the ALA are also being used for a multitude of purposes. Atlas of Living Australia tools provide capability in many areas across the spectrum of the information supply chain.

The Living Atlas platform supports many different systems – whether they be separate instances of the software suite, or hubs (different thematic interfaces over one common instance of the platform). Open APIs also allow others to independently access data and some data processing services.

The ALA is a strong supporter of citizen science and has part-nered with the Australian Citizen Science Association (ACSA) to provide the national citizen science projects catalogue. The Atlas of Living Australia also directly supports numerous projects collecting data through the BioCollect tool (http://biocollect.ala.org .au/acsa). The project finder exchanges project information with the SciStarter (www.scistarter.com) system in the United States and, through the PPSR-Core initiative, various catalogues of citizen science projects are being progressively connected to enable fast and simple discovery and access to projects of interest from a comprehensive list of projects from around the world.

Box 5.3. Case study — GBIF France

Marie-Elise Lecoq, GBIF France, Systems Development and Support

The Global Biodiversity Information Facility (GBIF) is an international open-data infrastructure for biodiversity data. GBIF encourages and helps participant countries to publish and share biodiversity data to support international biodiversity research, inform pan-national policy and improve management outcomes

(continued)

Fig. 5.3 Website Global Biodiversity Information Facility France

for biodiversity, in other words, better decisions to conserve and sustainably use the biological resources of the planet.

The Global Biodiversity Information Facility operates through a network of collaborating nodes which share skills, experiences and technical capacity. Its vision is, 'A world in which biodiversity information is freely and universally available for science, society and a sustainable future'. To achieve this, GBIF provides and endorses tools which help publishers share their own data using standards such as DwC (Darwin Core). Experiences and developments made by GBIF nodes increased the list of GBIF tools with a set of reusable ones. GBIF France decided to work with software developed by the community, especially the ALA platform (see box 5.2).

This platform was chosen because it is a powerful infrastructure that has already addressed a lot of GBIF France requirements, meaning that work was only needed to install the system and add national specificities (language, data, design, etc.). As a result, the French portal of GBIF France was established within a year and was later enhanced with the addition of the ALA spatial portal. GBIF France developed and optimised their performance to produce an attractive feature-rich portal within two years. Due to the efficiency of the development, GBIF France decided to participate more in growing and supporting the community around ALA modules.

Table 5.1 Countries currently using the Living Atlas platform

Country	Link to Platform
Australia	http://www.ala.org.au/
Argentina	http://datos.sndb.mincyt.gob.ar/
Brazil	https://portaldabiodiversidade.icmbio.gov.br
Costa Rica	http://www.crbio.cr/crbio/
France	http://portail.gbif.fr
Spain	http://datos.gbif.es/
Portugal	http://dados.gbif.pt/
UK	http://www.als.scot/ and https://nbnatlas.org/

The commitment of ALA, GBIF France, GBIF Portugal, GBIF Spain and several others to this community has multiple forms. Since 2013, a technical workshop held at least once a year has presented ALA modules to new users, to improve existing data portals and to learn from others' successes and achievements. For instance, GBIF France was the first outside the ALA team to install the spatial portal and gave feedback on this experience to the growing 'Living Atlas' community at the workshops. The meetings are motivating for new users because they can see that they can gain a powerful tool for themselves and for other participants with relatively little time and investment. Indeed, during training, technical teams get ideas from other projects and can also complete significant work on their own project. Community members have also shown the result of this collaborative work through presentations and posters at international conferences around the world. Finally, the international community around ALA have helped other institutions who do not have the technical competencies to implement their own data portal, especially in Africa.

Thanks to these engagements, seven data portals using the ALA platform were released between 2014 and 2016 (table 5.1), with several others currently in development and more investigating its use.

This ALA community is therefore helpful for organisations or associations who want to install a data portal but do not have the technical competence or staff to do so.

Box 5.4. Case Study – NBN, United Kingdom

Ella Vogel, NBN UK, Programme Development and Support

The UK National Biodiversity Network (NBN) has a long history of activity in biological recording and citizen science. In 2015 it undertook a review of its online data-sharing infrastructure and concluded that the current system was no longer fit to serve the growing needs of the Network. Three options were considered: (1) Develop a new platform from scratch; (2) re-engineer and enhance the existing platform to accommodate required functionality; and (3) adopt an existing platform to replace the old system. When the ALA open source platform (see box 5.2) was presented to the NBN Secretariat, it was clear that the most time and cost-efficient way to move forward was to adopt this infrastructure in the UK.

The pilot, NBN Atlas Scotland, was launched in 2016 as the precursor to the new core NBN Atlas. Implementation of the Living Atlas platform has enabled the UK to shift its attitude to data accessibility to being more open with improved data sharing both within the UK and globally. Previously, record sharing via the NBN Gateway was done under a bespoke NBN Data Exchange Format. Within the UK this worked well, but with a more global outlook it is important that common and interoperable formats are used. Data can now be shared both within the UK and internationally using com-

Fig. 5.4   Website National Biodiversity Network, UK

mon Darwin Core–based standards. The new system has also encouraged the use of creative commons licences, allowing datasets to be more easily used by others domestically and internationally in research, policy and planning at any scale.

Over the years, many questions have been raised about how to mobilise historic datasets; how to empower citizen scientists to collect biological records in a transparent, consistent and peer reviewed way so that their efforts are seen as equal alongside the work of professionals; how to provide access to biological records by network members; and how to combine datasets and data layers to undertake detailed analysis, without having to each have access to separate tools and different systems to perform each step. The Living Atlas infrastructure has provided solutions to these and many other issues and has given the NBN a clear direction for future development. With a global developer base to contribute to and learn from, there is stability in the future of the Atlas platform and endless opportunities for growth and development.

as facilitating change in the way that people think about the whole information supply chain and the value of their data beyond the project that they used to collect it.

Conclusion

It is not the aim of this chapter to pick 'winners' among the large pool of current technology solutions serving the citizen science community. Instead, it aims to highlight that 'best practice' in technology is a rapidly moving target and that at any given time there will always be a range of old and new technologies, features, capabilities and costing models among the wide array of tools available. However, within this environment, there are some fundamental considerations for citizen science projects when choosing appropriate infrastructure solutions to support their needs. These choices can determine the real value of a project's outputs to downstream scientific endeavours and supply chain outcomes.

Arguably, notwithstanding the direct and sometimes profound personal, social and environmental benefits of public participation in scientific activities, the most enduring element – where public contributions to science will likely have their greatest impact – is the information which they generate. However, to be of real value, this information must be accessible to the information supply chain in a timely manner and in a form which is suitable for use throughout.

Therefore, the application of standards in data collection, data transmission, and the descriptions of datasets and collection methods are critical to scientists and policymakers accepting and giving proper value and respect to, citizen science data and the enormous volunteer commitment made by citizen science participants worldwide. Well-designed IT infrastructures, which include in-built processes and rules to enforce standards and data quality, as well as mechanisms for standards compliant data sharing, can fulfil such requirements with minimal impact on users. Solutions that include such features should therefore be chosen over those that do not. Such market-based demand-driven choices will encourage all infrastructure providers to engage with the standards framework, which is critical to a functioning information ecosystem.

Acknowledgements

I acknowledge and thank my colleagues at the Atlas of Living Australia, Hannah Scott and Paul Box (CSIRO), for providing editorial comments during the drafting of this chapter.

6
Evaluating citizen science
Towards an open framework

Barbara Kieslinger[1], Teresa Schäfer[1], Florian Heigl[2], Daniel Dörler[2], Anett Richter[3,4] and Aletta Bonn[3,4,5]

[1] *Centre for Social Innovation, Vienna, Austria*
[2] *University of Natural Resources and Life Sciences, Vienna, Austria*
[3] *Helmholtz Centre for Environmental Research – UFZ, Leipzig, Germany*
[4] *German Centre for Integrative Biodiversity Research (iDiv) Halle-Jena-Leipzig, Germany*
[5] *Institute of Ecology, Friedrich-Schiller-University Jena, Germany*

corresponding author email: kieslinger@zsi.at

In: Hecker, S., Haklay, M., Bowser, A., Makuch, Z., Vogel, J. & Bonn, A. 2018. *Citizen Science: Innovation in Open Science, Society and Policy*. UCL Press, London. https://doi.org/10.14324/111.9781787352339

Highlights

- Evaluation concepts for citizen science are required both by policy-makers, to improve citizen science funding schemes and by project initiatives, to enhance their project management.
- Citizen science programmes should be evaluated along three dimensions of participatory science: (i) scientific impact, (ii) learning and empowerment of participants and (iii) impact for wider society.
- Evaluation and impact assessment should embrace the diversity and emerging nature of citizen science.
- An open framework for evaluation can be adapted and tailored to the specific goals of citizen science programmes.

Introduction

An exponential rise in citizen science projects is currently taking place (Kullenberg & Kasperowski 2016), bringing innovation potential for science, society and policy (Holocher-Ertl and Kieslinger 2015). There are indications that citizen science contributes to transformational change in science and society through the formulation of new research questions by both members of the public and the scientific community and through the joint discovery of solutions to regional (e.g., Lee, Quinn & Duke 2006), national and even global (Theobald et al. 2015) problems of societal and scientific relevance.

As citizen science can contribute to learning about the processes of scientific enquiry and to a deeper understanding of scientific outcomes (Riesch & Potter 2014; Bela et al. 2016; Richter et al. 2016; and see Edwards et al. in this volume), it may lead to improved understanding, uptake and implementation of transparent and responsive research in society. In this way, citizen science is an approach that encourages stewardship, fosters empowerment and contributes to Responsible Research and Innovation (RRI) (Sutcliffe 2011; Wickson & Carew 2014; and see Smallman in this volume). All in all, the innovation potentials of citizen science are in line with calls for open and responsible science (European Commission 2016d).

The growing appreciation of the power of citizen science has resulted in the establishment of new funding schemes for citizen science, such as OPAL in the UK (Imperial College London 2016), the TOP CITIZEN SCIENCE programme in Austria (Zentrum für Citizen Science 2016), or the new explicit citizen science funding scheme in Germany by the Ministry of Education and Research (BMBF). Associated with this development, context-adaptable evaluation criteria are required to assess the impact of citizen science programmes on science, society and policy. Evaluation criteria are needed to inform both proper citizen science funding support and effective project management. Evaluation should assess the value of citizen science for different outcomes and/or processes. This comprises a systematic assessment of both the effectiveness and efficiency of an activity or programme against a set of explicit or implicit standards and criteria. There are two aspects to evaluation: (i) outcome-based evaluation, which assesses the overall goals of activities or programmes and the benefits to participants and recipients of the results and; (ii) process-based evaluation, which identifies the operational strengths and weaknesses of activities or programmes.

This chapter presents a framework of evaluation criteria focusing on both the process and outcome level of citizen science projects. It is an open framework for evaluating diverse citizen science initiatives, based on an in-depth review of the characteristics and diversity of citizen science activities and current evaluation practices. These are applicable for projects ranging from grassroots initiatives to those led by academic scientists. The framework incorporates the scientific, social and socio-economic perspectives of citizen science and is aligned with the Ten Principles of

Citizen Science (see Robinson et al. in this volume). The indicators developed are intended to serve as a foundation for quantitative and qualitative data collection instruments.

Citizen science evaluation

There are currently no commonly established indicators for evaluating citizen science, and individual projects have the challenge of defining the most appropriate way to collect evidence of their impact. While some experts focus on the learning gains of participants (e.g., Phillips et al. 2014; Masters et al. 2016; and see Peltola & Arpin in this volume), others concentrate on the scientific gains and socio-ecological relevance (Jordan, Ballard & Phillips 2012; Tulloch et al. 2013; Bonney et al. 2014). Haywood and Besley (2014) made a first attempt towards an integrated assessment framework by combining indicators from science education and participatory engagement. The evaluation of the scientific impact of projects is challenging, since many approaches exist and many are criticised for their shortcomings (Allen et al. 2009).

Evaluation methods demonstrating impact on individual participants are common (e.g., Brossard et al. 2005), and include aspects like gains in scientific knowledge or skills as well as wider personal impact in terms of behavioural change, interest in science, motivation and ability to participate in science (Phillips et al. 2014). Personal development of participants is an important aspect of any citizen science project but evaluation is based only on personal learning outcomes and may miss out on other important aspects, such as wider societal impact. Behavioural changes, such as taking stewardship and civic action (Crall 2010; Phillips et al. 2014), point towards an assessment of such social implications. Shirk et al. (2012) therefore recommend a more holistic approach to project evaluation, accounting for impact on scientific knowledge and individual development as well as broader socio-ecological and economic impacts. Similarly, a more comprehensive approach to evaluation might operate on three levels – individual, programme and community – and stress the potential impact of citizen science on social capital, community capacity, economic impact and trust between scientists, managers and the public (Jordan et al. 2012).

Experts advise to define learning goals and expected learning outcomes at the beginning of a project to develop an appropriate and customised evaluation strategy (Jordan et al. 2012; Phillips et al. 2014; Tweddle et al. 2012). Otherwise, project evaluation risks not properly assessing the learning gains of individuals or documenting genuine impact (Skrip 2015). The use of a variety of evaluation methods is recommended, such as pre- and post-project surveys or examination of the correspondence between participants and project co-ordinators (Bonney et al., 'Public Participation', 2009). Evaluation also has a role in adaptive project management (Wright 2011). Continuously sharing experiences and lessons learnt with all stakeholders supports the social learning process and contributes to an iterative improvement of citizen science projects and programmes. This can be supported by iterative evaluation during the course of the project, allowing for flexibility and the possibility to counteract undesirable project developments (Skrip 2015; Dickinson et al. 2012).

Despite these contributions to evaluation, citizen science projects currently lack comprehensive evaluation frameworks that would allow for comparability across projects and programmes (Bonney et al., 'Citizen Science', 2009; Bonney et al. 2014; Crall et al. 2012). A recently published evaluation rubric (Tredick et al. 2017) tries to fill this gap in citizen science programme evaluation by including the main elements found in literature, but it still remains weak on the social implications of citizen science. Citizen science stakeholders continue to seek flexible evaluation strategies that adapt to specific project contexts (Schäfer & Kieslinger 2016) and initiatives have begun worldwide to build capacity (Richter et al. in this volume), guide citizen science development (e.g., Pocock et al. 2014b; Pettibone et al. 2016) and professionalise evaluation. The European Citizen Science Association (ECSA) has taken important steps by developing Ten Principles of Citizen Science (Robinson et al. in this volume) and the framework presented here aligns with these evaluation criteria.

Developing evaluation criteria for citizen science

The evaluation criteria presented in this chapter are the result of a review of existing projects and literature, as well as qualitative analysis including stakeholder consultation, expert interviews, and iterative adaptation and additional feedback loops with stakeholders. This was led by two working groups focusing on the social sciences and natural sciences, respectively, and the evaluation criteria have undergone a circle of refinement since this work began in July 2015 (see figure 6.1).

A narrative literature review included surveying the databases Scopus, Web of Science and Google Scholar as well as the library of the

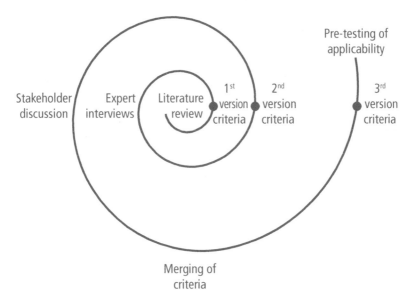

Stakeholder discussion Expert interviews Literature review 1ˢᵗ version criteria 2ⁿᵈ version criteria Pre-testing of applicability 3ʳᵈ version criteria

Merging of criteria

Fig. 6.1 Methodological approach to developing the evaluation framework

University of Natural Resources and Life Sciences, Vienna. Practical online evaluation guidelines were screened from citizen science organisations worldwide and websites that provide access to citizen science resources and projects (www.buergerschaffenwissen.de, scistarter.com, Citizen Science Central from the Cornell Lab of Ornithology, Centre for Ecology and Hydrology). Analysis of current evaluation practice focused mainly on areas in which citizen science projects differ from non-participatory scientific projects, such as communication, learning, technology participation and data management. The analysis was reinforced by 10 semi-structured expert interviews and expert consultation to gain feedback on scope, completeness, usefulness and applicability of the evaluation criteria and framework. The experts from Austria and Germany were selected based on their different approaches towards citizen science, covering practical as well as theoretical and evaluation-specific expertise, and with an even gender ratio. Further, a stakeholder workshop was conducted with 20 representatives of Austrian citizen science projects and four representatives of the funding body, the Austrian Federal Ministry of Science, Research and Economy to gain insight into the genesis of a citizen science project or programme.

Citizen science evaluation framework

Three core dimensions of evaluation emerged: 1) scientific dimension, 2) participant dimension and 3) socio-ecological and economic dimension (see table 6.1). For each of these dimensions, criteria are proposed at the 'process and feasibility' level as well as at the 'outcome and impact' level.

This framework can be applied for:

- Strategic planning and funding assessments of citizen science proposals;
- Monitoring progress during project duration; and
- Assessing impact at the end of a project.

In the course of the project lifecycle, the emphasis of evaluation would gradually shift from process and feasibility to outcome and impact. Process and feasibility ensures that projects prepare the groundwork for upcoming activities by engaging with concepts, methodologies and adaptive planning during their initial phase. Outcome and impact come into play when the first impacts on science, citizens and socio-ecological/economic systems can be measured.

Table 6.1   Citizen science evaluation framework

Dimension	Process and feasibility	Outcome and impact
Scientific	• Scientific objectives • Data and systems • Evaluation and adaptation • Collaboration and synergies	• Scientific knowledge and publications • New research fields and structures • New knowledge resources
Participant	• Target group alignment • Degree of involvement • Facilitation and communication	• Knowledge and science literacy • Behaviour and ownership • Motivation and engagement
Socio-ecological and economic	• Target group alignment • Active involvement • Collaboration and synergies	• Societal impact • Ecological impact • Wider innovation potential

Scientific dimension

Indicators at the *process and feasibility level* analyse the scientific grounding of the citizen science project. A clearly defined and genuine research question is the scientific basis of all future activities. It should be appropriate to citizen science approaches and meet the interests of participants (whether in terms of societal relevance or basic scientific curiosity). Good data quality control and validation processes are crucial success factors. Conceptual approaches such as research ethics, the proper management of (open) data as well as intellectual property rights issues need to be addressed from the beginning (see Williams et al. in this volume for more on these issues). Progress monitoring is also important; it should allow for flexibility and may lead to adaptive management during the project. New forms of sustainable collaboration between scientists, citizens and other societal actors and groups are also relevant here.

At the *outcome and impact level,* projects should be evaluated according to traditional academic standards, such as the generation of genuine scientific knowledge, captured in publications and possibly leading to new projects or collaborations. In addition, indicators should assess project impact on institutional or organisational structures and new forms of integrating traditional and local knowledge, thereby facilitating true knowledge exchange between science and society (see also Danielsen et al. in this volume).

Participant dimension

At the *process and feasibility level*, project design needs to include engagement and communication strategies. These should cater to different participant groups in terms of levels of engagement and interactive support measures and training to facilitate successful participation and collaboration (see Haklay in this volume). Working with civic society organisations may facilitate the participation of specific target groups and individuals with a genuine interest in the topic.

When it comes to assessing the *outcomes and potential impact* at the individual level, personal learning and development gains are key. Did participants develop new knowledge or skills, and does that increase their understanding of, and attitude towards, science? Did they enjoy the project and/or gain personal satisfaction from contributing to science and possibly to (local) policy development? Personal gains by individual participants may lead to changes in attitude and behaviour as well as an increased sense of ownership and empowerment, while the participation

Table 6.2 Evaluation criteria and supporting questions

Dimension	Criteria	Supporting questions
Scientific	***Process and feasibility***	
	Scientific objectives (Principles* 1, 2, 3)	
	Scientific goals	• Are the scientific goals sufficiently clear and authentic?
		• Is the scientific objective appropriate to citizen science?
		• Does the project adhere to the principle of joint knowledge creation in citizen science?
		• Does the scientific objective have relevance for society and does it address a socially relevant problem?
	Data and systems (Principles 2, 3,7,10)	
	Data quality and standards	• Does the project have clear processes defined to validate and guarantee high data quality?
		• Does the data adhere to common standards?
	Ethics, data protection, Intellectual Property Rights (IPR)	• Does the project have a data management plan, IPR strategy and ethical guidelines?
		• Are data ownership and access rights clear and transparent?
		• Is the data handling process transparent?
		• Do citizens know what the data is used for, and where it is stored and shared?
	Openness, interfaces	• Does the project have open interfaces to connect to other systems and platforms?
		• Is the generated data shared publicly and if so, under which conditions?
		• Is the project data appropriately archived for future analysis?

Table 6.2 (continued)

Dimension	Criteria	Supporting questions
Scientific	***Process and feasibility***	
	Evaluation and adaptation (Principle 9)	
	Project evaluation	• Does the project have a sound evaluation concept, considering scientific as well as societal outcomes?
		• Does the evaluation concept include indicators regarding the impact on individual participants and users of the project results?
		• Is evaluation planned at strategic points of the project?
	Adaptive project management	• Are project structures adaptive and reactive, including feedback loops for adaptation, and possibly a scoping phase?
		• Does the project have an appropriate risk management plan?
	Collaboration and synergies	
	Collaboration and synergies	• Does the project collaborate with other initiatives at the (inter-) national level to enhance mutual learning?
		• Does the project link to experts from other disciplines?
	Outcome and impact	
	Scientific impact (Principles 6, 8, 9)	
	Scientific knowledge and publications	• Does the project demonstrate an appropriate publication strategy, both in scientific and other media outlets?
		• Are citizen scientists recognised in publications and if so, can they participate in the dissemination of results?
	New fields of research and research structures	• Did the project generate new research questions, projects or proposals?
		• Did the project contribute to any institutional or structural changes?

(continued)

Table 6.2 (continued)

Dimension	Criteria	Supporting questions
Scientific	***Outcome and impact***	
	Scientific impact (Principles 6, 8, 9)	
	New knowledge resources	• Does the project ease access to traditional and local knowledge resources?
		• Does the project contribute to a better understanding of science in society?
Participant	***Process and feasibility***	
	Involvement and support (Principles 1, 4)	
	Target group alignment	• Does the project have an involvement plan that considers specifics of different target groups?
		• Are the options for participation and the degree of involvement diversified (e.g., gamification)?
	Degree of participation intensity	• Can citizens participate in various project phases?
		• Do citizens and scientists work as mutually respected partners in the knowledge generation process?
	Facilitation and communication	• Are support and training measures adapted to the different participant groups?
		• Are objectives and results clearly and transparently communicated?
		• Do citizens receive regular feedback?
		• How interactively is communication and collaboration between scientists and citizens organised?
	Outcome and impact	
	Individual development (Principle 3)	
	Knowledge, skills, competencies	• What are the learning outcomes with regards to new knowledge, skills and competencies for the participants?

Table 6.2 (continued)

Dimension	Criteria	Supporting questions
Participant	***Outcome and impact***	
	Individual development (Principle 3)	
	Science literacy	• Does the project contribute to a better understanding of science?
		• Does the project contribute to a better understanding of the scientific topic?
	Behaviour and ownership	• Does the project foster ownership amongst participants?
		• Does the project contribute to facilitating personal change in behaviour or political citizenship?
	Motivation and engagement	• Does the project raise motivation, self-esteem and empowerment amongst participants?
		• Are participants motivated to continue the project or involve in similar activities?
Socio-ecological and economic		
	Process and feasibility	
	Dissemination & communication (Principle 5)	
	Target group alignment & active involvement, two-way communication	• Does the project have a targeted outreach and communication strategy to reach a wide audience?
		• Does the project include innovative means of science communication and popular media, (e.g., art or hands-on experiences)?
		• Do citizens have the possibility for two-way communication?
	Collaboration and synergies	• Are collaborations planned with the media and science communication professionals?
		• Does the project leverage civic society organisations for communication and synergies?

(continued)

Table 6.2 (continued)

Dimension	Criteria	Supporting questions

Socio-ecological and economic

Outcome and impact

Societal impact (Principle 9)

Collective capacity		• Does the project contribute to the collective capacity of the participants in achieving common goals?
Political participation		• Does the project stimulate political participation?
		• Does the project impact on policy processes and decision-making (e.g., through agenda-setting or data contribution for policy evaluation)?

Ecological impact (Principle 10)

Targeted interventions, control function		• Does the project include objectives that protect and enhance natural resources and/or foster environmental protection?
		• Does the project contribute to higher awareness, knowledge and responsibility for the natural environment?

Wider innovation potential (Principles 9, 10)

New technologies		• Does the project foster the use or development of new technologies?
Sustainability, social innovation practice		• Does the project consider sustainability (environmental impact or sustained social relations) as part of the project plan?
		• Are the project results transferable to other contexts or organisations?
		• Does the project contribute to social, technical or political innovation?
Economic potential, market opportunities		• Does the project generate any economic impact or competitive advantages, (e.g., cost reduction, new job creation, new business models, etc.)?
		• Does the project foster co-operation for exploitation, (e.g., with social entrepreneurs)?

* Principles mentioned in this table refer to ECSA principles (Robinson et al. in this volume)

of young citizens may raise their interest in embarking on a science career (see also Edwards et al.; Makuch & Aczel; Harlin et al., all in this volume).

Socio-ecological and economic dimension

Appropriate dissemination and outreach activities need to be considered at the *process and feasibility level* to enhance the wider social, ecological and economic impacts of citizen science projects. Key stakeholders need to be engaged in a two-way dialogue to foster ownership and participation. Seeking collaborations with, for example, civic society organisations, tend to further enhance visibility and impact.

At the *outcome and impact level*, the wider societal impact should be assessed in terms of increasing civic resilience, social cohesion and social impact. Depending on the project, a focus on environmental or economic impact might be appropriate (see Owen & Parker; Schroer et al., both in this volume). The wider innovation potential of citizen science should be addressed against its contribution to societal transformation and sustainability goals.

Overarching assessment criteria can also be matched with supporting questions to qualify and detail potential evidence for each criterion (table 6.2). Such questions offer guidance for planning, monitoring and assessing citizen science projects, and have a reflective purpose, meaning that they should be tailored to specific projects or programmes. A mix of qualitative and quantitative assessment methods is recommended to collect the necessary data to answer these questions, such as online surveys, usage statistics, interviews, focus groups and so forth. The evaluation instruments need to be embedded in a solid evaluation plan tailored to each project, which may include concrete benchmarking of measurable targets to assess success during and after the project.

Discussion of the evaluation framework and its applicability

The presented framework touches one of the most relevant aspects of citizen science – how to evaluate citizen science? The developed open framework allows project managers and funders, the main target groups of this framework, to expand and adapt the evaluation criteria according to their specific needs. Adding the participant dimension on an equal level to the scientific and socio-ecological and economic dimensions indicates an expansion of focus from more traditional scientific projects. Empowering citizens and facilitating critical participation is on equal terms with scientific objectives, triggering a need for new research designs (Sieber & Haklay 2015).

Key decisions about framework implementation should be informed by a project's target groups and processes. It is also important to identify whether project evaluation will be performed by project members themselves, funding agencies, external experts or as a collaborative effort. Importantly, evaluation should be included in time and resource budgeting. Gathering evidence is resource-intensive and projects should seek a balanced approach in terms of measures and expected outcomes.

If funding organisations plan to apply such a framework of evaluation criteria, the definitions of citizen science and expectations towards it need to be clearly communicated (Eitzel et al. 2017). Support measures, including specific evaluation guidelines and methods for proper evaluation, will need to be developed, and can build on existing guidance (e.g., Pocock et al. 2014b; Pettibone et al. 2016) and the evaluation criteria framework presented in this paper.

The framework is intended to be comprehensive and its application needs tailoring and contextualising according to the spatial, temporal and socio-economic demands of the project or programme. Criteria need to be prioritised and may receive different weighting depending on project goals. While all Ten Principles of Citizen Science hold for all initiatives (Robinson et al. in this volume), some projects might have a special focus on social goals and succeed in creating greater societal impact, although they might not open new research fields or have economic potential. Nevertheless, all three dimensions – scientific, participant, and socio-ecological and economic – should be considered to benefit from the full potential of science-society collaboration. Synergies and trade-offs will need to be considered, and an initial clear set of criteria and evaluative scales adds transparency to the whole process. Recording and monitoring project experiences along this criteria framework is required to evaluate and demonstrate good practice examples that may inform the development of successful citizen science.

Overall, while a framework should be clear, adaptive capacity and openness is needed to embed learning and development in the project lifecycle. While evaluation should be comprehensive, it should not be static. In the course of a citizen science project, which often runs for years, the framework should allow for reflection on developments and contextual changes. In addition, long-term monitoring is necessary to capture a project's far-reaching impact.

Conclusions

This chapter has presented a citizen science evaluation framework that integrates three assessment dimensions: scientific advancement, citizen engagement and socio-ecological/economic impact. The evaluation criteria matrix and supporting questions can – and should – be tailored to different purposes.

For funding agencies, the framework could inform the development and selection of evaluation criteria for citizen science initiatives. For citizen science projects, the supporting questions can support holistic reflection on project strengths and weaknesses, as well as the potential for improvement both during project planning but also for adaptive project management and impact assessment. For scientific organisations, the three equal dimensions might enrich reflections on citizen engagement and impact on socio-ecological/economic systems. For civic society organisations, a closer look at the scientific perspective might offer opportunities to better exploit benefits from collaboration with science.

Thus the evaluation framework can be used as (a) a planning instrument for designing projects; (b) a mid-term and final self-evaluation for projects; and c) an external evaluation for funding agencies.

The presented framework needs to be transformed into a practical assessment tool for projects and initiatives, preferably through a mix of qualitative and quantitative methods, such as tailored online surveys, usage statistics, in-depth interviews or focus groups. It can assist in strategic planning, monitoring and impact assessment. It is hoped that these evaluation criteria will trigger further discussion on measures of success and evaluation for different project approaches and contextual settings within the wider citizen science community. Overall, a proper evaluation framework will help to professionalise the citizen science community, foster and guide targeted funding support and, ultimately, increase the desired impact of citizen science on science and society.

Part II
Innovation in science with and for society

7

Watching or being watched

Enhancing productive discussion between the citizen sciences, the social sciences and the humanities

Dana Mahr[1], Claudia Göbel[2], Alan Irwin[3] and Katrin Vohland[2]

[1] *University of Geneva, Switzerland*
[2] *Museum für Naturkunde Berlin, Germany*
[3] *Copenhagen Business School, Denmark*

corresponding author email: dana.mahr@unige.ch

In: Hecker, S., Haklay, M., Bowser, A., Makuch, Z., Vogel, J. & Bonn, A. 2018. *Citizen Science: Innovation in Open Science, Society and Policy.* UCL Press, London. https://doi.org/10.14324/111.9781787352339

Highlights

- The growing success and take-up of citizen science needs to be accompanied by increased reflexiveness in the field.
- Social science and humanities research shows that citizen science has a broad history and brings important alternative perspectives on the relationship between science and society.
- Better collaboration between citizen science and the social sciences and humanities, especially Science and Technology Studies (STS), should be facilitated to the benefit of all parties.

Introduction

Citizen science reshapes hopes for a democratisation of scientific knowledge production through the empowerment of grassroots initiatives to conduct research. At the same time, more and more professional scientists, scientific institutions and policymakers have started to engage with citizen science, often pursuing the benefits of fostering participatory research in terms of their own goals, which may differ from those of citizen scientists (see also Ballard, Phillips & Robinson; Haklay; Novak et al.;

Smallman, all in this volume). In this situation, it becomes important to reflect on citizen science, including the many and varied projects, methodologies and communities that make up this approach to science and technology, as well as its recent popularity and the side effects thereof.

Recent years have seen an increase in literature on citizen science from a growing and increasingly international (but mostly Western) networked community of practice (Kullenberg & Kasperowski 2016). Significantly, a journal has been founded to support discourse and reflections about citizen science, *Citizen Science: Theory and Practice*. These developments point to the potential for a growing (and shared) *reflexivity* of citizen science. Reflexivity is understood here as the generation and exchange of knowledge about how citizen science works, with the aim of better understanding and improving it. Such reflexiveness, however, cannot be limited to merely making more knowledge about citizen science available, but fundamentally requires critical engagement with the underlying assumptions of participatory research as well as the practical consequences of these assumptions. The social sciences and humanities have an especially important role to play here.

A reflexive perspective should consider how participants, the people who do the work in citizen science projects, could be explicitly acknowledged and invited to integrate their views and needs into the projects. However, the first issue of the *Citizen Science* journal appears to speak to the perspective of institutionalised science and the 'scientific outcome' of citizen science projects. For example, the most read articles cover topics including the 'credibility' of volunteered data (Freitag, Meyer & Whiteman 2016) and the 'effectiveness' of citizen science (Muenich et al. 2016). The democratisation and empowerment of volunteers, which could also be framed as valid goals for citizen science projects as 'the outcome for the people', are largely absent. Critical observations of this kind are important when working towards greater plurality and inclusivity in citizen science.

The success of citizen science and need to meet the expectations of various stakeholders (e.g., participants, researchers and policymakers) mean that citizen science practitioners in turn need to establish and continuously refine a self-reflexive culture. Within such a culture, topics like the power relations between amateurs and experts or the community impact of citizen science projects should be discussed with other practitioners and participants.

There is also a long history of scholars in the social sciences and humanities doing research *on* topics directly related to citizen science, even before the term 'citizen science' was coined in its contemporary usage (e.g., Irwin 1995). This scholarship typically reflects on the phenomenon from the perspective of the various academic fields which explore the shifting relationship between science and society. For example, historians have begun to ask how citizen science fits into the broader history of public participation in science, while sociologists and political scientists are concerned with how the phenomenon reshapes expertise and the demarcation of social spheres in democratic societies (Strasser et al., forthcoming). Such reflections from the social sciences and humanities offer important contributions to the field. Researchers in these fields might, for instance, work together with citizen science practitioners and participants to find and analyse pitfalls, and help identify and scrutinise the (sometimes implicit) biases that may occur while setting up a participatory endeavour. In this co-reflexive process, questions may arise, particularly around how to best manage access and remove barriers to research participation (e.g., at the level of language) and the manner in which the focus of science-public dialogue is framed (e.g., the kinds of questions that are – or are seen to be – important to the different parties to a citizen science project).

Despite the increasing number of venues for exchange and critical discussion among practitioners as well as the proliferation of research on citizen science, citizen science practitioners and scholars from the social sciences and humanities sometimes still appear to be disconnected. There is an often misleading, but perpetuated, self-understanding of these communities as being part of different intellectual spheres – here the natural sciences with their 'strict epistemologies' and there the more 'hermeneutical' humanities (*a longue durée* of C.P. Snow's 'Two Cultures' [Snow 1959]). This can make it difficult to find common ground for exchange and co-production, even when it comes to topics or projects where a joint endeavour could be promising. Setting up self-reflective and multi-perspective citizen science projects could be one of these endeavours and might hold the key to finally overcoming old distinctions, not only between 'experts' and 'laypeople', but also between the 'sciences' and 'humanities' (see Dobreva 2016; Crain, Cooper & Dickinson 2014).

This chapter has three aims: (1) to give examples from current social science and humanities research on citizen science; (2) to point out areas where joint ventures between these two communities promise to add value, illustrated by two case studies; and (3) to inspire further instances of co-operation by critically reflecting on the authors' own attempts to produce such an encounter. It is also hoped that making this possibly fruitful alliance accessible to the wider community of citizen science practitioners will stimulate further productive and critical engagement between the various communities engaged in citizen science.

Current research on citizen science

The first international European Citizen Science Association (ECSA) conference in Berlin (19–21 August 2016) aimed to give an overview of the current state of citizen science in Europe. From both a humanities scholar perspective and citizen science 'activist' perspective, it was evident that the citizen science scene is still in a phase of self-identification and development. While some, for example, the executive chair of ECSA in her welcome speech, addressed citizen science as a global movement which frames the 'idea of responsible citizenship and of responsible research' developing discursive and political power, others may treat citizen science more instrumentally as a tool for citizen involvement in the achievement of predetermined scientific and educational goals.

Many discussions focused on questions about how to make the best of the involvement of the public in terms of scientific outcome. Questions such as 'How reliable is the data produced by citizen scientists?', 'How can we measure "data quality"?' and 'How can we make citizens better "sensors" or better "observers"?' were important to many scientists, citizen science practitioners and policymakers. Likewise, the standardisation of such 'quality aspects' and citizen science in general, as well as the professionalisation of the field, were discussed. Other prominent topics included technology and learning outcomes (e.g., in schools) via citizen science. Citizen science was on the one hand framed as an additional 'scientific method' among others (that needs to follow an orthodox epistemology via 'universal' values like scope, data quality, fruitfulness, etc.) rather than as an 'opportunity for empowerment' (see also Wyler & Haklay in this volume). However, on the other hand, it has the potential to become both at the same time.

With its strong focus on developing 'policy' and 'standards', the community brought together at the ECSA conference framed citizen science in a way that did not focus on thinking about the societal and historical backgrounds of the phenomenon and corresponding theories. Additionally, the social sciences and humanities seemed to be rather absent from the main programme, which centred mostly on environmental sciences, citizen science technologies and methods, as well as the policy aspects of participatory approaches. Even if researchers from the social sciences and humanities do not necessarily do much citizen science themselves, their perspectives could enhance the field when considered and operationalised by practitioners and policymakers. Taking perspectives from the social sciences and humanities into account would benefit the citizen science community, for example, by bringing more knowledge about the sociology of citizen involvement or addressing some of the tensions and dilemmas involved in citizen science work.

Perspectives from Science and Technology Studies

Social scientists and scholars of the humanities played a part in the movement towards making science more participatory through the 1990s and 2000s, and have recently redeveloped their collective interest in the social structures, epistemologies and history of citizen science. Science and Technology Studies (STS), an interdisciplinary field comprising approaches from sociology, history, philosophy and other disciplines, is the most prominent field of investigation from which such reflective studies originate.

Current sociological and philosophical work on citizen science, for instance, discusses topics like the type and degree of participation and the agency of participants. Typical questions in the field include: How is participation framed by citizen science practitioners? How are volunteers engaged, and what is their motivation for partaking in citizen science? How does self-organisation function (e.g., Göbel et al. 2016)? Is citizen science part of a (serious) bourgeois leisure culture of the twenty-first century? Which endeavours and projects are framed as citizen science and why? A good example of this is the work from the research group around Lorenzo del Savio, Barbara Prainsack and Alena Buyx. In a current publication, they question whether crowdsourcing could also be framed as citizen science (del Savio et al. 2016). Furthermore, STS scholars Dana Mahr and Sascha Dickel (forthcoming) ask whether it is possible to enhance citizen science beyond 'invited participation' in a less linear way (with professional scientists 'on top' and participants 'at the bottom'), as Yochai Benkler's concept of commons-based peer production suggests (Benkler 2006).

From the perspective of historians of science, the emergence of citizen science is neither new nor surprising. It is embedded in the larger relational history of science, society and politics: from public experimentation in the eighteenth century (Shapin & Schaffer 1985), the large natural history networks of lay experts in the nineteenth century (Mahr 2014), the 'science for the people' and social responsibility of science movements of the 1970s, to the deliberative consensus conferences about environmental issues and participatory action research in the 1990s and 2000s (Irwin 1995; Mahr 2016). All these historically well-explored episodes prove that the demands of citizens to partake in processes related to science cannot be described as an exclusive phenomenon of the twenty-first century.

According to historical work, science almost always relies on lay expertise and lay assistance by members of the societies in which it unfolds. The scientific spectacles of the *Ancien Régime* testify to this as well as the networked activities of Darwin, Wallace and Mendel, or the masswork of volunteers collecting plant specimens for Carl Linnaeus and his binominal nomenclature (Shapin & Schaffer 1985; Golinski 1999; Bensaude-Vincent & Blondel 2008; Shapin 2010). The epistemological goal of this natural history–type of science was to unfold the book of nature by collecting and comparing huge amounts of data (Strasser 2011), an approach to research that provoked collaboration with various publics, for example, large-scale networks of volunteers conducting field observations in vast geographic areas for biogeographical research (Mahr 2014). In the nineteenth century, this resulted in a 'knowledge society' integrating scientific citizenship. Although the professionalisation of science had already begun at this time, the rising and confident bourgeoisie framed volunteer scientific work as a highly valuable and meaningful leisure activity. Therefore, thousands of laypeople-driven scientific societies emerged and fostered research that could keep up with the work conducted by professionals (Daum 2002). In sum, modern science was naturally considered as something that had tasks for almost everyone who was willing to participate. Science and society were inseparable.

This raises the question of why, in the early twenty-first century, science has become something that needs to be reconnected with society – why is modern science detached, estranged, unintelligible, not helpful on everyday issues and sometimes not even fully trustworthy (for example in the cases of nuclear research, GMO (genetically modified organisms) or pharmaceutical research)? Relatedly, why do many people hope to overcome this situation by participating in (or setting up) 'citizen science'? The answers to these questions are complex, but two factors are noteworthy: the rise of experimentalism in the twentieth century and the process of social differentiation. Experimentalism brought science from the field to the laboratory (Kohler 2002; 2006); in other words, from open spaces to closed ones, not accessible to everyone. Furthermore, experiments needed special – often expensive – equipment and required distinctive education. Social differentiation goes hand in hand with this since

Box 7.1. Case study 1. Who are the citizen scientists?

At the core of citizen science projects lies the belief that the making of science can be improved by extending participation in the research processes to a broader public. Whether they are called 'amateurs', 'the crowd', 'people' or 'citizens', unpaid participants are increasingly enrolled by scientists not just to discuss and learn science, but also to actively engage in the production of scientific knowledge. However, little is known to date about who these participants are, especially with regard to their education and professional backgrounds (but see also Haklay in this volume). The limited surveys which have been carried out tend to represent only the most active participants and do not represent the majority of participants.

A project by Jérôme Baudry, Elise Tancoigne and Bruno Strasser focuses on the identity of participants in distributed computing, where volunteers share their computer(s)'s power to advance data processing in several research areas. The project mines the online profiles of the dedicated BOINC platform (where projects include Seti@home, Rosetta@home and LHC@home, among others) as well as the users' data (e.g., points earned, country) to provide a richer picture of the demographics of volunteering in science.

the accelerated division of labour in the first half of the twentieth century finally led to the rise of professional 'scientists' and other 'experts' as distinct 'truth classes' (Mahr 2016). The old social contract was that science produces reliable knowledge while politicians make decisions for the good of society on this basis (Gibbons 1999). This succeeded as long as public trust in the expertise of experts remained (Beck 1991; Mahr 2016). Public clashes between experts exposing differences in underlying values and, with it, the knowledge they put forward, undermined this trust (Frewer et al. 2003). Today, discussion has turned to the role citizen science can play in a new social contract between science and society (Maasen & Dickel 2016; and see Smallman on Responsible Research and Innovation in this volume). The case studies of STS work on citizen science, discussed in the following two boxes, demonstrate the potential for enhanced and productive discussion between the two spheres.

Box 7.2. Case study 2. Citizen science between democratisation and economisation

Following a 'participatory turn', seeking to democratise science and technology (see for example Irwin 2006), new inclusive forums have been established on science- and technology-related issues over the last two decades. These spaces aim to promote mutual respect for different ways of reasoning and often portray public participation as free from strategic bargaining and manipulation. However, participatory approaches often lack reflection on, and remain disconnected from, their context of application. One important phenomenon here is the orientation of science and technology towards economic ends, which has been labelled 'economisation'.

To fill this gap, a project by Hadrien Macq studies public participation to assess the ways in which democratisation and economisation imperatives interact, conflict or complement each other, and how the design, process and outcomes of participatory exercises are impacted. He focuses on two domains and policy levels: the European research and innovation policy and the Walloon Region's digital strategy, which both promote political strategies relying on the creative potential of multiple societal actors to achieve economic goals. The project uses a two-step methodology to analyse the dynamics shaping participation in science and technology and its political-economic context across these policy levels. First, a critical discourse analyses if, and how, economisation influences the way participation is conceived by its sponsors. Second, participant observation and interviews with participation professionals and engaged parties assess the way the design, conduct and outcomes of participatory exercises are affected by the economisation rationale. Macq seeks to understand how the economisation of science and technology influences public participation, therefore providing a crucial platform for the theoretical and empirical investigation of the normativities of public participation in science and technology. In this respect, attention is paid to the reorientation of public participation in science and technology as conceived and promoted by the European Union under the Horizon 2020 programme. The recent promotion of citizen science as a priority within the new 'Open Science, Open Innovation, Open to the World' programme is scrutinised as part of the shift from public engagement in decision-making to public participation in innovation processes.

Citizen science studies session at the ECSA conference

With the aim of exploring links between citizen science practitioners and social science and humanities scholars, the authors, together with Anett Richter, organised a session at the ECSA conference in 2016. Initially perceived as quite a niche topic, we were surprised to discover the overwhelming resonance – the session received about one-fifth of all submissions for the conference.

The questions addressed can be summarised in four overlapping groups: (1) case studies by citizen science practitioners reflecting upon their own practices of doing and institutionalising citizen science, for example, Josep Perelló's 'brief story of the Barcelona Citizen Office: community of practice, the rules of governance, and the connection with citizens and public administration'; (2) surveys of the national landscapes of citizen science actors, disciplines and discussions, like Lisa Pettibone's 'What is citizen science today? A case study of current practice in Germany'; (3) studies of single systematic aspects of citizen science practice, such as Gitte Kragh's talk on 'Understanding motivations of citizen scientists'; and (4) generalising accounts that mobilise social science theory to offer reflective views on current practices as exemplified by Sascha Dickel's 'The (citizen-) scientification of society and the pleasures of research. Citizen science as science communication'.

The session format included two parallel streams of discussion with related presentations grouped per topic and at least two talks introducing different perspectives. A key lesson learned is that while many short presentations help to build mutual awareness, more time and focus is needed to explain underlying assumptions, a key in point for seriously exploring connections with substantially different points of view.

Conclusion

While citizen science practitioners are often highly reflexive of their own practices – as shown by the *Citizen Science* journal and work of citizen science associations – these initiatives would benefit from a closer relationship with the work of scholars in the social science and humanities, especially STS scholars, who critically engage with citizen science in their research on relationships between science and society. Moreover, the rising popularity of citizen science creates a growing need to work towards plurality and inclusiveness by collaborating in critical reflection on the practice of public participation in research, as well as on the standards and institutions forming within and around the community of practitioners. This also opens wider discussions concerning, for example, the relationship between citizen science and the 'knowledge politics' of contemporary societies.

This chapter provided a critical review of main topics of the ECSA conference to illustrate points of departure where more critical reflexiveness is needed. It argues that focusing on the scientific, educational and policy-relevant outcomes of citizen science, along with recipes to increase efficiency, is too narrow and risks treating participants as sensors rather than self-empowered citizens. This is especially concerning given calls for the standardisation of citizen science practice. In the brief overview of current research in STS, the chapter suggested that perspectives from the sociology and philosophy of science can help to scrutinise which forms of public engagement with science and technology are currently framed as citizen science (and thus receive higher attention of academic researchers and funders), which emancipatory aspects are sidelined, and how this can affect the knowledge generated. Historical studies contribute yet another level of reflexiveness by repositioning the current drive to reconnect citizens and science as part of a longer trajectory of changing relationships between science and society, in which lay participation continues to be a key part. The chapter argued that addressing such issues creates added value for both science and society. The authors' own attempt to produce an encounter between citizen science practitioners and scholars from STS was a first step to facilitating such productive exchange. While the workshop format can be improved, it initiated contacts between communities, ignited debates and increased the visibility of the social science and humanities scholars as a central part of citizen science.

There are numerous directions for further activities that promise to be productive for such endeavours. One example is the working groups of citizen science practitioner associations, such as ECSA. Here, citizen science practitioners and other researchers are invited to engage in cooperative projects, thus practising reflexivity in developing common frames of discussion and outputs that are meaningful for all parties. Another route is 'co-laborative' practice (Niewöhner 2016) where, rather than imposing a joint goal for working together from the start, exchanges happen on a more flexible basis with the primary objective of getting to know each other's knowledge practices and being open to where that might lead.

The authors hope this chapter might inspire others to seek new ground for debates surpassing the boundaries of their own disciplines, vocabulary and maybe even comfort zones. At the same time, peers need to challenge each other and bring about a more reflexive understanding of citizen science practices and how they can be explored, including the different motivations for advocating public participation in scientific research and where they might conflict within and between different stakeholder groups. Finally, shared spaces and tools are needed to identify, reflect and negotiate such goals.

Acknowledgements

The session, 'Citizen Science Studies. Engaging with the participatory turn in the co-production of science and society', was organised at the first international ECSA conference in 2016 and inspired this contribution. The authors thank all participants for their contributions and fruitful discussion – recognising that, due to the incredible interest, conditions were not optimal (especially noise levels) and that this has only been the starting points for many conversations that will need other venues and formats of exchange to unfold and bear fruit beyond mere mutual awareness.

With case studies provided by Jérôme Baudry, Elise Tancoigne, Steven Piguet, Bruno Strasser (all University of Geneva) and Hadrien Macq (University of Liège).

8

The value of indigenous and local knowledge as citizen science

Finn Danielsen[1], Neil D. Burgess[2,3,4], Indiana Coronado[5],
Martin Enghoff[1], Sune Holt[6], Per M. Jensen[3], Michael K. Poulsen[1]
and Ricardo M. Rueda[7]

[1] Nordic Foundation for Development and Ecology (NORDECO), Copenhagen, Denmark
[2] United Nations Environment World Conservation Monitoring Centre (UNEP-WCMC), Cambridge, UK
[3] Copenhagen University, Denmark
[4] World Wildlife Fund USA, Washington, US
[5] Herbario UNAN-León, León, Nicaragua
[6] Inter-American Development Bank, Managua, Nicaragua
[7] Universidad Nacional Autónoma de Nicaragua, León, Nicaragua

corresponding author email: fd@nordeco.dk

In: Hecker, S., Haklay, M., Bowser, A., Makuch, Z., Vogel, J. & Bonn, A. 2018. *Citizen Science: Innovation in Open Science, Society and Policy*. UCL Press, London. https://doi.org/10.14324 /111.9781787352339

Highlights

- International policies require land management to be informed not only by scientific but also by indigenous and local knowledge.
- A major challenge is how to use, and quality-assure, information derived from different knowledge systems.
- Possible data collection and validation methods include focus groups with community members and information collected on line transects by trained scientists.
- Both methods provide comparable data on natural resource abundance, but focus groups are eight times cheaper.
- Focus group approaches could increase the amount and geographical scope of information available for land management, while simultaneously empowering indigenous and local communities who generally have limited engagement in such processes.

Introduction

Countries that have ratified the Convention on Biological Diversity (CBD) are obliged to respect, preserve and maintain knowledge of indigenous and local communities (https://www.cbd.int). As part of the convention, the countries have agreed on a set of goals, the Aichi targets, which should be achieved by 2020. Aichi Target 18 states that, by 2020, traditional knowledge should be integrated in the implementation of the convention (https://www.cbd.int/sp/targets/). Moreover, the Intergovernmental Science Policy Platform on Biodiversity and Ecosystem Services (IPBES), which was established in 2012 and is in the process of completing the first global assessment of nature and its benefits to people, aims to bring different knowledge systems, including indigenous and local knowledge, into the science-policy interface (Diáz et al. 2015; United Nations Environment Programme 2016). Policy of this kind is one thing, but sometimes practice is another. How can the broad policy statements and the results of high-level global assessments be translated into practice in the 'real world'?

Citizen science encompasses a broad array of approaches that have in common that citizens are involved in one or more aspects of assessment and monitoring of the environment (Bonney et al. 2014; ECSA Ten Principles of Citizen Science, see Robinson et al. in this volume). In Western countries, citizen science programmes often involve community members only in data collection. The design, analysis and interpretation of the assessment results are undertaken by professional researchers (see discussion in Kennett, Danielsen & Silvius 2015). In tropical, Arctic and developing regions, experiments have been made to involve community members in all aspects of environmental assessment and monitoring, including programme design, data interpretation and use of the results for decision-making and action (Danielsen, Burgess & Balmford 2005; PMMP 2015; Johnson et al. 2016). Although there are still a number of scientific questions surrounding these approaches, and many programmes are still at an early stage of development, the new approaches show a great deal of promise.

This chapter summarises a recent case study which tested a simple approach to document and validate indigenous and local knowledge (ILK) from Nicaragua using focus group discussions, in comparison with scientific knowledge gathered from line transects (Danielsen et al., 'Testing Focus Groups', 2014). This approach provides the base evidence to support the inclusion of ILK alongside scientific knowledge. This example illustrates the issues that can arise from bringing ILK into science-based land management and the benefits that can be achieved. The conclusions also build on experiences from similar activities where ILK and community expertise in monitoring have been brought together with scientific approaches in different regions, providing valuable insights especially for tropical forest and Arctic regions, although some lessons will apply to a range of geographies (Brofeldt et al. 2014; Danielsen 2016; Danielsen et al., 'A Multicountry Assessment', 2014; Danielsen et al., 'Counting What Counts', 2014; Danielsen et al. 2017; Funder et al. 2013; Zhao et al., 'Can Community Members', 2016; www.monitoringmatters.org).

Indigenous and local knowledge

The world's approximately 370 million indigenous people include some of the world's poorest and most marginalised communities (United Nations 2009). To participate in decision-making, indigenous people need to translate their knowledge about their territories into a format through which they can be heard, for example in government land management plans (Dallman et al. 2011). Often, however, indigenous knowledge is not valued, or simply not available, in decision-making processes.

One challenge for the synthesis of information generated by different knowledge systems (Huntington 1998; Colfer et al. 2005) is that while scientific knowledge is validated primarily through peer review by other scientists, other knowledge systems have different validation approaches (Tengö et al. 2014). In other knowledge systems, for example, the concept of 'if it works, it is good' may count as an evidence (Tengö and Malmer 2012). Unidirectional scientific validation of other knowledge systems may therefore compromise the integrity and complexity of the knowledge (Bohensky & Maru 2011; Gratani et al. 2011) and promote power inequality between technocrats and communities (Nadasdy 1999; Bohensky, Butler & Davies 2013). Alternatively, validation of community-based knowledge through a respectful process of collaboration between scientists and community members could potentially facilitate mutual learning and empowerment.

Here, the term 'indigenous and local knowledge', or ILK, is used to emphasise that knowledge of resource abundance is closely linked with knowledge of resource management systems and the social institutions the management systems operate within (Berkes 2012). Indigenous and local knowledge, like scientific knowledge, implies a way of viewing the world. It is context-specific, hence may lose meaning when applied in other contexts (Stephenson & Moller 2009). In comparison, knowledge about resource abundance, bound by place and time, does not lose its meaning and is relevant to decisions about its management. Berkes (2012) used 'local knowledge' when referring to recent knowledge and 'indigenous knowledge' for the local knowledge of indigenous peoples, or local knowledge unique to a culture or society. To demonstrate how ILK on natural resource abundance can be used in environmental assessment processes, the below case study compares community-level focus group discussions against scientist-executed line transects.

Comparing ILK and scientific methods

One previous study has evaluated focus group results against direct counts of natural resources (Mueller et al. 2010). This compared assessments of species richness, diversity and height of grasses and trees by community members from a village in Niger, with direct counts made by scientists. The study found a good match on height and density for grasses and trees and tree species richness, but poor correlation on herb species richness and Simpson's D value for both trees and grasses. The study does, however, have a different temporal scale and different times for community members' focus group discussions and direct counts, preventing conclusions about the reliability of the focus group (Danielsen et al., 'Testing Focus Groups', 2014).

Case study location

The case study was undertaken in the Bosawás Biosphere Reserve in Nicaragua, inhabited by Miskito and Mayangna communities who use forest as their principal resource base (Koster 2007; Stocks et al. 2007). The area is a global priority for conservation (Miller, Chang & Johnson 2001). Conventional scientific knowledge is constrained by difficult access, rugged terrain and frequent heavy rains.

The research covered nine study sites located opportunistically, 2 to 15 kilometres from San Andrés and Inipuwás villages, within Bosawás Biosphere Reserve. All study sites are covered in dense evergreen tropical forest, which is used as a resource to different degrees. The area is inhabited by indigenous Miskito and Mayangna who practice subsistence agriculture and harvest non-timber forest products.

Methodology

Focus groups are not commonly used by biologists but are often part of social scientists' tool box. They involve group discussion on a particular topic, organised to improve understanding and involve participants carefully selected for their knowledge, or experience, of the topic. The discussion is guided, monitored and documented by a person from within the community and/or by an external person, sometimes called a moderator or facilitator (Kitzinger 1995). Line transect survey is a commonly used scientific method in ecology (Peres 1999; Luzar et al. 2011). It is a survey undertaken while moving on a path along which researchers count and record occurrences of the species of study (Bibby et al. 2000). The abundance assigned by the focus groups was compared to the abundance from the scientists' transects.

In this study, communities were contacted through a civil society organisation with long experience working with them. Researchers met the General Assembly of Miskitas in the two villages to obtain their advice and approval. Community members volunteered for the focus groups, based on their interest and experience with forest resources. During participatory planning workshops, members of the focus groups were involved in planning the process and deciding on the future use of the results (for more on models of participatory citizen science see Ballard, Phillips & Robinson; Haklay; Novak et al., all in this volume). This included scientists and community members agreeing on 10 resources important to the communities for food or other uses. They identified three plants, three birds and four mammal taxa to be monitored across nine sites and at the same time (three-month periods) by both the focus groups and line transects.

Focus group members included forest product harvesters, hunters, loggers, local park rangers, and both women and men. A volunteer group of 10–20 persons was established in each village to observe forest resources at study sites between discussions. From April 2007 to September 2009, these groups took part in two-to-three-hour meetings every three months. Community members had good knowledge of the forest (Koster 2007) and the resources studied were of interest to, and well known by, them.

The meetings were facilitated by a group of non-indigenous park rangers. Facilitators were selected based on their skills at communicating equitably between knowledge systems during meetings. There was no detectable political interplay between the facilitators and community members. The facilitators led community discussion on the abundance of different resources at each study site in the respective three-month period.

The following abundance categories were used (Danielsen et al., 'Testing Focus Groups', 2014):

1. Many resources: ≥10 individuals of the resource (e.g., ≥10 individuals of a plant species) were recorded in four hours of morning walks in the forest;
2. Some resources: One to nine individuals of the resource were recorded in four hours of morning walks in the forest;
3. Few resources: More than four hours of morning walks in the forest were required to record one individual of the resource, but the resource is still recorded regularly (≥4 times during the three-month period); and
4. Very few (or no) resources: Resource only recorded a few times (<4 times) during the three-month period.

During the focus group discussions, these categories were interpreted as 'many daily', 'daily', 'less than daily' and 'rarely'. Focus groups' validation was a careful process involving time and trust. Community members were in control of the process, agreeing what was right and wrong, and the facilitator assisted this process. Community members involved in focus groups had extensive experience of hunting and collecting forest products (see figures 8.1–8.3).

Line transect routes were established in the same month and year as the focus groups. Transects were surveyed for animals and plants by trained scientists.

The findings were returned to the communities so they could see how their observations connected with results from other methods, and could be used to promote indigenous and local input into reserve management. This two-way process helped underline that the study was not information 'harvesting' but a collaborative undertaking.

Outcome

The focus group discussions were unable to differentiate between what scientists considered 'very few', 'few' and 'some resources', but resources reported as plentiful ('many resources') were significantly different (more abundant) from all other categories for all types of resources.

The apparent inability of focus group reports to differentiate between the three categories of least abundance was caused by high spread out

Fig. 8.1 Tuno (*Castilla tunu*) has a fibre-rich bark. It is important for crafting clothing, bags and rope, among other things, in the Bosawás Biosphere Reserve, Nicaragua. The tree grows more than 25 metres tall and is rich in latex but, in contrast to the related species (*Castilla elastica*) also found in the area, the Tuno-latex does not have elastic properties. (Source: Sune Holt)

Fig. 8.2 Signs of the Nine-banded Armadillo (*Dasypus novemcinctus*) showing disturbed leaf litter, twigs and small holes, where ants, termites and other insects have been dug out. (Source: Sune Holt)

of the numbers (high standard deviation) within focus-group category 4 ('very few') and fairly even densities of focus-group category 3 ('few resources') and 2 ('some resources') (see figure 8.4). Reducing the number of abundance categories from four to three, by merging 'few resources' and 'some resources', delivered a clearer separation of densities for birds and plants, although not for mammals. Likewise, Spearman correlation coefficients for transect densities and focus-group categories were 0.43 ($P < 0.001$), 0.06 ($P = 0.32$) and 0.30 ($P = 0.04$) for birds, mammals and plants respectively, suggesting a stepwise reduction in densities (high, medium, low, very low) against focus-group categories (many, some, few, very few) for birds and plants, but not for mammals.

The cost of focus groups and scientist-executed line transects was estimated as the actual expenses incurred during the training and fieldwork at each site. Across all nine study sites, measurements through focus group discussions cost significantly less than scientists' transects ($P < 0.001; n = 9$).

Fig. 8.3 A Miskito community member recording his sightings and signs of plants, birds and mammals in the Bosawás Biosphere Reserve, Nicaragua. (Source: Sune Holt)

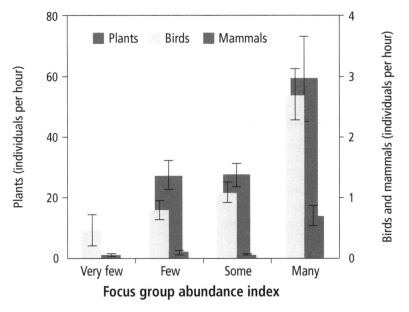

Fig. 8.4 Relationship between focus groups' statements of abundance of 10 plant, bird and mammal forest resources and the average abundance indices (number of individuals observed per hour, with SE) of the same resources obtained by trained scientists' transect walks between 2007–2009 at nine study sites in the Bosawás Biosphere Reserve, Nicaragua. Experienced community members' perceptions of forest resources, transmitted orally during focus group discussions, matched results from line transects by scientists. (Source: Danielsen et al., 'Testing Focus Groups', 2014)

Lessons for citizen science

The case study suggests that over a range of birds, mammals and plants, ILK documented and validated with focus groups provides similar abundance indices of wild species to trained scientists undertaking transects. The strongest agreement between focus groups and transects was for birds and plants, with lower agreement for mammals. This might be because mammals were mainly recorded by footprints and dung along transects, while birds and plants were directly observed, hence the number of mammals recorded on transects is subject to substantial individual interpretation.

Interestingly, focus group participants' understanding of individual abundance indices appears to vary between taxa. For instance, mammals recorded in the scientists' transects at 0.7–0.8 individuals per hour are considered 'many individuals' by the focus groups, whereas birds recorded on transects with the same density are considered to be 'few individuals' in the same focus groups. Focus groups are, thus, integrating community expectations, in other words, recording something as less abundant when fewer than expected are recorded given its identity, size (perhaps) or interest as food.

In the scientific knowledge system, reliability has two components: conformity to fact (lack of bias) and precision (exactness). The case study suggests that villagers' focus group assessments of abundance are similarly accurate (unbiased) to scientists' transects. The precision of the focus groups' assessments was not measured because abundance values from the focus groups are categorical, which hampers assessment of precision.

Focus groups involve interaction between group members (Gibbs 1997). Although the views of the most powerful members of the group might bias the results, observation in this case suggested that when potentially inaccurate information was provided by one or a few participants, after discussion, this information was generally corrected. Hence, the conclusion represented the group consensus.

The 'process' aspect of the focus groups was important to the community members. Focus group discussions were undertaken in an open learning environment, where participants had the right to vote and express opinions. They were the gatekeepers, detecting and deciding which data were complete and which were false or out of context, and should be discarded. The findings suggest that community members' ownership of the data and information and their control over the knowledge, validation process and application of knowledge were critical to their sense of empowerment (Stephenson & Moller 2009; Huntington 2011).

Central to approaches that facilitate exchange between knowledge systems is the concept that knowledge itself is power, which means that those who share knowledge should not lose power in the process (Nadasdy 1999; Gamborg et al. 2012; Tengö et al. 2017). The case study findings suggest that using focus groups to validate ILK about natural resources could increase the information available for measuring the status and trends of natural resources, while at the same time empowering indigenous and local communities. Guidelines already describe how to promote the use of indigenous knowledge (e.g., Tkarihwaié:ri Code; the Convention on Biological Diversity 2011) but to aid this process and

Box 8.1. Recommendations for how to increase the ability of community focus groups to provide natural resource abundance data which scientists would consider reliable (Danielsen et al., 'Testing Focus Groups', 2014; Danielsen 2016). Further recommendations for the participatory monitoring of biodiversity are available in the Manaus Letter (PMMP 2015).

1. Establish independent focus groups in multiple communities that know about resource abundance in the same geographical area (triangulation across communities).
2. Convene regular (e.g., annual) village meetings to present, discuss and interpret data, and obtain feedback from the entire community (triangulation across community members).
3. Facilitate the collection of auxiliary data, for example, through community members' direct counts of resources in the same area when possible (triangulation across methods).
4. Include focus group participants who are directly involved in using and observing natural resources (thereby increasing the number of primary data providers).
5. Use unequivocal categories for resource abundance.
6. Ensure that the moderator of focus group discussions has relevant skills and experience in facilitating dialogue.

increase the ability of community focus groups to provide natural resource abundance data which scientists would consider reliable, this chapter proposes a series of recommendations (box 8.1).

This approach should not, however, be rolled out uncritically – representatives of indigenous and local communities should decide whether focus groups on resource abundance can help them be heard. The UN Declaration on the Rights of Indigenous Peoples states that development must take place in accordance with their 'Free, Prior and Informed Consent' (United Nations 2008). Focus groups may also be a useful starting point from which broader regional and national monitoring and assessment programmes could be designed and implemented according to local conditions.

Conclusion: Implications for achieving management goals

The case study in this chapter has shown how ILK can inform land management policies and processes. Further, the authors have previously found that, for the same recurrent government investment in protected areas in the Philippines, far more conservation management interventions result from participatory natural resource monitoring approaches than conventional scientific ones (Danielsen et al. 2007). A large proportion of the interventions emanating from participatory monitoring addressed the most serious threats to biodiversity and led to changes in local policies with potentially long-term impacts.

In a meta-analysis of published monitoring results, the degree of involvement of local stakeholders in natural resource monitoring influences the spatial scale and speed of decision-making based on the monitoring data (Danielsen et al. 2010). The greater the involvement of local people in monitoring activities, the shorter the time it takes from data collection to decision-making. The most participatory approaches lead to management decisions typically taken three to nine times more quickly than decisions based on scientist-executed monitoring, although they operate at much smaller spatial scales. In contrast, scientist-executed monitoring typically informs decisions in regions, nations and international conventions.

Participatory monitoring of natural resources with the involvement of ILK depends on local people making a significant investment in monitoring. These approaches are therefore most appropriate: (1) where local people have significant interests in natural resource use; (2) when the information generated can impact management of the resources and the monitoring can be integrated within existing management regimes; and (3) when there are policies in place that enable decentralised decision-making (Danielsen 2016).

Promoting approaches such as those outlined in this chapter could provide an important set of results that, when published, could be used in the assessment work of IPBES as it seeks to fulfil its mandate to recognise and respect the contribution of ILK and bring it alongside scientific knowledge. The Intergovernmental Science-Policy Platform on Biodiversity and Ecosystem Services has an important catalytic role in promoting the use of new approaches to improve the capture of data and information, and bringing together material from different knowledge systems. This chapter has shown how social and natural science approaches can also validate the credibility of either approach (social and natural science), and allow more confidence in results used to make important decisions for the management of the natural world.

Acknowledgement

This chapter is a shortened version of a detailed report available in Danielsen et al., 'Testing Focus Groups', 2014.

9

Citizen engagement and collective intelligence for participatory digital social innovation

Jasminko Novak[1,2], Mathias Becker[2], François Grey[3] and Rosy Mondardini[3]

[1] *University of Applied Sciences Stralsund, Germany*
[2] *European Institute for Participatory Media, Berlin, Germany*
[3] *University of Geneva, Carouge, Switzerland*

corresponding author email: jasminko.novak@hochschule-stralsund.de

In: Hecker, S., Haklay, M., Bowser, A., Makuch, Z., Vogel, J. & Bonn, A. 2018. *Citizen Science: Innovation in Open Science, Society and Policy*. UCL Press, London. https://doi.org/10.14324/111.9781787352339

Highlights

- Digital social innovation shares the basic ideas of citizen science, as well as the common challenge of motivating and structuring citizen engagement. However, it is different in scope, focus, forms of participation and impact.
- Digital social innovation explores new models where researchers, social innovators and citizen participants collaborate in co-creating knowledge and solutions for societal challenges.
- There are critical issues and effective practices in engaging citizens as knowledge brokers and co-designers of solutions to societal challenges, which should inform the design and implementation of new projects and approaches.

Introduction

As citizen science matures, it finds itself part of a growing plethora of approaches democratising the processes of scientific enquiry and related modes of knowledge creation. Digital social innovation (DSI) and do-it-yourself (DIY) science are two examples that share citizen science's ideals and challenges of enabling citizen engagement (see also Mazumdar et al. in this volume). Considering typical challenges and types of citizen engagement models in DSI and DIY science, and how such platforms relate to approaches in participatory citizen science, may help the fields to learn from each other and inform new projects and approaches.

In citizen science (Bonney 1996; Cohn 2008), citizens are commonly involved in different types of activities in scientific projects, which are mostly led by professional scientists in institutional settings (Bonney et al., 'Public Participation', 2009; Shirk et al. 2012). The underlying assumption of science as the primary legitimate source of knowledge requires citizen participation to conform to the scientific process (Wyler & Haklay in this volume). More flexible forms of engagement relax this requirement by giving citizen participants more influence on the project design (e.g., in the choice of problems or outcome types) and empowering them to collaborate with different actors, among which scientists are but one kind (see also Ballard, Phillips & Robinson in this volume). This broadens the scope of projects, their goals and outcomes, and the types of activities performed by citizens. In particular, participatory citizen science and 'extreme citizen science' (Haklay 2013; Stevens et al. 2014) emphasise citizen involvement in core activities of the scientific process, such as problem definition, data analysis and interpretation (see also Gold & Ochu in this volume). These projects design tools for empowering participation from different societal groups (e.g., marginalised communities) in activities that would normally require scientific skills and knowledge. In doing so, they bring scientific enquiry to 'non-scientific' problems (e.g., problems important to the volunteers' communities) and 'non-scientific' knowledge (e.g., indigenous knowledge, local needs) (see Danielsen et al. in this volume).

Do-it-yourself science extends this to more informal, experimental methods and a broader range of outcomes: DIY scientists are people who create, build or modify objects and systems in creative ways, often with open source tools, and who share the results and knowledge (Nascimiento et al. 2014, 30). This includes non-specialists, hobbyists and amateurs, but also professional scientists doing science outside their traditional institutional settings. Many DIY science projects are private or community-based initiatives that use scientific methods combined with other forms of enquiry to explore techno-scientific issues and societal challenges (Nascimento et al. 2014; see also Mazumdar et al. in this volume).

This openness to different types of knowledge, outcomes and social settings is also part of the field of social innovation, which emphasises the societal impact of both scientific and practical knowledge creation. The concept of social innovation commonly describes novel solutions to social problems that are more appropriate than existing ones (e.g., more effective, efficient or sustainable) and that create value for society as a whole (Phills et al. 2008, 36). Many social innovations are increasingly based on the use of digital technologies, such as social networks, open data, open source hardware and software. Such digital social innovations are often defined as new solutions to societal needs developed through collaboration between innovators and target users, supported by digital technologies (Bria et al. 2015, 9). This resonates with an early view of citizen science as a science that addresses the needs of citizens and involves them in the scientific development process (Irvin 1995, xi). Such views of (participatory) citizen science and social innovation thus converge in the goal of producing knowledge that addresses societal needs.

A key commonality of participatory citizen science, DIY science and DSI is the focus on citizen engagement with different professional actors in a process of collaborative development and knowledge co-creation, in other words, a process of collective learning. Ideally, they all aim at engaging individual citizens and local communities in the entire process of scientific, exploratory or creative inquiry: from the problem definition and data collection, to analysis and interpretation, solution implementation and take-up. While exhibiting important differences in scope and focus, forms of participation and intended impact, all three approaches face similar challenges of motivating, enabling and structuring citizen engagement. They therefore explore various forms of collective intelligence that often require a lot of groundwork to be implemented (e.g., mobilising large numbers of participants) and can be overwhelming for a single project. A growing number of platforms aim at supporting citizen engagement in DSI by facilitating various forms of collective intelligence (for an overview see Bria et al. 2015).

Purposes and typologies of citizen engagement

Citizen involvement in social innovation is often valuable in its own right because it makes the development of solutions to societal problems more transparent to the people affected by them. There are also other common reasons for citizen engagement: citizens bring local knowledge about the problem and their needs; they can generate new solutions informed by their knowledge; and they bring different points of view, leading to more diverse perspectives on the problem (Davies et al. 2012a). When involved in the process, citizens are also more likely to accept the solutions. This is especially important to the many types of societal problems that inherently require citizens to change their actions or behaviour (e.g., public health, sustainable consumption) (Davies et al. 2012a; see also Schroer et al. in this volume). The benefit of this is emphasised by DSI that not only uses digital technologies as innovation enablers, but also makes the engagement of citizens in the creation of solutions a normative prescription (Bria et al. 2015). Many of these issues also echo the motivations for citizen engagement in citizen science (see Bonney et al., 'Public Participation', 2009). They are especially reflected in participatory approaches that involve citizens as equal partners with scientists, and that value different types of (non-scientific) knowledge from local, often marginalised, communities (Haklay 2013; Stevens et al. 2014).

Devising a DSI or a participatory citizen science project requires choosing appropriate forms of citizen engagement for the given purpose. Different typologies of engagement from both fields can inform such decisions (see table 9.1). With respect to the level of citizen influence on the project, Bonney et al., 'Public Participation', 2009, differentiate between projects where citizens collect and contribute data (contributory projects),

---

Box 9.1. Example of a digital social innovation project involving citizens as co-creators contributing local knowledge

Hybrid LetterBox

Hybrid LetterBox is an example of a project involving citizens as equal partners working with researchers in the development of novel solutions for local needs[1]. The project aimed at easing citizen participation in online discourses by connecting digital and the analogue channels of interaction. The Hybrid LetterBox is an 'augmented mailbox where anyone can throw a physical postcard that is automatically digitized, and uploaded to an internet platform to be spread and discussed' (Becker et al. 2015, 78). In developing the concept and prototype of the Hybrid LetterBox, the researchers initially collaborated with a group of elderly citizens, empowering them as co-designers. As the lead researchers Andreas Unteidig and Florian Sametinger describe in their project report, this helped them to discover new target groups, to better understand potential uses and to arrive at the final design of the original concept:

> The idea for the prototype emerged out of co-design workshops, since some of the predominantly elderly inhabitants

(continued)

of the neighborhood we worked with do not have access to digital media. This presented itself as a problem, since we were working on a local social network and particularly aimed at involving those who have not been active in the shaping of their neighborhood so far. We realized that we needed an interface that connects the digital and the analog world, and hence started working on the development of the early prototype together. [. . .] In the course of running the first tests and experiments with an early prototype in this neighborhood, many different groups – children, families, senior citizens – started using our technology in a broad range of ways: they formulated questions, ideas, they scribbled or contributed their thoughts in their respective mother tongue. It became clear that our target group is much bigger than we initially anticipated and that it proves useful in a variety of different contexts. Participating in discourses through the usage of our artifact proved attractive, also to those who are digitally well connected. (Becker et al. 2015, 84 and 88; see also Herlo et al. 2015).

Source: http://www.design-research-lab.org
/projects/hybrid-letter-box/

Fig. 9.1 Hybrid LetterBox. (Source: Matthias Steffen)

Table 9.1 Overview of typologies of citizen engagement

Design factor	Typology of engagement	Source
Control level	• Contributory, collaborative, co-created citizen science projects	Bonney et al., 'Public Participation', 2009
Cognitive complexity and type of contribution	• Crowdsourcing, distributed intelligence, participatory science, extreme citizen science	Haklay 2013
Type of contributed knowledge	• Information about present needs (understanding individual problems and needs, understanding larger patterns and trends) • Developing future solutions (co-developing or crowdsourcing solutions)	Davies et al. 2012a
Function	• Provision of information and resources • Problem-solving • Taking and influencing decisions	Davies et al. 2012b
Scale	• Small-scale vs. large-scale engagement	Davies et al. 2012a

projects where citizens help with the data analysis and may contribute to refining the project design (collaborative projects) and projects in which citizens co-design the project together with scientists, and are involved in all stages of knowledge creation (co-created projects) (see also Ballard, Phillips & Robinson in this volume). This framework can also be read as a map of different types of activities that are compatible with the chosen level of control over the knowledge creation process (see Bonney et al., 'Public Participation', 2009 for a detailed analysis). The typology proposed in Haklay (2013) can be read with respect to the level of cognitive engagement and type of contribution. Crowdsourcing resources (e.g., citizen sensors) are the 'simplest' form of participation, with little cognitive engagement and no citizen influence on the project design. Involving citizens in activities such as data collection and annotation is a way of harnessing their distributed intelligence ('citizens as interpreters'), whereas enabling them to contribute to the problem definition and data analysis leads to participatory science projects. In 'extreme citizen science', citizens are empowered to collaborate with professional scientists on many core aspects of designing the scientific project – from problem choice to the interpretation of results – and on ensuring the relevance to their local context. This modality also opens 'the possibility of citizen science without professional scientists, in which the whole process is carried out by the participants to achieve a specific goal' (Haklay 2013, 12). This matches DIY science and DSI, where citizens act as active co-creators and initiators of solutions to problems relevant to their social realities (see also Smallman et al. in this volume on Responsible Research and Innovation).

With respect to the type of knowledge generated, a social innovation project will typically involve citizens in gathering information about present needs and/or to participate in the development of future solutions (Davies et al. 2012a). This is often performed with ethnographic techniques, workshops or consultations for eliciting citizen knowledge of the problem, competitions for novel solution ideas, various testing and rating techniques for evaluating the suitability of different solution ideas or assessing the importance of different problem aspects. Co-developing new solutions in smaller groups is often performed through hands-on workshops and bootcamps involving citizens, scientists, technology and domain experts, while crowdsourcing is applied to extend the ideation process to (very) large groups of participants. From another functional perspective, citizens can help in the provision of information and resources (e.g., crowdsourcing data or donations), support problem-solving (e.g., competitions, co-design) or be involved in taking and influencing decisions (e.g., campaigning or participatory planning) (Davies et al. 2012b).

Methods and critical issues

Despite a large body of experience, citizen engagement remains a challenge, especially when it comes to harnessing more complex forms of citizen collaboration that go beyond data collection (Rotman et al. 2012). Digital social innovation and participatory citizen science projects have been exploring this challenge, but there is also a long tradition of precursors that provides helpful insights (see also Haklay; Mahr et al., both in this volume). Citizen engagement in knowledge brokering and co-designing is closely linked to the concepts of user-centred and participatory design, which both place the elicitation of user needs, feedback and ideas at the core of the solution design process (see also Gold & Ochu in this volume). While in user-centred design, the project design is defined by professionals (e.g., designers, technology experts), participatory design gives major influence to the users and stakeholders. It considers them as equal partners to the professional actors and makes co-creation activities a key element. Citizens as 'users' and stakeholders impacted by the problem and the solution being developed are involved through a range of methods, from needs and requirements workshops to focus groups and ethnographic studies to storytelling (see also Hecker et al. 'Stories' in this volume) and storyboarding (see box 2), games and co-operative prototyping, and to empowering lead users to experiment with, and adapt, solution prototypes in real-world settings (for an overview see Müller 2002). Such focus on joint learning and co-creation is closely related to co-created citizen science projects and to extreme citizen science. Similarly, the request for scientists to acknowledge and engage with the relationship of their work to a given social reality (Haklay 2013) resonates with the core ideas of participatory design.

A key issue for effective knowledge brokering and solution co-design is the creation of a shared understanding between the different worlds of citizens – their levels of knowledge and their lived reality – on the one hand, and those of professional scientists, domain and technology experts on the other. Methods such as concept visualisation, mockups, storytelling and prototyping can support this. Enabling effective joint exploration of the problem space and possible solutions includes the need to bridge information asymmetries and goal conflicts between different stakeholders (e.g., citizen volunteers, scientists, policymakers). This is frequently addressed through face-to-face interaction in physically co-located settings to further a sense of transparency and trust building. Supporting collaboration in such settings can benefit from adapting existing techniques and designing new tools for reducing information asymmetries, increasing transparency and reducing cognitive complexity, for example, through shared visualisations of multiple perspectives representing the views of different stakeholders (Novak 2009).

All such approaches come with a price: they require intensive engagement with participants and face-to-face interactions, often embedded in their day-to-day environments and across prolonged periods of time. Many studies have highlighted that participants are motivated by a wide range of factors, from identification with a project focus and goals to personal interest (e.g., learning new things), desire to help (e.g., helping science or society), shared values and beliefs (e.g., knowledge should be free), social recognition and reputation or simply fun and enjoyment (see for example Rotman et al. 2012; Raddick et al. 2013; Nov et al. 2011b;

Geoghegan et al. 2016). Recognition and regular feedback are key elements to ensuring continuing engagement and catering to changes in motivation (Rotman et al. 2012; Geoghegan et al. 2016). Regular social interaction (with scientists and other volunteers) is important (Geoghegan et al. 2016) but requires effort (e.g., regular face-to-face meetings and group activities). Even if locational constraints can be bridged by online interactions and mechanisms (e.g., crowdsourcing, continuous online feedback), online participation tends not to be fully representative. A few community members typically provide the majority of contributions, while others are 'passive' consumers, with a small portion of occasionally active participants (the 90-9-1 rule [Nielsen 2006]). Participation also tends to vary with time, requiring regular triggers of attention and dedicated community moderators to maintain activity dynamics over extended time spans.

Citizen engagement in co-creation activities (which are typically complex and demanding) thus risks reaching only a small portion of society. Engagement levels frequently change over time so activities with limited participants also risk failing to recruit new participants as existing ones become inactive. In fact, the transition of participants' roles (e.g., from passive to active) are an important mechanism of online participation (Preece & Shneiderman 2009). Successful community platforms tend to offer a range of different participation options requiring varying levels of

---

Box 9.2. Storyboards are often used in user-centred design to facilitate involvement of target users and stakeholders

Storyboards as a co-design technique

User-centred design techniques readily lend themselves to facilitating user involvement in co-creation and co-design processes for participatory citizen science or social digital innovation. Visual storyboards are an example of a technique commonly applied in system design practice. They are used to illustrate initial ideas about possible solutions and the ways they would be used in practice, in order to facilitate discussion about the actual problem, proposed solutions and new ideas with intended users and stakeholders. Below is an example from a project developing a platform for citizen engagement in water saving and sustainable water consumption (Micheel et al. 2014).

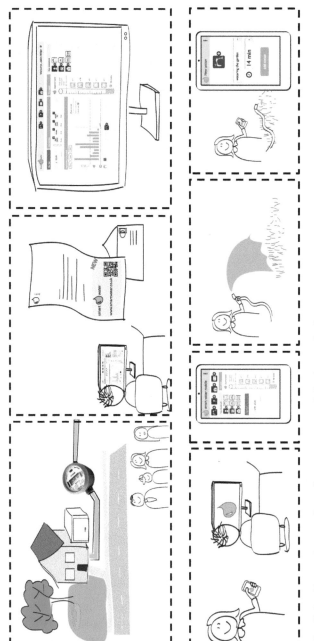

Fig. 9.2 Example storyboard as user-centred design technique

effort, allowing transition across different roles, based on participant's motivation, capabilities and situation through time (Anderson et al. 2012). A successful participation model will thus include simpler activities, with little complexity and cognitive effort (e.g., data collection) together with more complex activities, requiring more effort and/or more regular engagement (e.g., data analysis, solution co-design, evaluation and interpretation of results) (see Kieslinger at al. in this volume for more on evaluation). Face-to-face workshops or co-design sessions will be combined with online interaction and different options for the contribution of different types of knowledge, with some requiring more, others allowing less continuity of participation. Social recognition and reputation gained through regular feedback from the project can be combined with motivational designs using game-like elements to reward and make visible personal activity and achievements (Bowser et al. 2013; Iacovides et al. 2013). Joint exploration of the problem and solution space and co-creation of new knowledge will be facilitated by applying existing or developing new tools for alleviating information asymmetries between citizen volunteers and professional actors.

Online platforms

Effectively implementing such diverse and flexible models of citizen engagement is far from trivial. Beyond the issues identified above, other challenges concern the practicalities of implementation, such as choosing an appropriate engagement method for a given purpose, reaching the target groups and potential participants, disseminating the results of co-creation activities and supporting the uptake of outcomes and solutions. To facilitate this, (online) platforms designed for different types of citizen engagement and different forms of collective intelligence have been established (see Bria et al. 2015; Brenton in this volume). This section presents two cases studies: the CHEST platform for digital social innovation and the Open Seventeen citizen science challenge.

CHEST Enhanced Environment for Social Tasks

In the European project CHEST[2], citizens, social innovators, scientists, technology experts and other stakeholders collaborated in the participatory development of innovative solutions to societal challenges enabled by digital technologies. The CHEST online platform provided different tools and supporting measures including seed funding schemes, crowdsourcing

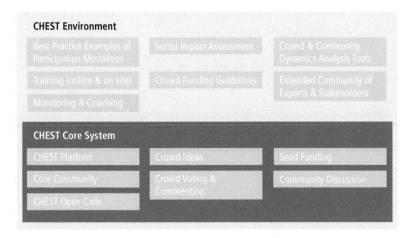

Fig. 9.3 Architecture of the CHEST Enhanced Environment for Social Tasks

tools, on-site/online coaching and training, and best practice guidelines for knowledge co-creation (figure 9.3). The project was carried out by three main partners, Engineering – Ingegneria Informatica SpA, European Institute for Participatory Media and PNO Consultants – extended by a network of 18 supporting partners and enlarged by 23 new partners through open calls (Chest 2016).

CHEST supported 35 ideas and 28 projects over three years, such as a platform exploring the use of Blockchains in product supply chains to foster transparency and sustainable consumption; a low-cost crowd-based traffic sensing device and analysis tool; a solution for self-monitoring and sharing of air pollution data; apps supporting people suffering from eating disorders or mental health; and many others. Such projects have actively involved 36,000 citizen participants in the different stages of the innovation process (table 9.2): They have provided knowledge (e.g., on social needs and solution ideas) and resources (e.g., placing traffic sensing devices in their homes), participated in problem-solving (e.g., analysing traffic and air pollution), co-designed solutions and influenced decision-making (e.g., voting on ideas to be funded, influencing local planning). Citizen engagement in different forms of collective intelligence has been facilitated at two main levels: several crowdsourcing schemes and instruments have been implemented at the platform level (e.g., crowd voting, commenting and monitoring), while coaching and training has been provided for the selection and implementation of appropriate citizen engagement methods at the individual project level.

Table 9.2 Overview of the main citizen engagement methods in the CHEST platform for digital social innovation

	Stages of the social innovation process		
Problem identification and selection	Development of new solutions	Evaluation and monitoring	Uptake and scaling
Idea competition, crowd commenting, crowd voting	Crowd commenting, user-centred and participatory design	User-centred evaluation, crowd monitoring	CHEST extended community and crowd

Problem identification, idea generation and selection

The bottom-up selection of societal problems and the generation of solution ideas has been supported through three competitions with monetary rewards: (1) call for ideas outlining a solution to an important societal problem requiring further exploration (e.g., of technical feasibility or potential social impact), with 35 proposals awarded €6,000 each; (2) call for projects developing an initial idea into a product or service ready for deployment, with five winners awarded up to €150,000 each; and (3) call for prototypes turning a solution into a functional prototype evaluated with target users, with 23 winners awarded up to €60,000 each (see Ficano 2014).

The call for ideas implemented an open innovation design where all the submitted proposals were publicly visible and could be commented on by a crowd of volunteers (e.g., critique, improvements). The submitters responded to comments and engaged in collaborative idea refinement. The submitted ideas were also voted upon by the public (after a registration process) and the submissions with the highest number of votes were selected as winners. The recruitment of the crowd of volunteers was supported by a Europe-wide dissemination campaign, resulting in nearly 5,000 registered crowd members. The call for projects and the call for prototypes also implemented a competition design, but the selection of proposals was performed by an expert jury (including researchers, technology experts, social innovation experts, civil society, public institutions and media representatives).

The call for ideas generated 1,141 comments by 956 participants (19 per cent of total crowd) and 28,851 votes by 4,886 participants (98 per cent of total crowd) over 21 weeks. This is a high engagement rate compared to much lower rates of typical online community participation (1–10 per cent active users), suggesting that the voting worked as a low-effort

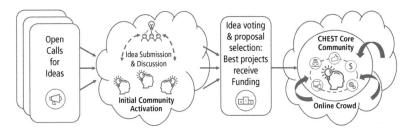

Fig. 9.4 CHEST bottom-up problem selection and solution generation process

activity motivating engagement. A visual network analysis performed on the voting and commenting activity has shown that many users commented on and endorsed different ideas, rather than supporting only one idea for which they may have been mobilised by the entrants (see Becker et al. 2015).

Project implementation

All 28 projects applied different methods for citizen engagement in knowledge brokerage (e.g., providing information and knowledge about the specific societal problem and citizen needs), resource provision and co-creation (e.g., co-designing solutions, co-analysing data, testing prototypes). This was facilitated through group and individual coaching (on-site, online, email) and training materials. The vast majority of the projects (79 per cent) involved citizens in the main co-design process: from the identification of the specific needs and requirements for a given problem, through co-developing solution ideas to evaluating the suitability of the developed solution concepts and prototypes. Only a smaller number of projects also involved citizens in the (re)definition of the problem to be addressed (18 per cent) (table 9.3).

Engagement methods used by most projects included on-site workshops (93 per cent, see for example figure 9.6), traditional interviews (71 per cent) and surveys (64 per cent). More "sophisticated" methods, such as lead user involvement in experimenting with the prototypes (figure 9.7), piloting (i.e., testing prototype solutions in prolonged real-world usage) and continuous online feedback were also used, though to a lesser extent (14 per cent, 21 per cent and 39 per cent respectively, see figure 9.5). The most popular were combinations such as on-site workshops with interviews (seven projects), surveys with on-site workshops and online continuous feedback (three projects) and surveys with on-site

Table 9.3 Citizen involvement in individual project phases

Project level citizen engagement in CHEST	
Citizens involved	31.047
Target groups*	79
Project phase	No. of projects
Problem (re)definition	5 (18 per cent)
User needs & requirements	25 (89 per cent)
Solution design and implementation	24 (86 per cent)
Test/Evaluation	28 (100 per cent)

* The target groups varied from project to project (depending on their specific goals) and ranged from children, youth and schools to elderly people, people with eating disorders, refugees, citizens in general and many others.

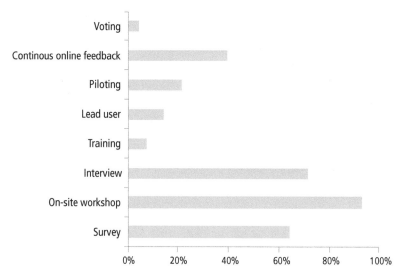

Fig. 9.5 Citizen engagement methods applied by CHEST-supported projects

workshops, interviews and online continuous feedback (three projects). Due to the small sample, no statistical correlations between the used method mix and the project evaluation rating (see next section on assessment process) could be established. However, it sticks out that the top three rated projects regarding the suitability of developed solution used a method mix of three or more methods. Moreover, the project with highest

Fig. 9.6 Co-design workshop in the TransforMap project. (Source: transformap.co)

Fig. 9.7 End-user test session in Project99/AyeMind. (Source: We Are Snook Ltd)

evaluation (4.94 on a 1–5 scale; see box 9.1) used the second highest number of methods of user engagement (5 methods), including lead user involvement and piloting.

A combination of offline and online activities in implementing the above methods was the most effective engagement strategy by the number of participants and the diversity of target groups, as well as in the number of tools developed to alleviate information asymmetries (table 9.4).

Box 9.3. Lead user and piloting methods in the Magenta TrafficFlow project

Magenta Traffic Flow

This CHEST-supported project for participatory traffic monitoring and management implemented a Living  Lab approach in Florence (Italy) to co-design its solution. Starting from day one it involved a small group of initial users to gather feedback, assess and setup the technology developed in the project. Existing grassroots communities (e.g., Ninux, Fab Lab) were also involved in the co-design processes through on-site workshops and online feedback.

Fig. 9.8 Sensor for traffic monitoring

Participants set up the privacy-preserving traffic monitoring points in their homes and tested the sensor and tool in real-world use. They provided input regarding sensor requirements, privacy and the design of the analysis tool. The sensors collected more than 50 million data points, classified in terms of their location, size of the vehicle, speed and type. All data has been made available in the open data portal of Florence and has been used in participatory traffic planning sessions.

Source: http://www.magentalab.it/

Table 9.4 Strategies of implementing methods of citizen involvement in CHEST

	Projects	Target groups	Citizens involved	Info. asymmetry tools
Offline involvement	12	38	277	53
Online involvement	2	6	110	19
Offline and online	14	61	31,000	203

Monitoring and project evaluation

CHEST also used a crowdsourcing model to collect citizen feedback on project progress and success, throughout the project cycle. The results of citizen assessment were provided as a feedback to the projects (and were not visible publicly), rather than as control instance for the funders. This allowed projects to assess their progress and take corrective actions. Using the CrowdMonitor tool developed for this, citizens assessed projects by rating them on a 5-point Likert scale with respect to three main aspects: the solution approach ('The project implements an appropriate solution to the addressed social problem'), the project progress ('The project is likely to reach its goals'), and the regularity of project updates ('The project informs regularly about its progress'). The CrowdMonitor has been available for 6.5 months during which a total of 521 different users made 580 assessments of the 28 projects funded by CHEST, totalling 1,738 responses to individual questions[3]. Most assessments were positive or very positive (82 per cent) with a minority of undecided (13 per cent) and a small portion of negative votes (5 per cent). Most negative and undecided votes related to the regularity with which the projects informed about their progress.

Such rating patterns suggest that the crowd assessments can be considered credible, though probably skewed by votes from avid project supporters. The use of CrowdMonitor for continuous feedback rather than a final verdict of a project's success is likely to have contributed to more realistic feedback. This is supported by the assessment of projects based on predefined social impact key performance indicators (KPIs) by the CHEST consortium, which were even more positive than the crowd results.

While the crowdsourcing model worked well in this case due to the voluntary engagement of participants (based on interest in the topic and/ or results of a given project), and low-effort feedback on project progress, critical issues can arise, ranging from the relationship between participant motivations and the quality of contributions, to ethical concerns such as the relationship between benefits accruing to participants and those accruing to the project leaders. These critical issues in different models of crowdsourcing and citizen science have received increasing attention and should be carefully considered when applying crowdsourcing methods (see e.g., Harris & Srinivasan 2012; Gilbert 2015; Resnik, Elliot & Miller 2015; Bowser et al. 2017).

Citizen Cyberlab and the Open Seventeen Challenge

At Citizen Cyberlab (CCL)[4], researchers from different backgrounds experiment with new forms of public participation in research, encouraging citizens and scientists to collaborate in new ways to solve major challenges. The lab is a partnership between the European Particle Physics Laboratory (CERN), the UN Institute for Training and Research (UNITAR) and the University of Geneva. In September 2015, the United Nations adopted Agenda 2030, which includes a set of 17 Sustainable Development Goals (SDGs) that aim to end extreme poverty, fight inequality and injustice, and tackle climate change over the next 15 years. The Open Seventeen Challenge[5], launched by the Citizen Cyberlab in 2015, is based on the understanding that some of the datasets best able to monitor progress towards the SDGs are local in nature, and can thus be better generated and collected by individuals and organisations representing civil society. The Open Seventeen Challenge involves three other partners: GovLab (the Governance Lab at New York University)[6], the advocacy group ONE Campaign[7], engaged in actions to end extreme poverty and preventable diseases, and SciFabric[8], which develops open source crowdsourcing tools.

Approach

In traditional citizen science, the involvement of professional scientists helps to address issues of data quality due to wide variability in the skills and expertise of participants. However, modern technology means that even those without research experience can in theory set up a participatory initiative using open source hardware sensors and software platforms that automate statistical validation procedures. This is particularly true for social and civic projects in which participants are asked to collect data or contribute to data analysis.

The Open Seventeen Challenge provides step-by-step coaching in the design and implementation of crowdsourcing projects led by non-professionals to increase their chances of success and impact. This includes both technological and social aspects. The Challenge recurs every six months, and involves the following elements:

- A project pitching phase: Candidates identify open data relevant to an SDG (e.g., photos, scanned documents, video clips, tweets), define a crowdsourcing project with clear, measurable outcomes, and then submit their idea. A maximum of 10 projects judged viable, or having a good potential of becoming so, are selected.
- Online coaching sessions: Sessions use a web conference platform and specifically designed online tools for project development. Over three months, the partner organisations help refine the project concept, including how to use crowdsourcing and ensure data quality with CCL, and how to optimise social impact at the community and policy levels with GovLab.
- Technical implementation and promotion: The projects set up a prototype crowdsourcing app on an open source platform, web or mobile, with the help of SciFabric[9] and are then promoted through their networks and at international events, benefiting in particular from the ONE Campaign's[8] strong international following and social media savvy to raise awareness.

Results and challenges

In 18 months, the Open Seventeen Challenge has issued three calls and coached more than 25 projects in diverse areas. From the first two calls, partners coached 10 projects, including crowdsourcing for a street guide to sustainable businesses, a platform to facilitate access to generic medicines for specific diseases in Latin America, projects to crowdmap sexual violence in India, tracking water policies in Nigeria, mapping the resources in a mega-slum of Mexico City, and other initiatives enabling SDG monitoring led by civil society.

In the most recent call, the Open Seventeen Challenge invited citizens to tackle specifically SDG 11, which is about making cities inclusive, safe, resilient and sustainable. The projects participating in the ongoing coaching sessions include mapping food markets in cities; sampling and monitoring air quality in Santiago, Chile, and Geneva, Switzerland, with wearable open source air detectors; and monitoring the international reconstruction work in Gaza.

While traditional sources of official data remain important, such data can also be expensive to generate and leave large data gaps in areas where traditional data gathering methods are not applicable. The next

> **Box 9.4. Using the crowd to map rural services in Crowd2Map Tanzania**
>
> *Crowd2Map Tanzania*
>
> In this project, a teacher reached out wanting to map rural Tanzania. Through Open Seventeen, she learned about the open source application Epicollect (http://www.epicollect.net/). With contacts from the partners' networks and the help of the crowd, Crowd2Map Tanzania was set up. The project has already mapped hundreds of services in Tanzanian villages and hosted an international mapping day. Data are now open and publicly available on OpenStreetMap.

step for the Open Seventeen Challenge will be to connect the grassroots initiatives to official government data producers and inter-governmental institutions, to ensure that crowdsourcing of open data by the public becomes a valuable resource in achieving the SDGs. As the executive director of UNITAR, Nikhil Seth, recently stated in a co-signed correspondence piece in *Nature,* 'governments will need to support projects that promote public participation in measuring progress towards the SDGs. National statistics offices must develop best practices for integrating crowdsourced data' (Flückiger & Seth 2016, 448).

Conclusions

Digital social innovation and participatory citizen science share the goal of engaging citizens with scientists and other professional actors in the collaborative development of different types of scientific, professional and practical knowledge, related to social needs. Ideally, individual citizens and local communities collaborate in the entire process of scientific, exploratory or creative inquiry: from the problem definition, through data collection, to analysis and interpretation, solution implementation and take-up. Successfully realising such types of engagement requires supporting different types of motivations and participatory activities, and appropriate methods for different purposes and project stages. In addition to existing experiences in the fields of citizen science and DSI, the methods and lessons from user-centred and participatory design provide actionable insights into how this might be successfully achieved.

Online platforms for collective intelligence can facilitate practical implementation by providing an initial community, access to crowdsourcing resources and (particularly important) coaching and monitoring support. Lessons from literature and case studies discussed in this chapter suggest that successful platforms will offer a range of different participation modalities with varying levels of effort, allowing citizens to switch between different types of engagement based on their motivation, capabilities, needs and resources. This should include both simpler activities, requiring little effort and little continuity, and more complex activities, requiring more effort and/or more regular engagement. Face-to-face workshops or co-design sessions can be effectively combined with online interaction such as continuous online feedback and well-known innovation methods such as lead user involvement and citizen experimentation in real-world piloting. Incorporating regular feedback to the participants and different mechanisms of social recognition are important for supporting the continuity of engagement. Coaching, training and monitoring support (online and offline) are essential enablers, but are resource and effort intensive. Crowdsourced approaches can provide one part of the solution (e.g., for continuous project feedback and monitoring). Other possible solutions could include better support for peer-exchange between different projects, recruitment of scientists and other professionals as volunteer mentors, or a community-driven massive open online course (MOOC) on designing and implementing DSI and participatory citizen science projects.

Notes

1 The Hybrid Letterbox project was partially supported by the European Commission within the CHEST project, itself partially funded by the EC, grant agreement No. FP7-ICT-611333, http://chest-project.eu (see case study presented in this chapter).

2 The Collective Enhanced Environment for Social Tasks (CHEST) project was partially funded by the European Commission (grant agreement No. FP7-ICT-611333, http://chest-project .eu) within the Collective Awareness Platforms for Social Innovation and Sustainability (CAPS) programme: https://ec.europa.eu/digital-single-market/en/collective-awareness.

3 A few users did not reply to all three questions.

4 http://citizencyberlab.org/

5 http://openseventeen.org/

6 http://www.thegovlab.org/

7 https://www.one.org

8 https://scifabric.com/

10
Creative collaboration in citizen science and the evolution of ThinkCamps

Margaret Gold[1] and Erinma Ochu[2]

[1] *European Citizen Science Association (ESCA), Berlin, Germany*
[2] *University of Salford, UK*

corresponding author email: mg@margaretgold.co.uk

In: Hecker, S., Haklay, M., Bowser, A., Makuch, Z., Vogel, J. & Bonn, A. 2018. *Citizen Science: Innovation in Open Science, Society and Policy*. UCL Press, London. https://doi.org/10.14324/111.9781787352339

Highlights

- Creative collaboration events foster co-creation, co-design and collaborative thinking at key points in the citizen science research cycle. They can help to grow science capital and thus deliver on the principles of citizen science.
- Such events can be held at any or all stages of the project lifecycle, from initial development to sharing outcomes.
- The hybrid ThinkCamp event format is well-suited to citizen science and can diversify participation, support knowledge sharing and engage a wider audience in the development of new ideas and projects.
- ThinkCamps can support engagement with policymakers to bring community-based citizen science initiatives into the fold of existing scientific activities that inform policy and civic action.

Introduction

The global aim of citizen science is to actively engage the public in the scientific process, with an emphasis on the importance of being open and inclusive, and a desire to facilitate creativity, learning and innovation throughout (see also Hecker et al. 'Innovation' in this volume). Initiators of citizen science projects are increasingly encouraged to engage more diverse participants to grow 'science capital' and deliver the benefits of science outcomes to as wide a population as possible (see also Edwards et al. in this volume).

While citizen science is traditionally driven and initiated by researchers who then reach out and engage citizens to help them solve research challenges, more communities are becoming active in devising and leading their own citizen science projects (see Ballard, Phillips & Robinson; Mahr et al., both in this volume). This provides an opportunity for practitioners to support grassroots community involvement throughout the entire research process: from defining the problems and framing the questions, through designing and launching the project, to collecting and making sense of the data – including writing academic papers, sharing findings widely, and taking action in their community (see also Novak et al. this volume, on digital social innovation approaches; and Kieslinger et al. in this volume, on outputs from citizen science projects).

This chapter discusses how to harness the potential of creative collaboration through ThinkCamp events – an 'unconference' style event with an open and creative environment designed to foster co-creation, co-design and collaborative thinking at key points in the citizen science research cycle. It draws on the authors' experiences of running (and participating in) creative collaborative events and explores their potential to support inclusive, co-creational approaches to citizen science. Finally, it makes specific recommendations for project initiators, event organisers and policymakers.

Science for all: The case for creative collaboration

The role of the 'citizen' in citizen science has been strongly emphasised since the mid-1990s, when the term 'citizen science' was first coined (Bonney et al., 'Public Participation', 2009; Irwin 1995). More recently, Schäfer and Kieslinger (2016) plea for even more diversity in citizen science to further close the divide between society and science, and recommend a wider range of approaches including 'the emergence of new forms of collaboration and grassroots initiatives'. (Schäfer & Kieslinger 2016, 1)

Citizen science project initiators are encouraged to pursue collaborative and democratic methods that involve the public in all aspects of citizen science, as in 'extreme' citizen science (Haklay 2013) where, 'Approaching and coaching communities to express their needs has the potential to generate very innovative projects that not only contribute to knowledge making but also to true social change – this is part of a wider approach of participatory action research' (Cunha 2015). Extending this approach to also influence policy by engaging policymakers provides another political dimension to citizen science. We propose that Think-Camps might offer a way to facilitate this in practice, an approach that contributes to the field of participatory democracy (See Smallman in this volume).

The value of cross-disciplinary collaboration across traditional organisational boundaries is well recognised in business (Mattessich & Monsey 1992), in scientific research (Hara et al. 2001) and in facilitating radical innovation within industries (Blackwell et al. 2009). The role that cross-disciplinary collaboration can also play in citizen science, to broaden and deepen the role of citizens, is becoming increasingly clear:

> We thus ask ourselves how may the combination of insights from artist-designers, natural and social scientists, change the status and indeed the experience of engaged citizens beyond the denomination of mere 'data drones'?. . . . it is perhaps here that interdisciplinary collaboration becomes most relevant, allowing us to be more inventive with people and with technology . . . In this way the conventional parameters of what is expected of public participation and what counts as monitoring can be potentially shifted.
>
> (Hemment et al. 2011, 63)

The concept of creative collaboration arose in the business world in an effort to embrace a more grassroots approach, where collaboration is:

> an act of shared creation and/or shared discovery: two or more individuals with complementary skills interacting to create a shared understanding that none had previously possessed or could have come to on their own. Collaboration creates a shared meaning about a process, a product, or an event.   (Hargrove 1997, 33).

These characteristics of creative collaboration – endeavouring to achieve shared value and create something new – are well-suited to citizen science, where the process is as important, if not more so, than the outcome (Freitag 2013). This diversity of input also improves the effectiveness of the approach and the quality of the outcomes of citizen science: 'Incorporating diverse ways of knowing into the analysis of a given issue increases understanding of the issue and offers solutions better tailored to the full context' (Freitag 2013, 2).

Growing science capital

One lens through which to view the role that citizen science plays in society is the concept of science capital, which looks at the level and depth of exposure that communities, families and individuals have to science knowledge and scientific thinking (see also Edwards et al. in this volume). Science capital is related to social capital and cultural capital in that it encompasses all the science-related knowledge, attitudes, experiences and resources that one acquires through life (Archer et al. 2015), and may lead to the pursuit of a career in science (Edwards et al. 2015). Citizen science projects can have a tangible impact on growing science capital by designing recruitment and engagement efforts to reach as broad a spectrum of people as possible, with an emphasis on involving children, young adults, and families with low science capital (Edwards et al. 2015; see also Makuch & Aczel; Harlin et al., both in this volume).

Organising creative collaboration events around community-specific issues that impact people's lives directly gives participants the opportunity to a) mingle with scientists to broaden their understanding of what science entails and what scientific careers look like; b) direct a line of scientific enquiry towards outcomes for their communities, incentivising active involvement and fostering ownership; and c) co-create new citizen science projects with a genuine local impact.

This approach builds on the spectrum of public involvement goals established by the International Association for Public Participation (IAP2), which begin with information sharing and build up to collaborative acts of partnership across the decision-making process, such that the final decision is in public hands (Ramasubramanian 2008). It is important to recognise the potential power dynamics inherent in community-based participatory research (Banks et al. 2013) and citizen-led digital innovation (Whittle et al. 2012), and to ensure these events present the opportunity to foster scientific citizenship among all participants (Irwin 2001).

Indeed, the first of the Ten Principles of Citizen Science is: 'Citizen science projects actively involve citizens in scientific endeavour that generates new knowledge or understanding'. (ESCA 2015; Robinson et al. in this volume). This major central theme of inclusiveness and involvement is re-emphasised in the third principle, 'Both the professional scientists and the citizen scientists benefit from taking part', and again in the fourth principle, 'Citizen scientists may, if they wish, participate in multiple stages of the scientific process'. Delivering on these principles in practice requires building in opportunities for collaboration between citizens and scientists throughout the project, from initiation to conclusion.

Ideally, this allows citizens to define at the outset what research questions are most relevant for them and their immediate environment, and how they can benefit from the process and outcomes (Sanders & Stappers 2008). Creative collaboration events provide a space to bring these principles to life, curating around the potential chaos of many voices.

Creative collaboration events

Management books are full of good advice about how to nurture creative collaboration within organisations (Hargrove 1997), or how to open the innovation process to a wide range of beneficial partnerships (Chesbrough 2003). These formalised methodologies are well-suited to a commercial context with either a shared profit motive or the desire to develop innovative new products and services, but are less useful for garnering public participation.

New online models of co-creation, collective intelligence and deliberation that foster scientific agency and democratic participation are emerging (see for example, Miah 2017; Saunders & Mulgan 2017), but the reality in citizen science is that individual participants can be widely spread demographically as well as geographically, with unequal access to the internet (see Haklay in this volume). Face-to-face events have therefore evolved to embrace the principles of citizen science and are designed to support creative collaborations locally, while also being compatible with cross-border citizen science by dispersing such events across a wider range of locations.

Creative collaborative events can also be held throughout the lifecycle of a citizen science project, when formulating research questions, designing the project, co-designing any tools, launching the project and sharing and celebrating the outcomes.

Creative collaboration events are often known as 'unconferences', a term dating back to the 1998 announcement of the XML Developers Conference in Montreal, Canada (Bosak 1998). Their original purpose was to be more participatory than the classic 'sit-and-listen' formal conference, and to facilitate in-depth conversations and knowledge sharing. Unconferences are participant-driven, often with no set agenda beyond an opening statement, and they are frequently based on the Open Space Technology technique developed by Harrison Owen in the mid-1980s

(Owen 1993). Today, there are several common types of events in this category: Open Space, BarCamps and hack days and hackathons. These are considered in more detail below.

Open Space events

Open Space Technology – oddly named, as it is more properly an approach or technique – brings order to chaos by relying on individual participants' ability to self-organise when a safe and welcoming space is provided for them. In essence, people are brought together around a defined subject and then provided with the space to raise the issues that matter most to them, thus setting the discussion agenda for the rest of the event. When people gather in an Open Space group, the 'law of two feet' applies – any individual not contributing or getting anything out of the break-out group should move to another group.

Open Space is most effectively used within organisations, communities or groups of people who have a strongly shared goal because it relies on participants taking ownership of any actions arising from the sessions. It works best when high levels of complexity, diversity, conflict (real or potential) and urgency are present (Owen 2008). A useful repository of resources for organising Open Spaces can be found on the Open Space World website (http://openspaceworld.org).

BarCamp events

The BarCamp format was inspired by the Friends of O'Reilly Conference, known as FooCamp (Tantek 2006), created by O'Reilly Media founder, Tim O'Reilly, at the turn of the millennium. The defining feature is a whiteboard or brown-papered wall on which participants draw up their own agenda for the event. As Tim O'Reilly recalls,

> We did the very first Foo Camp in 2003. It was in the middle of the dotcom bust, and we had a lot of empty space. It was really for fun, a thank-you to all the people who had given us the gift of their time, attention and ideas over the years. The output is not what we learn but what they learn. It goes back to creating more value than you capture. I love helping people make new connections.
>
> (O'Reilly in Levy 2012)

BarCamps similarly enable the spontaneous creation of the agenda and session content at the event itself, by way of a scheduling wall where participants post and announce their sessions. 'There are no spectators', the BarCamp philosophy goes, 'there are only participants' (DeVilla 2011). BarCamp gatherings are increasingly widespread globally, including science-themed BarCamps (http://lanyrd.com/2011/scibar camb/) and citizen science–themed BarCamps (https://wikimedia.de/wiki /Wissenschaft/csbarcamp; and http://buergerschaffenwissen.de/bar camp).

These events support self-directed learning and knowledge sharing, and can strengthen a sense of community. They are not usually designed for prototyping or the development of new ideas, and rarely lead to action planning beyond the event. A useful repository of event information and resources for organising BarCamps can be found at the official BarCamp wiki (http://barcamp.org).

Hack day and hackathon events

Finally, hack days and hackathons stem from formalised approaches to collaboration and co-creation that began to move beyond the realm of open innovation and open research and development (R&D) in the early 2000s (Chesbrough 2003), and into the realm of open source communities and technology organisations. The open source community pioneered 'outside-in' creative collaboration events to produce code and develop new functionality and features, and created a space that went beyond idea generation and information sharing. OpenBSD and Sun invented the hackathon event format in 1999 to enable a high-intensity collaborative coding effort around a shared code base (http://www.openbsd.org/hackathons .html). A more free-flowing hack day format was introduced by Yahoo! in 2006 to engage with their external developer community, enhance internal product development and support the creative application of their developer tools and software development kits (Dickerson 2005; Dickerson 2006).

As with BarCamps, hack days and hackathons continue to grow in popularity as a creative outlet for developers and a way for organisations to engage with a wider community of participants than usually possible. They have now expanded beyond their initial software developer orientation into fields such as civic engagement (https://www.bathhacked.org/), science (http://sciencehackday.org/), health (http://nhshackday.com) and museum engagement (https://museumhack.com).

Hack days are usually focused on the technology community and those with technology skills so are particularly well-suited for prototyping new ideas on the fly, testing prototypes for new citizen science mobile or web applications (Sanders 2008), and inviting the creation of new tools for citizen science based on existing software, technology platforms or devices. An excellent best practice guide to organising hack days can be found in the Hack Day Manifesto (http://hackdaymanifesto.com/).

The ThinkCamp approach to creative collaboration

The ThinkCamp methodology was first developed by the Mobile Collective (Gold 2011) to provide an open and creative environment for developing new products and services at the cross-section of different fields, such as mobile technology and health services. It came from the observation that a new generation of health care professionals were technology savvy and saw opportunities around them, but did not have the developer skills to act on them; at the same time, many in the technology community were passionate about health care provision based on personal experiences, but had no direct channel to make a positive impact.

The event format was born out of the desire to combine the improvisational creativity of the hack day with the self-organising principles of Open Space Technology. The ThinkCamp methodology also incorporates the interdisciplinary approaches to open innovation of the 'Fuzzy Front End' of R&D (Rubinstein 1994; Sanders 2008), which optimises creative problem-solving by taking the process outside the walls of a single organisation (Rochford 1991).

ThinkCamps invite participants from a diverse range of disciplines, skill sets and experiences to collaborate on addressing problems, rising to challenges and taking advantage of new opportunities. A key goal is to lower the bar for non-technical participation so that people without coding skills who might not feel comfortable at a hack day are able to join teams and make a significant contribution. This format evolved further during the EU-funded Citizen Cyberlab project (http://archive .citizencyberlab.org/) to provide a space for offline community-building and creative problem-solving, where scientists and citizens could meet to devise new projects or further develop the Citizen Cyberlab toolkit (Gold 2012). Although participants do not require computer programming or other technology skills, they can still contribute to the development of new technology features and functionality in the role of 'user as co-designer' (Sanders 2008), and provide inputs to prototyping at the event.

The hybrid ThinkCamp event format is uniquely suited to the context of citizen science, where external voices are valued. Supporting the sharing of knowledge among diverse participants and building bridges to engage a wider audience in the development of new ideas and projects helps to deliver on the principles of citizen science throughout the project lifecycle.

The evolution of the ThinkCamp format

The first iteration of the event format was the MC ThinkCamp mHealth organised by the Mobile Collective in June 2011, to address opportunities and challenges in health care by applying mobile and web technologies in innovative ways. Seventy-five participants attended, primarily mobile developers, technologists and health care professionals. The event opened with two keynote talks to provide context and inspiration for the discussions alongside demonstrations of current mobile app initiatives in health care. The Mobile Collective team then facilitated the creation of the agenda in the Open Space style and provided support for the working groups that emerged organically. Nine working groups formed, eight of which presented outcomes at the end of event and two of which continued after the event (Gold 2011).

The engagement and interest among the participants was high, with many indicating that they wished to stay involved in the further development of the ideas that emerged. However, it was not possible to 'own' the projects as event organisers and few participants were in a position to take on product development outside the scope of their day job. Bringing this format to citizen science, with the aim of supporting grassroots public involvement, therefore means ensuring project ownership is in place to take things forward.

ThinkCamps start with short presentations to set the scene and provide context as inspiration or to present the challenges for the day. If the agenda is to be set by participants, then the event can unfold as described above, which requires little prior planning, but relies on participants pursuing the ideas generated after the event. If the agenda is to feature pre-defined challenges, organisers invite the submission of ideas beforehand, work with challenge 'owners' to present them in a way that invites collaboration, and structure challenges so they can be reasonably tackled within the time allotted. A challenge can be a problem within an existing project or technical platform, a new technology, a new opportunity, an idea for a new project and so on.

Challenges need to be presented by the owner – the person with insight into the problem or opportunity, who is inviting participation but can also take ownership of any ongoing actions beyond the event itself, either by incorporating them into existing processes, or taking the lead on new initiatives. After the challenge presentations, participants self-select which working groups to take part in, facilitated by the challenge owner. As with any Open Space, the 'law of two feet' applies so participants should always feel free to move among discussions to those they are learning from, contributing to or enjoying. The event culminates in a 'show and tell', where each group presents their challenge and the outcome of their work or discussion, closing with a request for participants to indicate any desire to stay involved.

Challenge-driven ThinkCamps for citizen science

The next two iterations of the ThinkCamp format took place as part of the international Citizen Cyberscience Summit conference series in London in 2012 and 2014 (CCS12 and CCS14). Although billed as a hack day for ease of communication and to attract external participants with technology skills for prototyping, the format followed that of the mHealth ThinkCamp but was more deliberately curated with a range of pre-defined challenges connected to the themes of the conference and presented by challenge owners. The goal was to open the event beyond the traditional conference community of practitioners (primarily citizen science practitioners from research institutions and academic organisations) to harness the knowledge and skill sets of a wider audience for creative problem-solving to the benefit of current and future projects.

This included inviting members of the regional hack day and DIY science communities (see Novak et al. in this volume), inviting volunteer participants from the citizen science projects represented at the conference, posting event information to Meet-Up groups related to the challenges (meetup.com), sharing information with grassroots organisations in related fields, making event registration public on Eventbrite (eventbrite.co.uk) and promoting on the event discovery platform Lanyrd (http://lanyrd.com/2014/citizen-cyberscience-summit/).

The challenges were framed to address problems in the field, define and develop the next step for existing projects, respond to challenges in practice and take advantage of new opportunities. Each challenge represented a different stage of the project cycle, from ideas for new projects to the furthering of existing projects. Both events opened with the challenges being presented in 'elevator pitch' style (a persuasive sales speech that takes no longer than an elevator ride), in front of a wall of posters for each of the challenges (see figure 10.1a). Participants were then

Fig. 10.1 The Citizen Cyberscience Summit ThinkCamp 2014, London England. *Image A* – Ian Marcus of the Centre for Research and Interdisciplinarity, Paris, introducing the SynBio4All Challenge during the 'Elevator Pitches'. *Image B* – Jesse Himmelstein of the Centre for Research and Interdisciplinarity, Paris, working with fellow participants on the RedWire.io Challenge. (Source: Margaret Gold, CCBY)

invited to join relevant working group tables based on their own experience, skills, personal interests or ability to make a contribution (see figure 10.1b).

The CCS12 conference featured 13 challenges and approximately 50 participants, and led to a number of projects moving forward with new ideas and fresh participation. A range of interesting prototypes were demonstrated at the end (see figure 10.2c) and an audience vote was taken on various prizes to be won (see figure 10.2d).

> For me, several highlights of the conference included the impromptu integration of different projects during the summit. Ellie D'Hondt and Matthias Stevens from BrusSense and NoiseTube used the opportunity of the PLOTS balloon mapping demonstration to extend it to noise mapping; Darlene Cavalier from SciStarter discussed with the Open Knowledge Foundation people how to use data about citizen science projects; and the people behind Xtribe at the University of Rome considered how their application can be used for Intelligent Maps – all these are synergies, new connections and new experimentation that the summit enabled. (Haklay 2012)

Building on this success, CCS14 featured a fresh set of 14 challenges and approximately 60 active participants, with five challenge outcomes presented at the end. Not only did the collaborations result in a wide range of projects being moved forward, but a number of new initiatives came out of the connections made.

Fig. 10.2 The Citizen Cyberscience Summit Hack Day 2012, London, UK: *Image C* – Leif Percifield of Newell Brands presenting the outcomes of the Air Quality Egg challenge at the Show & Tell. *Image D* – Louise Francis of Mapping for Change and UCL ExCiteS taking the audience vote, with a noise metre held aloft, for prizes to be won. (Source: Cindy Regulado, CCBY)

> The Cyberscience Summit Hack Day 2014 was a great experience for us at the Lightyear Foundation. We met many people, particularly Rick Hall from Ignite! From this meeting grew an idea for a Lightyear-Ignite! collaboration on Lab_13 Ghana: a pupil-led science space based at a school, based on similar projects in the UK. Following this we raised the funds, recruited volunteers, and in April 2015 launched the pilot at the Agape Academy in the Bosomptwe district in Ghana, which has already worked with 29 local schools and over 600 students. None of this would have happened without the Cyberscience Summit Hack Day!
>
> (Gavin Hesketh, UCL/Lightyear Foundation)

Workshops ran in parallel, which fit well with the hands-on theme and often provided relevant know-how but took time away from the Think-Camp itself. It takes about half a day for participants to embed themselves in a challenge, so where possible, a citizen science ThinkCamp should be a two-day event, with stronger connections between the workshops and the challenges. CCS14 also had a Citizen Science Cafe, based on the World Cafe format for hosting large group dialogue (http://www.theworldcafe .com/). This was introduced the evening before the ThinkCamp and brought 50-plus volunteers from various citizen science projects together with the organisers and scientists behind the projects. This was an important recognition of the value of the volunteer community and a chance to meet like-minded people, as well as providing project owners with valuable feedback and insights. Unfortunately, almost none of these external attendees participated in the ThinkCamp the next day, perhaps due to the relative ease of attending an evening event over a full-day weekend event.

However, these events demonstrated that participants who had attended the full three days of conference sessions (keynotes, talks and workshops) came to the ThinkCamp with a range of new ideas and were eager to apply them in a new context, enhancing the discussions around the presented challenges. As the conference organiser reflected, after a day of 'listening' and a day of 'talking', the third and final day of the summit was about 'doing' (Haklay 2014).

The citizen science ThinkCamp at ECSA 2016

The most recent iteration of the challenge-driven ThinkCamp format was at the first international European Citizen Science Association (ECSA) conference in Berlin in 2016, as a full-fledged citizen science ThinkCamp to which the local Berlin DIY science, bio-hacker and maker communities were invited (see also Mazumdar et al. in this volume). Organised together with Lucy Patterson, who is co-organiser of Science Hack Day Berlin and the Berlin Science Hacking Community, the event was held on the third and final day of the conference and was structured as a day of collaboration, sharing and the exchanging of ideas (see box 10.1 below). To reduce barriers to attendance, the event was free for non-conference participants, held in a ground-floor space for ease of access and on a Saturday so that taking time off work would not be necessary. Participants were also encouraged to attend any of the mainstream conference sessions happening in parallel with the ThinkCamp for free.

Box 10.1. Citizen science ThinkCamp, ECSA Conference 2016

Why: To engage with local Berlin grassroots science and maker communities as part of the conference, collaborating on opportunities and addressing challenges in citizen science.
When and where: May 21, 2016, Berlin, Germany
Event wiki: https://sites.google.com/a/gold-mobileinnovation.co .uk/ecsa2016—citsci-thinkcamp/About-the-Think-Camp /home

---

Who: Over 75 participants, of which 12 attended from outside the conference – participation was encouraged from local Berlin DIY science, bio-hacker and maker communities as well as volunteer participants in the citizen science projects represented at the conference.

What:

1. **The ECSA Inclusiveness Challenge** – How can we ensure that ECSA becomes an inclusive organisation?

2. **The WeCureALZ 'Engaging Diversity' Challenge** – Help us design unique and effective strategies to engage and retain diverse communities.

3. **The CitSci Communities of Europe Challenge** – Mapping the citizen science communities of Europe: How and why should we do this?

4. **The Overleaf Collaborative Writing Challenge** – How can Overleaf support collaborative writing between academics and citizen scientists?

5. **The Museum Data Visualisation Challenge** – How can the visualisation of observation data gathered in the field be made more engaging and dynamic for participants?

6. **The HealthSites.io 'CitSci for Health' Challenge** – What Citizen Science projects become possible with the health facilities geodata being mapped on the HealthSites.io platform?

7. **The Motion-sensing Camera Trap Challenge** – Help us to design and build a DIY camera trap for citizen scientists around the world.

8. **YOUR Citizen Science Challenge** – Two challenges were proposed spontaneously by participants on the day: 1) How can we apply citizen science to the issues faced by refugees? and 2) How can we make sure that citizen science projects are interoperable?

Outcomes: Of the seven pre-defined challenges, four are still actively being worked on at the time of writing, and two may lead to new collaborations. The two spontaneously presented challenges led to fruitful discussions and new connections made between the participants.

A key innovation at the ECSA ThinkCamp was to host a 'Citizen Science Disco' the evening before, which featured a series of talks from the local DIY science, hacker and artistic communities to provide them with an important voice that might otherwise have been missed. This set the scene for the ThinkCamp challenges the next day, where the goal was to collaborate with the broadest local audience possible (Patterson 2016).

Outcomes of the citizen science ThinkCamp challenges

Benefits to the projects and project owners who presented a challenge at the ThinkCamp included making new contacts, the exploration of project goals and audiences, insights into engaging audiences and new practical solutions. Having project leaders present to lead discussion was key to ensuring results and ownership of new actions, and this also worked particularly well for the spontaneous challenges where challenge owners were motivated by the projects presented and opportunities to collaborate.

Participants in the ECSA Inclusiveness Challenge session (see the challenge poster in figure 10.3 image E) agreed that citizen science has the potential to be a transformative approach and make research more inclusive, but that work needs to be done to achieve this. Three main areas of focus were defined during the discussion, with a range of main

Fig. 10.3 The ECSA citizen science ThinkCamp 2016, Berlin Germany: *Image E* – The ECSA Inclusivity Challenge poster. (Photo credit: Margaret Gold, CCBY). *Image F* – The ECSA ThinkCamp participants in working groups alongside the related challenge posters. (Source: Florian Pappert, CCBY)

points for attention and action items being picked up by ECSA head-quarters in partnership with synergistic activities such as the Citizen Science COST Action. These points were worked on further at the Doing-it-Together Science (DITOs) European Stakeholder Round Table on Citizen and DIY Science and Responsible Research and Innovation (RRI) (Göbel 2017).

For the WeCureALZ (now 'EyesOnALZ') Engaging Diversity Chal-lenge, 'the ThinkCamp had a huge beneficial impact . . . across many dimensions – a testament to the preparation, participants, and format' (Pietro Michelucci, Human Computation Institute), including renaming the project and associated game, a new approach to designing game lev-els, removal of the 'test phase' at the start of the game to lower barriers to entry, and consideration of accessibility factors for an older audience, such as larger fonts, buttons and full-screen video elements (Ramanauskaite 2017b).

The facilitator of the CitSci Communities of Europe Challenge, Jose Luis Fernandez-Marquez of the Citizen Cyberlab and University of Geneva, reported that the ThinkCamp brought new contacts, which will be beneficial to the DITOs project they are participating in, as well as establishing a number of key functional requirements:

I was especially surprised with the interest of the EC [European Commission] in these kind of maps. Initially the goal of the map was outreach – to increase the visibility of CS [citizen science] projects over Europe, allowing citizens to easily find new CS projects. How-ever, the information we were gathering was very useful for the EC to evaluate CS projects, their impact, to see what happen with the CS projects over the long term (especially those funded by the EU). Also, CS project owners were very interested in the map. They wanted to see the different technologies each of the projects is using. They mentioned as an example, that there are more than 10 CS pro-jects tracking foxes in cities, and they implemented the apps every time from zero.

The owner of the Overleaf Collaborative Writing Challenge was unfortu-nately unable to attend the event, but another participant at the conference volunteered to lead the discussion. A detailed discussion ensued, which identified the potential for a small research project and generated the enthusiasm to take it forward. However, lack of ownership or further investment might hinder development. The Healthsites.io 'CitSci for

Health' Challenge suffered a similar fate, with the project owner unable to attend at the last minute and no volunteer facilitator available. Consequently, this challenge failed to form a group of participants.

Participants in the Museum Data Visualisation Challenge discussions spent time defining who museum audiences are, and what their motivations and interests might be for museum data, before bringing that back into recommendations for the digital representation of data.

The Motion-sensing Camera Trap Challenge attracted a mix of participants with hardware hacking and DIY science skills, who further defined the challenges to building your own camera trap, and examined three alternative pieces of kit by taking them apart and making notes on the challenge Etherpad (Hsing 2016). Work on this challenge was moved forward beyond the event by posting a challenge to the broader DIY science community on the Hackaday.io platform (Ramanauskaite 2016), and running an open workshop session at the annual Mozilla Festival in London in November 2016.

The importance of encouraging and providing space for external participants to raise issues that matter to them, in order to draw on the wide range of experience and skills in the room, was again evident at the ECSA 2016 ThinkCamp. The two spontaneous challenges (see box 10.1) both led to fruitful conversations. Spontaneous challenge owners are often uniquely placed to act on any outcomes beyond the event because it is inspired by something directly relevant to them, and they gain the support of new contacts.

Best practice recommendations for creative collaboration events

Creative collaborative events can foster co-creation, co-design and collaborative thinking at all points in the citizen science research cycle. Challenges that are well-suited to creative collaboration have represented the full spectrum of the project lifecycle, from ideas for new projects and the beginning phases of newly funded initiatives, through mid-project improvements and impetus for new directions, to the creative application of existing tools and platforms in new ways, and finally to the representation of data upon research conclusion.

Additionally, by taking the time to reach out to a wider group of potential participants, particularly those connected to the subject matter of the challenges as well as those traditionally under-represented in citizen science research, means that more diverse experiences and viewpoints are brought to the table.

To meet their potential to support inclusive, co-creational approaches to citizen science, the following steps are recommended for project initiators and event organisers.

Before the event

1. Resource:
   a. Budget for a part- or full-time community manager, and the support of grassroots community spaces in funding applications for your events;
   b. Consider accessibility, travel and dietary requirements of participants in advance and budget for these costs.
2. Ownership: Invite pre-event challenge submissions and encourage attendance by the challenge owner. This is key to attracting participants and to following up on actions post-event. Consider how contributions will be recognised and accredited by the project owners.
3. Outreach: Actively reach out to a diverse range of participants and be sensitive to removing barriers to attendance, including time of day and physical location. Explore ways to give local people a platform and a voice, particularly those who would not call themselves citizen scientists.

During the event

4. Context: Set the context and find ways to make it relevant to what people already know.
5. Equity: Create the space to value and share knowledge and experience between all participants on an equal footing for mutual benefit. This might require self-regulation from some participants to ensure everyone's contributions are valued.
6. Representation: Build elements into the event programme that actively allow other voices to be heard such as World Cafe–style dialogue or guest talks.
7. Spontaneity: Invite, encourage and support spontaneous contributions from participants.
8. Innovate: Embrace serendipity, failure and unexpected outcomes to enable innovation.

9. Openly evolve: Document, evaluate and reflect on your events to share and help evolve creative collaboration approaches further. Prototypes, videos, reports and code can all be posted online. Be sure to credit everyone and get prior informed consent.

After the event

10. Connect: Provide forums or facilitate connections through which people can stay in touch and updated on progress (but which they can also opt out of).

Event organisers need to consider the fact that many people outside the existing community of citizen science practitioners do not necessarily identify with the label 'citizen scientist' (Eitzel et al. 2017; Lewandowski et al. 2017), even when they may be participating in activities that fit the academic definition.

> Problems include: The fact that not all communities are included in the conversation: not everybody identifies themselves with the same labels we use. That means we have the responsibility to be aware of these communities and reach out with them.
>
> (Ramanauskaite 2016)

A diverse range of voices contributing to the ThinkCamp process is possibly as important as the outcomes of the event itself: 'Citizen science does not replace this definition of "best available science" but adds a new dimension. The broader definition includes wider participation, broader impacts to society, and chances for many perspectives to add their voices to the final analysis' (Freitag 2013, 2).

Serendipity must be embraced when designing and running any variation of creative collaboration event, where participants are being invited to shape or entirely drive the agenda. Although the organiser can structure ThinkCamp events to support a certain desired outcome, once the event begins, control is handed over to those who are in the room – it is their event now, and they will take it in the direction that meets their needs, satisfies their curiosity or resolves their desire to seek a particular solution.

Harrison Owen advises strongly on the importance of letting go all control of Open Space, and defines the four principles as '1) Whoever comes is the right people. 2) Whatever happens is the only thing that could have. 3) Whenever it starts is the right time. 4) When it is over it is over'

(Owen 1993, 31). ThinkCamps should therefore not be resourced and funded with strict 'performance criteria' in mind, such as defined outputs or attendance numbers.

In fact, challenges spontaneously proposed by participants at the event itself should be actively encouraged and supported. Successful creative collaboration events can bring the virtuous circle of 'informal, unstructured and social' learning (Jennett et al. 2016, 15) to life in face-to-face interactions between participants and scientists:

> It is important to provide enough creative space where grassroots initiatives can flourish side-by-side with more established forms of scientific knowledge production and a platform where the community can meet and exchange ideas so as to establish fertile grounds for the broader dissemination and uptake of this collaboration between citizens and scientists. (Schäfer 2016, 10)

Further, with the advent of the DIY science, open science and maker movements, it is important to consider how to foster and build capacity, support the crossover of knowledge and know-how, and share with creative citizens participating in these spaces, as there is much to be learned from different groups (Patterson 2017). Holding project-funded events in grassroots community locations such as Fab Labs and Hackerspaces is one tangible way to provide these communities with much-needed financial support. (Patterson 2017; Ramanauskaite 2017a)

Indeed, the recent Arizona State University Maker Summit brought the maker and citizen science communities together to share insights, tools and best practices (Prange, Lande & Cavalier 2018). The Learning Outcomes and Next Steps report from this event can be found online at https://makersummit.asu.edu/. All such approaches to generate insights and foster cross-pollination by bringing these communities together through creative collaboration are welcomed.

Theoretically, creative collaboration events for citizen science can be situated alongside other creative research methods. For example, within media and communication studies, Gauntlett (2011, 4) considers 'making' as a way of connecting ideas, to other people and to the social and physical environment:

> This rarely seems to be a matter of 'making what I thought at the start,' but rather a process of discovery and having ideas through the process of making. In particular, taking time to make something, using the hands, gave people the opportunity to clarify thoughts or

feelings, and to see the subject matter in a new light. And having an image or physical object to present and discuss enabled them to communicate and connect with other people more directly.

It is here that creative collaborative events involving diverse participants might add considerable value to what citizen science offers, in terms of making sense of the world around us, our relationships to it and to one another.

Conclusion: Towards participatory democracy

The evidence and outcomes of ThinkCamps, which were designed to open up formal academic conferences to participation from a wider community, point to the value of embedding such events more deeply within citizen science projects. They are a valuable tool with which to foster co-creation, co-design and collaborative thinking during the citizen science research cycle.

Experience evolving these creative collaboration event formats to embed them within citizen science demonstrates the potential to deliver on the promise of science capital, and the principles for diversity and inclusion within citizen science as set out within the ECSA Ten Principles (see Robinson in this volume). More in-depth evidence and further research is required but it is important to consider how these approaches might be of value to supporting democratic participation in science policy by bridging the science-society gap.

The citizen-led approach to a shared understanding of both the problem and the solution, with event-based support for co-creation throughout, has clear implications for how policy could be formed in areas where science has a vital role to play, such as biodiversity management, air and water pollution, and fracking. For example, the pan-European DITOs project sets out to involve citizens in both bio-design and, critically, to contribute to policy on environmental monitoring.

Furthermore, as more communities become active in devising and leading their own citizen science projects, there is an opportunity for policymakers to not only play a key stakeholder role in the project life-cycle, but also to support such grassroots efforts by ensuring that there is a pathway to action and funding:

> Strategic policy-making needs to consider inclusive programme designs and funding mechanisms. . . . When we talk about funding,

agencies should consider a funding programme for citizen science projects that aims to collect the manifold experiences from the different project typologies of this ever evolving research methodology and that creates visibility for the potentials of citizen science for researchers and the public.   (Schäfer 2016)

Sociologists have outlined both the possibilities and practical and ethical challenges of deliberative democratic methods to engage citizens in public policy-making (Irwin 2001; Árnason 2012; Saunders & Mulgan 2017). The field of participatory democracy and the concept of the 'participatory turn' (Bherer et al. 2016) provides guidance as to how creative collaboration events could further bridge the gap between science and society, by scaling this approach to engage citizens, scientists and policymakers together. This has implications for funding bodies and how they select the initiatives which they support:

> A clear challenge to design a programme that allows participation of "grassroots" initiatives, which are driven by civil society organisations or by independent citizen scientists, therefore presents itself. . . . In the long run, citizen science should not be seen as separate from other research areas but as an integral part of existing scientific activities comparable to science communication. Thus the involvement of citizens could become one of the selection and evaluation criteria in existing funding schemes.   (Schäfer 2016)

Those planning their own future citizen science projects, or practitioners seeking to support grassroots initiatives for scientific enquiry, should therefore consider not only introducing such events as a tool for inclusion and co-creation, but also deliberately engaging with policymakers to bring community-based citizen science initiatives into the fold of existing scientific activities that inform policy and civic action. Policymakers should also be encouraged to consider the recommendations for running creative collaborative events as a process to facilitate a range of expertise contributing to and influencing decision-making.

Finally, given that citizen science projects often use the internet, and that participative democracy needs to draw on wider contributions, it will be important to consider and evaluate effective, equitable and accessible ways and tools to foster contributions to the co-design, analysis and reporting of citizen science projects. This might also help with tracking follow-up actions and contributions, sharing methods, innovations and progress more widely.

11

Integrating citizen science into university

Daniel Wyler[1] and Muki Haklay[2]

[1] University of Zurich, 8001 Zurich, Switzerland
[2] University College London, UK

corresponding author email: wyler@physik.uzh.ch

In: Hecker, S., Haklay, M., Bowser, A., Makuch, Z., Vogel, J. & Bonn, A. 2018. *Citizen Science: Innovation in Open Science, Society and Policy*. UCL Press, London. https://doi.org/10.14324/111.9781787352339

Highlights

- Universities are an integral part of citizen science activities.
- Universities gain breadth and strength in research by adopting and supporting citizen science, which consolidates their position and recognition in society, brings new resources and increases public trust in universities.
- Universities contribute to citizen science by providing professional infrastructure, knowledge and skills; ethical and legal background; educational facilities for present and future citizen scientists; sustainable teaching; and funding.
- University engagement in citizen science faces a number of challenges, which can be managed through project planning and the support of funders and policymakers.

Introduction

Research universities are usually seen as the place of the highest level of learning and home of cutting-edge science and innovation (Altbach 2011). They have been successful in training professionals and researchers and in establishing a sustainable research culture in many fields of knowledge, as well as providing fertile ground for the birth of new fields. Moreover, they create and maintain strong infrastructures for carrying out research, such as dedicated laboratories or global research networks. Along with this, scientific research has become significantly focused and successful scientists are often specialists in a highly professionalised environment (see Mahr et al. in this volume). This has led to a certain isolation of universities and research from society at large. Although universities are engaged with their local communities, their emphasis on the universalist value of science and globalised purview create the impression that they are detached from society (Bond & Paterson 2005). Previous policy on the integration of universities and society has focused on their economic mission – leading to innovations, creating companies and educating a knowledge-based workforce (e.g., the Triple Helix model of Etzkowitz & Leydesdorff 2000) – while their wider societal mission has been somewhat neglected (e.g., Harkavy 2006; Bond & Paterson 2005). To counteract these developments, universities have strengthened their relationship with civic society, for instance by broadening their science communication activities, engaging with schools or becoming sources of policy advice. Furthermore, they have launched campaigns such as the 'digital days' in Switzerland (https://digitaltag.swiss), which inform citizens about the digital revolution (see also Smallman in this volume, on responsible research). Furthermore, links and collaborations with industry are developed. Indeed, a dichotomy between academy and society is unsustainable, as both need each other (see also Mahr et al. in this volume).

Increasingly, laypeople are engaging in science and research activities (for recent overviews, see Haklay 2015; Bonney et al. 2014). There are many reasons for this, including the rise of societal concerns (such as climate change and demographic ageing), novel research projects that need a large number of participants and a certain mistrust of academic (official) research. Public participation in research has been further facilitated by the availability of modern communication technologies as well as the increased level of education across society (Haklay in this volume).

For universities, this rising interest and involvement of citizens in scientific activities should be viewed as a boon, despite some critical voices (e.g., Editorial 2015). It allows universities both to expand their research into new areas, and to strengthen their ties with society. Involving a large number of motivated people allows the investigation of problems previously beyond the means of a single researcher or a limited number of researchers. Involving citizens in research also makes it more compelling and acceptable to the public, thereby strengthening the position of universities in society (see also Hecker et al. 'Innovation' in this volume).

The involvement of lay people challenges established ways of doing academic research as well as the self-image of universities and their role within society. This requires new ways to properly and productively manage large and diverse groups of people with different levels of knowledge and access to necessary tools. There might also be legal and ethical questions that need to be taken up. This chapter addresses these issues based on a citizen science initiative of the League of European Research Universities (LERU) (League of European Research Universities 2016).

Citizen Science and universities

Many amateur scientists have contributed significantly to science over the years (Cooper 2016), notably in astronomy, archaeology, linguistics, zoology or botany. They were usually linked to universities and research institutions, as was the case with Charles Darwin. As the complexity and specialisation of science increased, opportunities for citizens without professional credentials to participate decreased, especially in the second part of the last century (Mahr et al. this volume). In recent decades, there has been a revival of citizen science activities, often outside universities. Indeed, citizen science activities are initiated and organised in many ways. For citizens and policymakers there is an increased need to find answers to great societal challenges such as environmental pollution, declining biodiversity or ageing. The growth in citizen science has been made possible by the rise of information technologies like the internet and smartphones (see Mazumdar et al. in this volume; Schroer et al. in this volume), open source software, digital fabrication technologies, online social network platforms and science 'kits' making science available to many people (see Volten et al.; Ceccaroni & Piera, both in this volume). There are also co-ordination efforts, with influential associations (e.g., Citizen Science Association [CSA] in the United States, European Citizen Science Association [ECSA], Australian Citizen Science Association [ACSA]) promoting best practice and knowledge sharing between practitioners and projects.

In view of these developments, universities need to find their own position, both to maintain their strong and recognised position in research, and to further the integration of science and society. Universities can also benefit from positive impacts on their research, teaching and even access to (financial) resources.

1. **Opportunities to increase the breadth and strength of research.** Citizen science projects substantially expand a university's research scope. Furthermore, involving citizens may mean results are higher

quality and more relevant to the research questions (e.g., Watson & Floridi 2016). This applies to projects which need skills beyond simple machine learning, or health projects where the citizens are themselves both the subjects and researchers. In these contexts, citizens provide essential knowledge to a project, enhancing its relevance.

2. **Teaching and student motivation.** Through participation in citizen science activities, students feel more engaged in their learning by participating in genuine scientific investigations where they are contributing to world knowledge (e.g., Oberhauser & LeBuhn 2012; Harlin et al. in this volume).

3. **Bridging the gap to society.** Increased university involvement in society means research can be tuned to local needs, thus integrating better with its community, addressing the gap between the universalist mission of universities and their obligations to the community within which they reside (Bond & Paterson 2005). This contributes to a science- and education-friendly environment, and complements other efforts for an informed society (see Edwards et al. in this volume). Through citizen science activities, universities can strengthen bonds with their localities, local communities, schools and local governments.

4. **New funding and resources**. Experience shows that citizen science projects open up new sources of potential funding to universities (Silvertown 2009), for instance because many funders and charitable foundations are interested in contributing to projects that benefit society or promote public engagement with science. In some cases, one may consider crowdfunding to further engage the public as research stakeholders although crowdfunding should be seen as a way to reach out and engage stakeholders more than as a mechanism to raise funding. Such projects can also bring new resources, as in the practice of volunteer computing in which people volunteer their unused computing resources or even build specialised computing devices to assist scientists and provide scientists with access to significant computing abilities (Cooper 2016; Curtis 2015).

Citizens can participate in university research at different levels. As in other citizen science activities, they often act as data collectors and more recently, also due to advances in software technology, as data analysts. Truly participatory projects can be even more rewarding as citizens are involved at all levels, from planning the research goals to analysing and communicating the results, such that projects benefit from the breadth of participants' knowledge, skills and enthusiasm (see also Novak et al. in this volume; Gold & Ochu in this volume). With many European societies passing the 40 per cent mark in terms of securing tertiary education (see also Haklay in this volume), the exchange between universities and citizens becomes easier and the potential pool of participants with an interest in specialised topics of science increases.

Universities' unique role in citizen science

Universities can benefit fundamentally from citizen science but including universities in citizen science activities also brings major advantages to citizen science projects and other stakeholders: Universities can support citizen science activities by using their assets, which include access to high-quality professional and research skills (see Volten et al. in this volume), existing infrastructure (including citizen science platforms), and interdisciplinary networks and collaboration.

1. **Professional infrastructure:** Research at universities can rely on existing well-developed state-of-the-art infrastructure such as computer facilities and well-equipped libraries, with knowledgeable staff that can assist with the interaction with the public, and in curation of data that emerges from citizen science activities in an open way that allows for public access.
2. **Professional knowledge:** Universities have experts in diverse fields who can provide essential support in statistics, computer science, legal and ethical knowledge, quality assessment, communication activities and more, which may yield higher quality and more trustworthy results.
3. **Teaching facilities and students:** Universities as teaching institutions can help instruct participants in research practice; and can bring large numbers of students to participatory science projects.
4. **Sustainable teaching:** Including citizen science in their teaching means that universities can educate their students to have a lifelong understanding of science and scientific decision-making, including engagement with research (see also Edwards et al. in this volume on the concept of science capital).
5. **Funding opportunities:** Universities have well established funding channels and networks that they can put to use for citizen science projects.

6. **Education of teachers:** Teachers have a critical role in public acceptance of research and the scientific method. Including citizen science as part of university-based teacher training courses supports teachers in raising awareness of science and encouraging pupils to engage in such projects (see Makuch & Aczel in this volume).

Projects at universities

There are general advantages of university participation in citizen science, but there are also particular types of citizen science projects that benefit from strong relationships with research universities.

1. **Large and complex data:** Projects where large and complex datasets are necessary, for example in sociology, astronomy or biodiversity, where standard existing automated data analysis tools may not reveal aspects that humans can perceive. Related to these are studies in which the data generated by individual citizens play a key role, for example in health-related studies requiring participants to regularly record their own biomarkers or behaviour.

2. **High-end technology:** Designing and developing sensors or software relies on the high technical and computing skills available at universities (see also Schroer et al. in this volume about the role of expert facilitators). The international ties of universities further enlarge the potential for accessing appropriate expertise. Universities also have an established tradition of archiving and in-depth research of past advances, allowing new projects to be grounded in, and expand on, available knowledge.

3. **International collaborations:** Universities are ideal for research over large geographical areas requiring distributed observations to provide evidence, for example about the movement of a species, the evolution of natural phenomenon or the impact of diseases. Relatedly, this also applies to studies that rely on the specialist knowledge and experience of many individuals, such as in linguistic studies on the distribution and history of languages.

4. **Interdisciplinarity:** Since research universities provide expertise in many areas, they can support interdisciplinary research projects. This is a particular strength for citizen science projects where diversity is natural to groups of citizens.

5. **Highest standards:** Finally, universities are well suited to projects intended to form the basis for far reaching, important (political)

decisions, which therefore require the highest standards of quality or research, including when it comes to ethical and legal considerations (see Shirk & Bonney; Nascimento et al., both in this volume). Here, the social position of universities and their internal procedures and structures are particularly valuable.

Below, this section turns to a few (well-known) citizen science projects where the connection to academia proved important for success and impact.

1. FoldIt (FoldIt 2017): The scientific goal of the project FoldIt is to decipher and understand the structure of proteins. Knowing a protein's structure is the basis to understanding how it works and how it can be targeted with drugs. FoldIt transforms scientific questions into games and was developed by highly skilled computer scientists at the University of Washington. This implies a strong interdisciplinary collaboration between biologists, computer scientists and others. By playing the FoldIt game, citizens reveal the protein's form. One of the best known outcomes from this project was resolving the structure of a protein which plays a key role in the growth of HIV (Khatib et al. 2011).

2. Zooniverse (Zooniverse 2017): Zooniverse is a large platform that supports citizen science projects in several areas and which grew out of the project Galaxy Zoo (Galaxy Zoo 2017) at Oxford University. Through the combined work of professional scientists, engineers, science funders and large institutions, the Sloan Digital Sky Survey and others have collected a huge number of photographs of galaxies. To analyse them, many citizens worked on the images, which brought new understanding of the factors that determine the formation and growth of galaxies. Among other findings, it has led to the surprising discovery of an unusual luminous gaseous structure called Hanny's object (Cox et al. 2015). Before its discovery by Dutch teacher Hanny van Arkel, it was unknown.

3. CASE (CASE 2017). The Centre for Ageing and Supportive Environments is an interdisciplinary centre for participatory research on ageing and health in Lund. It brings together researchers from medicine, engineering and social science with elderly citizens to develop concepts for healthy ageing. The established research team of Lund University has been able to attract considerable funding from the Swedish Research Council for Health. Collaboration

between three departments in the university means the research is interdisciplinary in nature, and it also involves extensive international collaborations and a graduate school which trains future experts on healthy ageing. This project also benefits from the ethical and legal support available in universities and which is particularly important in research on topics of this kind. Citizens have participated from the beginning of the project, for example, by identifying the priorities, helping design the project and more.

4. CLRP (Chintang Language Research Program) is focused on linguistics and aims to support the in-depth analysis of Chintang, a language of the Kiranti subgroup of Sino-Tibetan spoken in Eastern Nepal (CLRP 2017). The grammar, lexicon and language use are to be documented, but an additional major goal is to analyse how children learn the language. This programme is one of few to document the acquisition of an endangered language in a remote area by focusing on the natural development of children, without restricting children's conversation partners as is usual for most longitudinal studies in advanced industrialised societies. CLRP is a collaboration between the Psycholinguistics Laboratory of the University of Zurich and institutions in Nepal. The project requires the international ties and background in language acquisition studies available at the university. The intensive exchange and collaboration with native speakers makes this project highly participatory.

Nine challenges for universities

Contemporary citizen science activities have been running at universities for several years. To date, these are mostly scattered projects connected to individual scientists, which make up a small part of universities' total research effort and where citizen participation is often limited to providing and generating data. The trends towards open science, promoting stronger citizen participation and strengthening universities relationship with society, however, demand a more concentrated effort. Indeed, some universities are attempting to expand citizen science activities, including by initiating larger projects, often together with other institutions, and establishing platforms for citizen science. In some fields, such as health research or linguistics, the focus of research itself may shift to issues that require strong citizen participation.

Nevertheless, citizen science projects often meet with scepticism from established researchers who voice a series of concerns. These include doubts about the quality of the research (Editorial 2015), questions on ethical and legal aspects, and concerns about the degree of citizen participation and influence over the project. Researchers may fear that established research projects that do not include citizens could be curtailed to satisfy calls for citizen science. There are concerns about how to assess the quality and impact of citizen science projects and whether the established criteria are fit for purpose. In addition, funding organisations may lack appropriate mechanisms to support such projects, which may make them less attractive for researchers; as a result, citizen science is often funded by private organisations or political bodies, or viewed as part of public engagement in science and not as part of the main body of research (compare with Sforzi et al. in this volume about how museums are using citizen science).

There are therefore a number of challenges and corresponding requirements for citizen science projects at universities (see also ECSA 2015, ECSA Ten Principles of Citizen Science). While some of these may apply to citizen science more generally, the challenges are particularly strong at universities, which are embedded in a tight network of evaluations, competition and observation.

1. Reaching an adequate share of citizen science projects

In adopting more citizen science projects, universities must work towards a reasonable balance between highly successful traditional research practices and citizen science projects. If citizen science is to enjoy the acceptance of the academic community and funding agencies, universities need to ensure that citizen science projects conform to the usual academic standards. Universities need to set up appropriate infrastructures and build personnel capacities for such projects (see Richter et al. in this volume); furthermore, citizens and university members need to be made aware of the opportunities for citizen science and its potential, in particular for interdisciplinary research.

2. Maintaining quality and impact

Research at universities must satisfy high standards of quality. This is usually assured by peer review, careful handling of data and evaluation criteria. Citizen science projects are no exception and must undergo the same scrutiny as other projects (see Kieslinger et al. in this volume). Where existing evaluation tools are incomplete, the academic community should provide tools in agreement with established procedures. While studies have shown that properly collected citizen science data are not inferior to those collected by experts (e.g., Danielsen et al., 'A Multicountry Assessment', 2014), the large variability of citizen science data may require special attention, above all in cases where statistical validation by standard sampling techniques may not be an option. Concerning impact – usually weighted by publications – broader metrics should be considered, keeping with emerging standards of open science (San Francisco Declaration on Research Assessment [DORA 2012]).

3. Improve openness and transparency

Citizen science can be viewed as part of the open science movement, which includes open access to publication, open source software and open data standards. In fact, recent EU policy (European Commission 2016c), which strongly supports open science practices, highlights also citizen science. Obviously, citizen science projects are greatly enhanced by open science practice. Therefore, institutions and their researchers running citizen science projects should be encouraged to adhere to open practices. Furthermore, full transparency of the research objectives, research protocol and analysis techniques ensure the trust of participating citizens and the full documentation of the quality and reproducibility of results (see also Williams et al. in this volume).

4. Strengthening learning and creativity

Ideally, citizen science projects are designed to enable the public to learn about science through active participation as well as to develop their talents, creativity, skills and responsibilities. This requires pedagogical and presentational skills. Developing adequate pedagogical content, such as courses on citizen science, should be planned when designing citizen science projects (see Harlin et al. in this volume).

5. Optimising organisation, communication and sustainability

Governance issues are important for success, especially for larger and more complex projects which are designed to be long running (e.g., long-term ecological monitoring). The diversity of the participants' qualifications and the typical shorter term presence of graduate students or postdocs at universities pose special challenges. Health science projects, which involve many stakeholders including health services providers, require steering groups and may necessitate ethical reviews. In many cases, a dedicated community manager with long-term employment may be needed. There is also a need to establish online forums and monitor their content, encourage and publish blogs by scientists, organise periodic electronic chats and face-to-face video discussions to allow participants to ask questions and voice concerns about a project and to avoid negative phenomena, such as online harassment.

6. Establish suitable credits and rewards

Properly acknowledging and rewarding participant contributions is important; there are many options depending on the level and intensity of citizen participation, from internal letters of recognition and motivational rewards to an acknowledgement in a scientific publication or naming of a discovery. In all cases, permission should be sought before releasing names or private information about participating citizen scientists.

7. Increasing funding for citizen science projects

Citizen science projects can be difficult to fund as they are not always accepted as scientifically sound and there may be a perceived absence of 'star-appeal' (can one get a Nobel Prize for a citizen science project?). Other difficulties include the fact that citizen science research often involves long-term monitoring, in contrast to the more project-oriented research at universities, which makes comparison difficult. In this situation, it is important to extend rules of funding, both inside universities and in funding agencies, to include citizen science activities and thus provide equal opportunities. One important basis for this is establishing a visible citizen science entry point at universities (League of European Research Universities 2016). An example is the new centre for citizen science in Zurich where the University of Zurich and the Eidgenössische Technische Hochschule Zürich (ETH) are funding rooms and collaborators to support citizen science projects and at the same time develop answers to the nine challenges listed here, such as the necessary ethical and legal rules.

8. Developing ethical and legal procedures

Involving citizens in research projects raises ethical and privacy issues not present in purely academic projects where the researchers are members of the institution that oversees the projects. Both professional and citizen scientists have certain rights and obligations within the project. They may include issues such as ownership of data gathered or analysed by citizen scientists and intellectual property produced (see also Williams et al. in this volume). A written code of conduct should ensure that all parties are aware of their rights and should define appropriate procedures to handle disputes. The legal status of participants in citizen science projects and of the generated knowledge may require special attention.

9. Balance between society and researchers

A broader issue is the balance between greater public participation in the research process and the interests of individual researchers. A consent-based approach is suggested so that, where relevant, researchers may introduce ways for citizens to participate in the operational phases of a research project. Particular care applies to health-related research where the citizens may act as researchers and test persons at the same time and may share personal health-related data collected before the start of a study. In this situation, adequate informed consent must be sought and researchers should provide clear terms and conditions to participating citizen scientists, consistent with both open science and personal privacy requirements. Where useful to the project, citizens may be involved in decision-making aspects. Where appropriate, they should retain control over personal data they have shared, including beyond the end of the project.

These considerations may not be exhaustive, but they provide actionable advice based on experience from existing citizen science projects and are based on recent trends in the organisation of citizen science projects. As the interaction between humans and computer changes (for example in progress in machine learning), citizen science projects will evolve in their methodology. Therefore best practice should be continuously reassessed in the light of technological advances and societal change. This also requires universities, funders and policymakers working together with citizens. Table 11.1 makes some recommendations for each stakeholder intended to guide their engagement with citizen science.

Table 11.1 Recommendations to universities, funders and policymakers

Universities	Funders	Policymakers
Recognise citizen science as an evolving set of research methods, as well as having societal and educational benefits. Develop rules where needed to carry out citizen science projects;	Recognise a wide range of success criteria when supporting citizen science projects, including but not limited to traditional measures of scientific quality;	Encourage independent studies to evaluate the reliability of citizen science and help ensure projects use evidence-based methodologies, recognised by scientific institutions;
Consider creating a single point of contact and co-ordination for citizen science within the institution, to advise scientists and ensure liaison with national and regional citizen science initiatives;	When evaluating citizen science projects, ensure adequate funding for community management, platform development and other non-research functions characteristic of citizen science;	Develop clear guidelines for legal, ethical, commercial and privacy issues that arise in citizen science, and encourage productive participation of citizens if possible;
Provide suitable training programmes for researchers interested in citizen science;	Promote the use of open science practices in citizen science projects, such as open-access publication, open data standards or the use of open source software;	Encourage long-term collaboration between research universities and non-governmental organisations to ensure that citizen science is sustainable.
Raise researcher awareness of criteria for successful citizen science, including community management, pedagogical practices, open science standards and social, intergenerational and general diversity policies issues;	Set clear legal and ethical criteria for data privacy according to existing laws, such as personal data control.	
Ensure that proposals to funding bodies for citizen science projects include long-term commitment to infrastructures;		
Establish and support data repositories, in line with other research projects with long-term scientific or societal benefits;		
Ensure that project participants comply with ethical, legal and privacy regulations relevant to the scope of a given citizen science project, and have access to professional advice for this purpose;		
Adapt research evaluation and reputation systems to include metrics that can characterise projects with a high societal impact, such as successful citizen science projects, and develop ways of assessing citizen participation;		
Support the public-facing mission of research libraries in universities, in supporting citizen science activities and assisting in data management and curation.		

Source: League of European Research Universities 2016.

Conclusion and outlook

This chapter is concerned with the contribution of citizen science methods to research at universities. It argues that universities should adopt citizen science as part of the movement to open science practices, which in turn can enrich and enlarge the research scope of universities to previously inaccessible areas. Universities face a number of challenges in engaging with citizen science projects, which can be addressed with corresponding project planning. This can be further facilitated by funders and policymakers, so that universities can progress towards citizen science becoming part of their regular academic research activities, making them eligible for regular funding. While citizen science is important for communities and policy, exploiting its full potential rests on the accumulated and established scientific knowledge at research institutions like universities, where the future generation of researchers are also educated.

Part IIa
Case studies

12
Citizen science on the Chinese mainland

Chunming Li[1]

[1] *Chinese Academy of Sciences, China*

corresponding author email: cmli@iue.ac.cn

In: Hecker, S., Haklay, M., Bowser, A., Makuch, Z., Vogel, J. & Bonn, A. 2018. *Citizen Science: Innovation in Open Science, Society and Policy*. UCL Press, London. https://doi.org/10.14324/111.9781787352339

Introduction

Public engagement and direct contribution to scientific activities in China are limited. Recently, advances in low-cost sensors (Volten et al. in this volume) and information technologies (Novak et al. in this volume), as well as an increase in the level of education across the population (Haklay in this volume), mean that more citizens have become involved in citizen science projects. This set of case studies demonstrates that China is witnessing activities across the spectrum of citizen science – from bird watching to air quality monitoring and from biological observations to volunteer computing.

Bird watching

Bird watching is a popular activity in China as in other countries. Nowadays, about 24 bird watching societies have been founded, forming a network across the Chinese mainland. Participants can find their contact information via a website (http://www.chinabirdnet.org/network.html), which provides details on local programmes.

These societies have proposed targeted bird watching projects, such as the China Coastal Waterbird Census, initiated by The Hong Kong Bird Watching Society (HKBWS) in September 2005. In this project, participants conduct monthly surveys of 12 permanent and 3 irregular sites along the eastern coast of the Chinese mainland to study the distribution, migration and seasonal changes of waterbirds and contribute to the conservation of China's biodiversity and Important Bird Areas (HKBWS 2015, ii).

Participants can use either use Birdtalker (an online record submission system), email or regular mail to submit their records (see Mazumdar et al. on multiple methods).

The China Ornithological Society (COS) collects, compiles and reviews bird watchers' observations and publishes the annual China Bird Report in Chinese and English. The China Bird Watching Database has also been established based on the China Bird Report (2003–2007). Over five years, the database compiled 30,936 records covering 1,078 species, representing more than 80 per cent of all bird species in China, which reveal bird distribution and changes (Li, X. et al. 2013, 649).

Plant classification

The Institute of Botany, Chinese Academy of Sciences, established a plant classification programme with citizens' help in 2007. Participants take photographs of plants and upload them to Chinese Field Herbarium (CFH) (http://www.cfh.ac.cn), a plant information collection and classification platform (Zhang et al. 2013, 747). There were over 10,000 registered users and more than eight million plant photographs, including one million with a GPS location, as of May 2017. More than five million photographs had already been identified and classified by this time (CFH 2017).

Air quality monitoring

FLOAT Beijing was a community-driven air quality monitoring project developed in 2012 using air-quality sensors and kites. The project founders held workshops with Beijing residents to demonstrate to them how to build the sensors. In addition, open online tutorials allowed more people to become involved.

The colour of an LED on the kites changes with the air quality condition. The kite sensors are based on the Carnegie Mellon Air Quality Balloons project (Maly 2012), which used Figaro's volatile organic compounds sensor (TGS 2620) and diesel/exhaust sensor (TGS 2201) (Kuznetsov et al. 2011). The minimum detection of TGS 2620 is 50 parts per million (ppm). The minimum carbon monoxide (CO) and nitrogen dioxide (NO_2) detection of TGS 2201 is 1 ppm and 0.1 ppm, respectively.

While China's ambient air quality standards (GB 3095-2012) shows that the limit of average concentration in one hour of CO and NO_2 is about 8.73 ppm and 0.11 ppm based on 25°C and 1 atmosphere, respectively, which means TGS 2620 cannot be used to monitor the CO and it would be difficult to monitor NO_2 with TGS 2201. Although TGS 2201 can be used to monitor the CO, its accuracy should be observed closely because this semiconductor sensor is designed for automobile ventilation control and also easily affected by air temperature and humidity due to its materials.

Water quality monitoring

Xiangjiang Watcher is a citizen science project initiated by Green Hunan, an environmental nongovernmental organisation (NGO) in China. The project engages participants to monitor the water quality in the Xiang River Watershed, Hunan Province. It has attracted more than 60 participants from 23 local cities and counties along the river, including industrial workers, farmers, students, professors and public officials. Participants periodically conduct basic tests of water quality at the monitoring sites, record any environmental changes in the watershed, photograph companies that are secretly discharging pollution into the river and advocate for solutions to the pollution (Yan 2012).

Using Weibo (the largest social network in China) means that participants can disseminate the pollution information on the internet, which has brought the issue to the attention of the environmental protection department and put pressure on the polluting companies. Participants hope to reduce pollution by using information technology for advocacy.

Computing for Clean Water

Computing for Clean Water (C4CW) is a scientific computation project launched by Chinese scientists in 2010, in collaboration with the Citizen Cyberscience Centre in Geneva. Participants can contribute their computing power to the project using desktop client software supported by IBM's World Community Grid (https://www.worldcommunitygrid.org/).

Researchers at Beijing's Tsinghua University use computing power from more than 50,000 participants to extend simulations to probe flow rates of just a few centimetres per second, significantly reducing computing time. This increased computing power enabled the study of

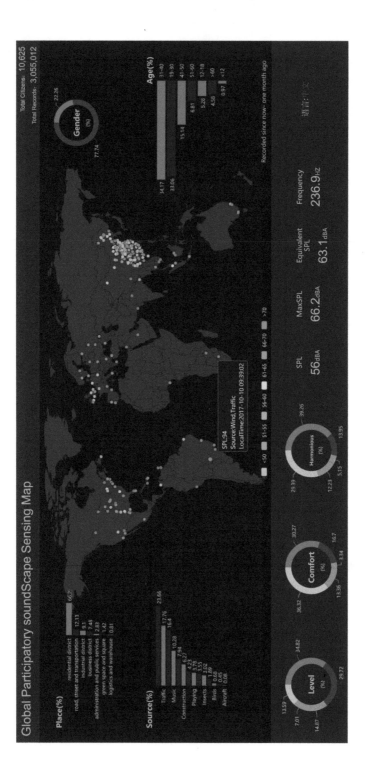

Fig. 12.1 Participatory soundscape sensing online analysis and visualisation website

the characteristics and working conditions of real nanotube-based filters (Ma et al. 2011, 1), which is useful for designing better low-cost, low-pressure water filters and making water purification cheaper and more accessible (Drollette 2012).

Soundscape evaluation

Participatory soundscape sensing (PSS) is an ongoing, worldwide soundscape investigation and evaluation project, initiated by the Research Center of Digital Urban Environmental Network at the Institute of Urban Environment, Chinese Academy of Sciences. The project collects soundscape data with the help of public participation and mobile phones equipped with SPL Meter software (see: http://www.citi-sense.cn /download).

The first version of PSS was launched in 2011 (Li, C. et al. 2013, 262) and collected little information, such as sound pressure level, sound frequency, GPS location and subjective feeling. The latest version was updated in March 2016 and has more useful functions, such as land use and sound source identification, soundscape evaluation (subjective evaluation of sound level, sound comfort level and sound harmony characteristics), online data analysis and visualisation (figure 12.1). The data collected supports the analysis of the temporal-spatial characteristics of soundscape, offers a high-quality evaluation model and facilitates the optimisation of urban sound environment policy.

13

The European citizen science landscape – a snapshot

Susanne Hecker[1,2], Lisa Garbe[1,2], Aletta Bonn[1,2,3]

[1] Helmholtz Centre for Environmental Research – UFZ, Leipzig, Germany
[2] German Centre for Integrative Biodiversity Research (iDiv) Halle-Jena-Leipzig, Germany
[3] Friedrich Schiller University Jena, Germany

corresponding author email: susanne.hecker@idiv.de

In: Hecker, S., Haklay, M., Bowser, A., Makuch, Z., Vogel, J. & Bonn, A. 2018. *Citizen Science: Innovation in Open Science, Society and Policy*. UCL Press, London. https://doi.org/10.14324/111.9781787352339

The increased importance of open science in the European Commission research policy makes it important to understand and analyse the development of the field. The Open Science Monitor of the European Commission is being developed to meet this need (European Commission 2017). In 2016, the authors conducted the first large-scale explorative survey of the European citizen science landscape to help establish a baseline for the monitor.

The survey focused on five major areas of interest, including the types of citizen science projects being undertaken, their perceived impact and added value, challenges, current funding schemes for citizen science, and project outcomes. Data was collected through an online survey in October and November 2016, predominantly with closed question formats to facilitate participant response and to cover as many projects as possible. This provided reliable and quantifiable basic information about different citizen science projects across Europe. The data is available upon request. This snapshot covers the main findings.

Geographical scale of projects

The survey attracted responses from 174 co-ordinators of citizen science projects. Most of the respondents are either from Central (40 per cent) or Western Europe (32 per cent), with only a few respondents from Southern

(16 per cent), Northern (10 per cent) or Eastern (1 per cent) Europe (see figure 13.1). Major activities across Europe were recorded from the UK, Germany and Austria, which may also reflect the fact that, at the time of the survey, the citizen science communities in these countries were most connected and thus the survey might have gained more traction here.

In terms of the scale of the projects, many initiatives cross local and even national boundaries. Most of the projects are at the national (41 per cent) or global (19 per cent) level. A smaller number of projects is being carried out at the regional (14 per cent) or European (12 per cent) level.

Project focus and leadership

The disciplines of the projects range from archaeology and engineering to zoology. However, there is a clear focus on projects within the life sciences (76 per cent) including ecology, environmental sciences and biology (see figure 13.2). This is in line with Kullenberg and Kasperowski's (2016) meta-analysis of citizen science studies, which also found environmental sciences and ecology to be at the forefront of citizen science research (see also Owen & Parker in this volume). Almost half of the surveyed projects are coordinated by a scientific organisation (45 per cent), followed by educational organisations (14 per cent) and non-governmental organisations (11 per cent).

Level of engagement

More than two-thirds of the projects are contributory or collaborative (see figure 13.3; the categories are based on those developed by Shirk et al. 2012 – see table 13.1; see also Haklay; Novak et al. both in this volume). Thus, most citizens are mainly involved in data collection and sometimes in the project design or data analysis.

Regarding the length of the projects and involvement of participants, more than 40 per cent of the projects involve citizens continuously during the research process (see figure 13.4), which may last several years.

The number of people engaged in citizen science projects varies widely. The average number of citizens engaged continuously, over a long period, is about 1,800, while the number of those who engage occasionally averages at about 7,900 per project. It is estimated that at least 1.2 million people participated once (or more) across the 174 projects sampled in the survey.

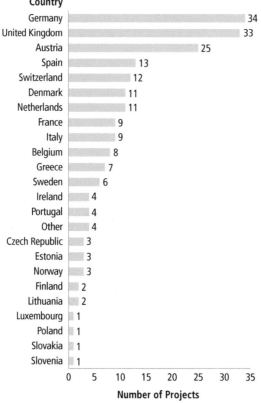

Fig. 13.1 Distribution of projects from the European Citizen Science Survey 2017

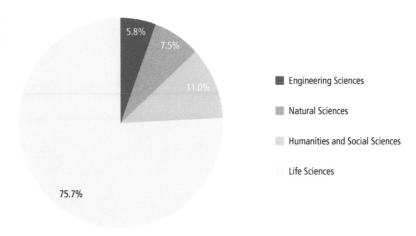

What is the primary discipline of the project?

- 5.8%
- 7.5%
- 11.0%
- 75.7%

Engineering Sciences

Natural Sciences

Humanities and Social Sciences

Life Sciences

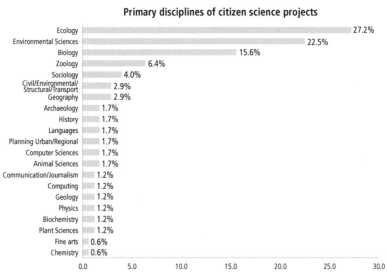

Primary disciplines of citizen science projects

Discipline	%
Ecology	27.2%
Environmental Sciences	22.5%
Biology	15.6%
Zoology	6.4%
Sociology	4.0%
Civil/Environmental/Structural/Transport	2.9%
Geography	2.9%
Archaeology	1.7%
History	1.7%
Languages	1.7%
Planning Urban/Regional	1.7%
Computer Sciences	1.7%
Animal Sciences	1.7%
Communication/Journalism	1.2%
Computing	1.2%
Geology	1.2%
Physics	1.2%
Biochemistry	1.2%
Plant Sciences	1.2%
Fine arts	0.6%
Chemistry	0.6%

Fig. 13.2 Primary discipline of citizen science projects

Outputs and funding

The most common outputs of the projects are contributions to media (78 per cent of projects; see Hecker et al. 'Stories' in this volume), social media (72 per cent), conferences (72 per cent) and publications of the data (71 per cent) (see figure 13.5). Other common outputs include articles

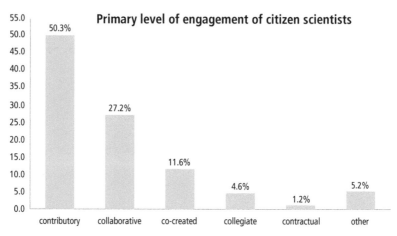

Fig. 13.3 Level of engagement in European citizen science projects (according to Shirk et al. 2012)

Table 13.1 Different types of participant engagement

Contributory	Scientists generally design projects to which members of the public primarily contribute data.
Collaborative	Scientists generally design projects to which members of the public contribute data but also help to refine project design, analyse data and/or disseminate findings.
Co-created	Scientists and members of the public work together and participants are actively involved in most or all aspects of the research process.
Collegiate	Citizens run projects with no professional scientist involvement.
Contractual	Communities ask professional researchers to conduct a specific investigation for them and report on the results.

Source: Shirk et al. 2012

in publicly accessible journals (61 per cent), public events (53 per cent), reports for participants (52 per cent) and teaching materials (48 per cent). Less common are contributions to newsletters (40 per cent), policy briefs (22 per cent) and articles in non-public journals (21 per cent) or guidebooks (15 per cent).

Around 25 per cent of the projects receive either no funding or less than €10,000 funding (see figure 13.6). Many projects (43 per cent)

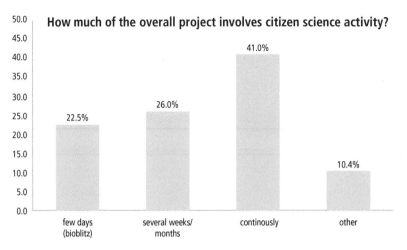

Fig. 13.4 Citizens' involvement within citizen science projects

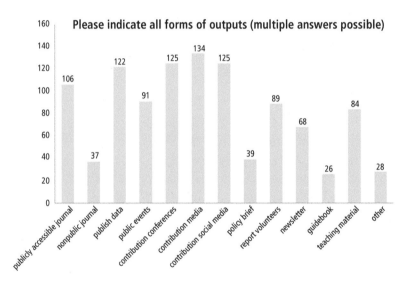

Fig. 13.5 Outputs of citizen science projects

receive between €10.000 and €250,000 in funding; while approximately a third (31.8 per cent) of the projects receive substantial funding of over €250.000, with 14 per cent receiving more than €1,000,000. Overall, most of the funding is from national research funds, nongovernmental organisations or EU research funds. Projects often have several sources of funding.

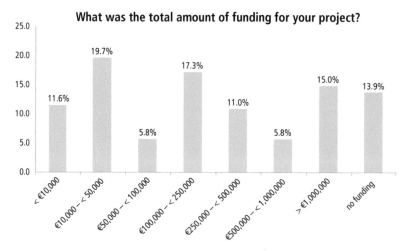

Fig. 13.6 Funding received by citizen science projects

Fewer than half of the respondents (38 per cent) agree (fully or partly) that the amount of initial funding is appropriate; only a minority of project co-ordinators (15 per cent) agree that the amount of long-term funding is appropriate.

Challenges and added value

When asked about challenges in citizen science, a clear majority of the respondents highlighted insufficient funding (75 per cent) and concerns over data quality (70 per cent) (see figure 13.7; and see also Williams et al. in this volume on data quality). In addition, there were concerns about the recognition of citizen science in co-ordinators' professional fields, with a lack of appreciation in academia (60 per cent of respondents) and of integration in education (68 per cent) the most pressing. The fact that citizen science projects are time consuming (65 per cent) was also considered a challenge.

The main added value for the majority of the respondents is the generation of large datasets (75 per cent). Around half of the respondents also value citizens providing expertise (47 per cent). Respondents strongly disagreed that citizen science saves time (84 per cent) or money (76 per cent) (See also Danielsen et al. in this volume). Seventy per cent of the respondents do not think that citizen science raises new research questions and only 30 per cent think that it produces knowledge other than

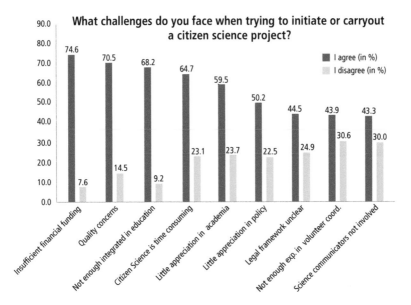

Fig. 13.7 Challenges for citizen science projects

scientific data. Slightly less than half of the respondents think that citizen science makes research more relevant (45 per cent).

This is slightly offset by the question on impact where respondents claim enhanced science-community interaction (77 per cent) and education (75 per cent) as the most important impacts of their citizen science projects (see figure 13.8). Enhanced community-policy interaction (40 per cent), enhanced science-policy-interaction (49 per cent), perceived behavioural change (43 per cent) and enhanced evidence (47 per cent) are also important perceived impacts.

Citizen science project leaders were also asked about their perception of the policy impact of their projects and where they perceive the project to have the most impact in the policy decision-making process. Forty-three per cent of the contributors stated that their project had a policy impact, whereas 50 per cent said that it currently had no policy impact but could have in the future. Only 7 per cent of the respondents did not think that their project had an impact or could have impact in the future.

Overall, respondents saw the possible influence of their project at all steps of the policy decision-making process, with the strongest potential linked to issue identification and measurement of effectiveness, which corresponds to the steps of agenda-setting and policy evaluation in the policy cycle (Howlett & Ramesh 2009) (see figure 13.9).

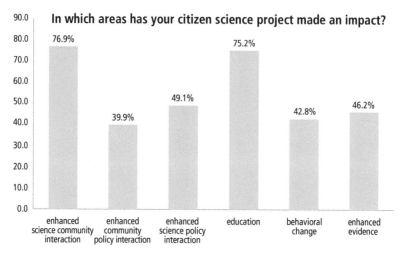

Fig. 13.8 Areas of perceived impact of citizen science projects

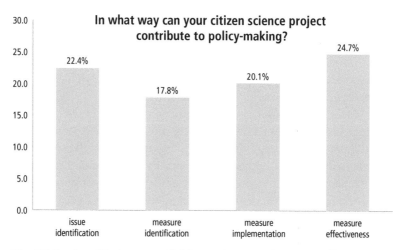

Fig. 13.9 Possible impacts of citizen science projects on policy decision-making

Conclusion

The survey results show that the European citizen science landscape is currently dominated by contributory and collaborative projects that are mainly related to the life sciences. Unlike the citizen science landscape in the United States or Australia where projects share English as a common language, communication in the citizen science projects in the EU is mostly carried out in the respective national language of the home country of the projects. Enhancing the interoperability of projects through adaptation into different national languages may facilitate greater participation, but these survey results may reflect participants' preference to be involved in national projects. National interfaces for citizen science projects may facilitate international contributions and thereby enhance scientific results (e.g., the Living Atlas of Australia – Brenton et al. in this volume). However, as pointed out by Ballard et al. (in this volume), the different spatial scales of projects may serve different purposes with respect to scientific and socio-political goals, which may require smaller scales of interaction.

Respondents indicated that only one-fifth of the projects publish their results in non-publicly accessible journals while project data was published in some form by the majority of projects (72 per cent) and the results were communicated at conferences. This may either reflect the early stage of many projects or a current lack of capacity to publish scientifically, since the number of citizen science publications in general is rising (Kullenberg & Kasperowski 2016). It may be important to provide scientific training suited to citizen science projects as well as avenues to make data available for scientific analyses by others, so that they can also be published in scientific journals and thereby advance science (see Richter et al. in this volume).

Regarding social innovation, most projects were understood as having an impact, although only half of all projects saw their contribution as being to policy-making (Haklay 2015). This potential may not yet be fully realised by the primarily scientific co-ordinators, while the European Commission and Environment Protection Agencies view this as an important facet of citizen science (Nascimento et al.; and Owen & Parker, both in this volume). It will therefore be important to tailor citizen science projects so that they can contribute to ongoing policy processes without compromising their creativity. Early interaction with local or national agencies may help to develop the project design so that outcomes can be useful to promoting innovation in policy. Overall, it will be important to monitor developments in citizen science communities over time, and to observe the advances and maturity of the European citizen science landscape.

If you are interested in the raw data, please feel free to contact the authors.

Acknowledgements

The authors gratefully acknowledge the survey participants across Europe who shared their time and expertise, and the attendees of the First International ECSA conference in Berlin 2016 and the Conference of the Citizen Science Association in St. Paul, US, 2017, who were an inspiration to this research. Thanks also to the institutions that circulated the invitation to participate in the survey, the Leibniz Institute for the Social Sciences GESIS in Germany for advice on the survey design and Gi-Mick Wu for statistical advice.

14

Stakeholder engagement in water quality research: A case study based on the Citclops and MONOCLE projects

Luigi Ceccaroni[1,2] and Jaume Piera[3]

[1] *Earthwatch, Oxford, UK*
[2] *1000001 Labs, Barcelona, Spain*
[3] *Institut de Ciències del Mar, Barcelona, Spain*

corresponding author email: lceccaroni@earthwatch.org.uk

In: Hecker, S., Haklay, M., Bowser, A., Makuch, Z., Vogel, J. & Bonn, A. 2018. *Citizen Science: Innovation in Open Science, Society and Policy.* UCL Press, London. https://doi.org/10.14324/111.9781787352339

Impact, policy and governance

The participatory engagement of decision-makers, including policymakers, is one of the most important components of the planning and development of a citizen science initiative (Nascimento et al. in this volume). Meaningful engagement depends on the ability of the civic educators involved in citizen science to build a healthy, lasting and trusting relationship with decision-makers and local communities. The approaches developed by citizen science initiatives are intended to define and develop this process of engagement with decision-makers, often in the domain of environmental monitoring (see Owen & Parker in this volume).

These approaches can be summarised in six steps:

1. Stakeholder mapping;
2. Understanding and engaging decision-makers;
3. Working with decision-makers;
4. Developing a participatory-science approach;
5. Decision-makers acting as advisers;
6. Developing co-management approaches.

When attempting to advance science, foster a broad scientific mentality and/or encourage democratic engagement (which allows society to deal rationally with complex modern problems; Ceccaroni, Bowser & Brenton 2017), stakeholder mapping and knowledge of influential actors and institutions, including their perspectives and interests, is needed (Sclove 2010). In particular, technology assessment in citizen science involves engaging a group of lay citizens who are representative of the general population but who (unlike political, academic and industry stakeholders) are generally under-represented in technology-related policy-making (Tomblin et al. 2015). Much like policymakers for science aspire to cultivate a research enterprise that generates 'usable' research in the service of complex issues like biodiversity, calls for public engagement with science and technology demand equal attention for processes that articulate 'usable public values' representing not only stakeholders and interest groups, but also the knowledge and experience of a diverse public (see Smallman in this volume).

More than a dozen European nations plus the European Parliament established their own technology assessment agencies from the mid-1980s

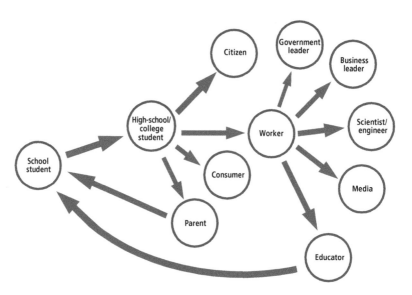

Fig. 14.1 Understanding and engaging decision-makers and the public in their lifecycle. Public engagement with science motivates the design of programmes for people in many different roles who make choices that help shape socio-technological futures and also influence the choices of others. (Source: Worthington et al. 2012)

onward. Most remain in operation as vital contributors to science and technology policy discourses and developments in their respective countries. In addition to strong analytical capacities, these agencies have pioneered promising methods for citizen participation (Worthington et al. 2012). Designing programmes to engage the public with science involves a reconceptualisation of audiences as not only learners, but as decision-makers in society, followed by a commitment to understanding and engaging these decision-makers through their lifecycle (figure 14.1; and see Edwards et al. in this volume).

Engaging stakeholders from citizen volunteers to decision-makers

Citizen science begins with practitioners determining the types and level of engagement, including what practitioners want to achieve by engaging decision-makers. Appropriate implementation at this stage should provide a solid foundation for moving towards more sophisticated and mature levels of engagement with decision-makers.

Different techniques and tools to develop a participatory-science approach are useful at different stages of project development. For example, a scientific advisory board can be involved during the stakeholder mapping process (understanding and engaging decision-makers) and also during problem identification (see also Haklay; Novak et al. both in this volume). It is also crucial to get decision-makers involved from the earliest planning stages to (a) get commitment from them and (b) find out the different points of entry for using the knowledge produced in the decision-making process. At these early stages, it is important to focus on building trust and engaging decision-makers. This can be also achieved by creating an advisory body of decision-makers or engaging in a co-operative management approach, starting when the citizen science co-ordinators are beginning to work with decision-makers. These approaches will become even more useful in the later steps of the project's development.

Engagement between decision-makers and citizen science practitioners is expected to evolve and aspires to move decision-makers and communities to a point where they will ultimately take some level of responsibility for jointly managing the citizen science initiative. Citizen science practitioners may have already worked with some of the decision-makers, and this can provide a concrete foundation on which to build a mature citizen science initiative in terms of decision-makers' engagement.

As the types and levels of decision-makers' engagement become more complex and responsive, so do the monitoring and decision-making tools and frameworks required for making multi-stakeholder participatory decisions. In fact, citizen science projects select the type of engagement they want with different stakeholders at different times, so the engagement model is not always progressive, but rather provides a range of possibilities depending on the context. Citizen science projects operate in parallel to decision-makers' and other stakeholders' activities, and there is often a core of primary stakeholders that does not change over time but may, at times, include or exclude other stakeholders.

As others have demonstrated, the identification and integration of the major stakeholder groups involved in citizen science – decision-makers; educators; developers; volunteers; civil society organisations, informal groups and community members; academic and research organisations; government agencies and departments; participants, including corporate-programme participants; formal learning institutions such as schools; and businesses and industry – is important (Mazumdar et al. 2017; Göbel, Martin & Ramirez-Andreotta 2016). According to the Citclops European project (see below), the main objectives of stakeholder engagement are (1) to improve citizens' understanding of environmental observations and monitoring, and (2) to enhance community decision-making and co-operative planning. These objectives are usually achieved through the use of technology, the organisation of events and the definition of plans and actions for public involvement.

About the Citclops and MONOCLE citizen science projects

The FP7 European Commission–funded project Citclops (Citizens' Observatory for Coast and Ocean Optical Monitoring), which started in 2012, is one of the largest citizen science projects worldwide, at a total cost of €4,743,458. Citclops introduced an innovative concept for water quality monitoring to help oceanographers and limnologists monitor natural waters, with a strong focus on long-term data series related to environmental sciences. In this context, forging connections among citizen science, the appropriate use of supporting tools and technology, and policy was key to achieving successful outcomes.

Citclops developed several new sensor systems based on existing optical technologies. These respond to a number of scientific, technical and societal objectives, ranging from more precise monitoring of key environmental descriptors of the aquatic environment (water colour, transparency and fluorescence) to the improved management of data collected with citizen participation and engagement (see also Volten et al. in this volume). Requirements have been translated into engineering specifications, leading to the development of new solutions based on citizen science. Sensors have been tested, calibrated, integrated on several platform types, scientifically validated and demonstrated in the field. Cost-efficiency has been improved via the implementation of several innovations, such as greater interoperability of sensors and data, and multiplatform integration.

In 2015, co-operation was established between organisations involved in Citclops, sailing race organisations, skippers, scientists and citizens to ensure continued ocean observations and use of the tools and sensors developed by Citclops (rebranded EyeOnWater) after the European Commission funding period.

In 2018, the Horizon 2020 European Commission–funded project MONOCLE (Multiscale Observation Networks for Optical monitoring of Coastal waters, Lakes and Estuaries), at a total cost of €4,999,863, started to extend Citclops tools, creating a network of in-situ sensor systems that links citizen observations to other types of earth observations. MONOCLE innovates and develops technologies related to sensors, platforms and data handling, to increase coverage and lower the cost of in-situ sensors in inland and coastal water bodies. The ecosystems related to these water bodies are particularly vulnerable to direct anthropogenic impacts, and they are of high economic importance and crucial to sustainable food, energy and clean water supply. At the same time, these water bodies represent areas of low performance in present earth observation (EO) capability. The MONOCLE system reduces uncertainties in EO by characterising atmospheric and water optical properties. MONOCLE deploys new and improved sensors on autonomous platforms (buoys, ships, drones), and further fills information gaps by developing low-cost complementary solutions for citizen scientists. This provides essential reference observations needed to further improve and grow EO-based water quality services. MONOCLE is requirement-driven and implemented by sensor and platform developers, sensor-data infrastructure experts, EO scientists and citizen scientists. Also, a service-oriented data storage, processing and visualisation infrastructure based on open data standards will integrate MONOCLE seamlessly with existing platforms.

Stakeholder mapping in Citclops

To guide the development of the project and related technologies, Citclops project leaders conducted a stakeholder mapping with a scenario-based approach, identifying key stakeholders, a unifying concern, the current situation at the beginning of the project and the ideal situation at the end of the project. This stakeholder mapping helped shape stakeholder engagement and all subsequent phases of the project's implementation.

Main stakeholders: Citizens; decision-makers; academic and research organisations

Concern: Citizens are concerned about the water quality in their marina and learn through enquiries to their regional environmental protection agency that water quality monitoring stations are not granular enough to represent their marina.

Situation at the beginning of the project: Citizens purchase a water quality sensor they find online and recruit additional neighbours to place the sensors in the marina. These sensors collect data for six weeks. According to citizens' research, the data demonstrate a violation of national water quality standards. Citizens share these data with their regional environment protection agency contact, who informs them that the data cannot be used because of quality-assurance issues. They are frustrated and left wondering if the regional environment protection agency is hiding something.

Situation at the end of the project: Through a quick online search, citizens find the Citclops/EyeOnWater's website that is supported by the European Commission. The citizen science project makes it easy for them to join. Citclops uses a number of established standards, tools and technologies to monitor various aspects of water quality. The citizens regularly meet up for training, collecting data and to share their concerns with project representatives from academic and research organisations (see 'Stakeholder engagement' below) and decision-makers. The citizens, the academic and research organisations, and the decision-makers develop mutual respect and work together to discover and address community concerns. Data collected by citizens are uploaded to the Global Earth Observation System of Systems Portal (GEOSS) repository and taken into account by decision-makers. Citizens feel more empowered about the environmental monitoring of their marina.

Stakeholder engagement in Citclops

Strategies to promote public engagement, involvement and participation in decision-making processes include information dissemination (e.g., to increase environmental awareness); information exchange; and inclusion of a variety of stakeholders in all project phases. Stakeholder engagement in Citclops often consists of events held in public spaces that are accessible to schools, local citizens, policymakers and tourists and thereby act as multipliers to approach citizens and engage them in environmental stewardship. At these events, citizen science measurement principles are introduced with hands-on materials and technology, for example, with a colour wheel with different standardised water colours, a Secchi disk to measure water transparency and smartphone apps. Visitors are informed about the citizen science initiative and encouraged to engage in environment monitoring with smartphones. Elements from these events including the training materials and supporting technologies are then transferrable to other citizen science initiatives, as well as to other media, such as websites. In Citclops, different stakeholders are engaged at different phases as discussed below (see figure 14.2).

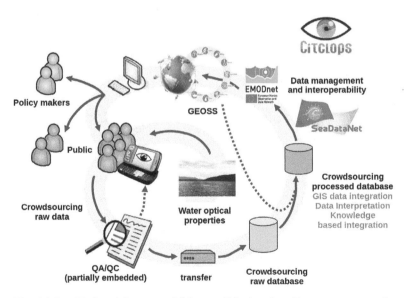

Fig. 14.2 Citclops' data acquisition, validation (quality assurance and control, QA/QC), processing and delivery. (Source: Authors)

- Collection of water's optical properties (raw data): citizens;
- Quality assurance: citizens and automatic (AI-driven);
- Data transfer: citizens and automatic;
- Hosting in databases: automatic;
- Data processing (GIS integration, interpretation, integration): academic and research organisations, and automatic;
- Data management and interoperability: automatic; and
- Information delivery: citizens, academic and research organisations, and decision-makers.

Conclusions and recommendations

Citclops's research provided several insights and recommendations about the engagement of a specific type of stakeholders: the decision-makers. To produce a good transfer of results from the research world to the public world and to achieve implications of the public world, civic educators and decision-makers should provide a good scheme for data gathering together with a clear explanation of the meaning of the data. Decision-makers and decision support systems should take into account that citizens are mostly triggered to observe when their natural system looks 'wrong' (algal blooms, off-colour, smell). Decision-makers and decision support systems should be aware of and take into account citizens' values' priority. For example, in the coastal environment, this priority is, generally (1) visible presence of plants (sea grass); (2) visible presence of macro algae; (3) visible presence of animals (charismatic megafauna); and (4) cleanliness. Decision-makers should formalise that citizens can measure for fun (curiosity), for science and for legal cases, and should differentiate the engagement approach for each case. Decision-makers should clarify if, to make better decisions in a specific context, it is useful to try to get citizens to measure more than once on the same spot, and, if this is the case, they should improve current apps, which are generally not good in the visualisation of concentrated, overlapping observations. Citizen science projects should be able to straightforwardly show officials and decision-makers the level of accuracy and precision that can be achieved by citizen observations and DIY instruments. Projects like Citclops demonstrated that citizen science monitoring can be easily extended beyond the initial scope towards covering the needs of decision-makers, for example from monitoring water fluorescence to monitoring the *Marine Strategy Framework Directive* (MSFD)'s descriptor 5 (eutrophication). Sessions on governance during specific events can give decision-makers insights on citizen science and on how citizen actions can influence decision-making on the environment and other topics. In addition, websites are important tools decision-makers can use to transfer information to, and engage, the public. In the best cases, they are also an excellent means to (1) exchange ideas and information between the public, scientists and decision-makers, and (2) collect qualitative and quantitative feedback related to citizen science tools, events and activities. Finally, and most importantly, when citizen science data about water monitoring, such as the ones produced by the Citclops and MONOCLE projects, are uploaded to open-access servers, processed and archived remotely, and resulting information can be accessed by decision-makers (e.g., at local administrations), these can use the information to improve the management of the aquatic environment.

Acknowledgements

The research described in this paper is partly supported by the Citclops and the MONOCLE European projects (FP7-ENV-308469 and H2020-SC5-776480). The opinions expressed in this paper are those of the authors and are not necessarily those of Citclops or MONOCLE projects' partners or the European Commission.

15
Global mosquito alert

John R.B. Palmer[1], Martin Brocklehurst[2],
Elizabeth Tyson[3], Anne Bowser[3],
Eleonore Pauwels[3], Frederic Bartumeus[4,5,6]

[1] Universitat Pompeu Fabra, Barcelona, Spain
[2] European Citizen Science Association, Europe
[3] Woodrow Wilson International Center for Scholars, Washington DC, US
[4] Centre d'Estudis Avançats de Blanes (CEAB-CSIC)
[5] Centre de Recerca Ecològica i Aplicacions Forestals (CREAF)
[6] Institut Català de Recerca i Estudis Avançats (ICREA)

In: Hecker, S., Haklay, M., Bowser, A., Makuch, Z., Vogel, J. & Bonn, A. 2018. *Citizen Science: Innovation in Open Science, Society and Policy.* UCL Press, London. https://doi.org/10.14324/111.9781787352339

An exciting recent development in citizen science has been the emergence of a variety of projects to fight disease-vector mosquitoes. These projects have shown that citizens can play an important role in alleviating the global burden of the diseases these mosquitoes transmit, but the projects are mostly limited to a handful of countries and have yet to benefit much of the world's most heavily mosquito-affected regions. The Global Mosquito Alert Consortium (GMAC) seeks to change that. The initiative is bringing diverse citizen science projects together to tackle disease-vector mosquitoes worldwide.

The problem of disease-vector mosquitoes

The re-emergence and global spread of vector-borne diseases during the past two decades has given mosquitoes a prominent place on the international public health agenda (WHO 2014). Dengue has skyrocketed, reaching 100–390 million cases annually (Castro, Wilson & Bloom 2017). Outbreaks of chikungunya and Zika since 2005 have infected millions of people, with Zika triggering a global health emergency due to its rapid expansion and its link to microcephaly and other neurological complications (Petersen & Powers 2016; Weaver & Lecuit 2015; Christofferson 2016; WHO 2016b). Malaria incidence and mortality have decreased

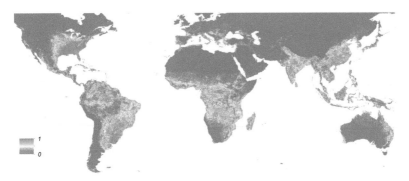

Fig. 15.1 Predicted global distribution of *Aedes albopictus*, mosquitoes that can serve as vectors for dengue, chikungunya, Zika and other viruses. Colours indicate probability of occurrence (from 0 blue to 1 red) at a spatial resolution of 5 km × 5 km. (Source: Kraemer et al. 2015, CC0)

since 2000 but the disease continues to affect enormous numbers of people, with over 200 million cases and over 400,000 deaths each year (WHO 2016a; WHO 2014). Moreover, recent upticks in malaria suggest stalled progress and the need to revitalize efforts by gathering real-time data with greater geographic precision (WHO 2017; Gates 2018).

These diseases and others, all transmitted by mosquitoes, place massive burdens on society – particularly the poor (Bhatt et al. 2013; Stanaway et al. 2016). Invasive vector species like the Asian tiger mosquito, *Aedes albopictus*, have spread quickly around the globe (see figure 15.1), and the World Health Organization has issued a strong warning that governments and development agencies must act quickly to improve vector control before this 'alarming situation' further deteriorates (WHO 2014). In contrast to treatments and vaccines aimed directly at vector-borne diseases, the WHO has concluded that targeting mosquitoes and other vectors provides an 'excellent, but underutilised opportunity' to fight these diseases and address the poverty and inequality that they cause (WHO 2014).

Citizen science as a scalable and flexible solution

Citizen science offers a highly effective strategy for tackling disease-vector mosquitoes. Traditional methods of mosquito surveillance and control are costly and often implemented in unco-ordinated patchworks at a time when public sector budgets are under increasing pressure (Hadler et al. 2015). Citizen science, in contrast, can be highly scalable, connecting the mosquitoes' human hosts into massive, active networks. These networks can act as effective mosquito sensors across large geographic scales, providing early warning and mosquito prevalence estimates comparable in quality to those from traditional methods (Palmer et al. 2017; and see Danielsen et al. in this volume). Citizen science projects that already provide information on disease-vector mosquitoes at national or supranational scales include Mosquito Alert (http://www.mosquitoalert .com), active in the Mediterranean Region; Muggenradar (https://www .naturetoday.com/intl/nl/observations/mosquito-radar), active in the Netherlands; and Zanzamapp (http://www.zanzamapp.it), active in Italy, among many others.

Further, engaging the public in vector monitoring through citizen science has numerous benefits beyond enhanced data collection. Participation in citizen science often leads to enhanced topical knowledge or knowledge of the scientific process (Edwards et al. in this volume). In

Fig. 15.2   Tiger mosquito photograph submitted by an anonymous participant through Mosquito Alert. CC BY 4.0. Participants in many mosquito-related projects may submit photographs along with their reports of mosquito detections to help researchers validate the reports. Other projects allow participants to submit specimens.

disease vector–monitoring, citizen science may also help motivate and improve co-ordination between individuals and families, so that they can implement effective protective and preventive measures such as interventions on mosquito-infested private property that public authorities cannot easily access (Oltra, Palmer & Bartumeus 2016).

Citizen science is also flexible, encompassing a wide array of approaches. Different approaches may be better suited to particular regions, depending on social, economic and ecological factors. Citizen science can easily adapt to offer this variety. The mosquito-centred citizen science projects that have been emerging across the globe are unique in terms of goals and methods (Kampen et al. 2015; Vogels et al. 2015; Waterhouse et al. 2017; Yong 2017; Mukundarajan et al. 2017), though many share common practices like the collection of photographs or specimens as vouchers for species identification (figure 15.2).

Global Mosquito Alert

Despite the apparent scalability of networked citizen science, existing mosquito-related projects are mostly limited to a handful of countries. This appears to result from two basic challenges. First, the need to communicate effectively with participants and work closely with local public health and vector control authorities adds an inherently local aspect to vector-mosquito citizen science. Specific vectors, especially invasive species, will differ from one municipality to another, as will the authorities responsible for vector monitoring, management and control. Second, projects have struggled, thus far, to find funding sources that are sufficiently large and sustainable to create the infrastructure needed for both long-term local implementation and global interoperability.

The GMAC initiative took shape as a way to address these challenges. After initial discussions between existing projects, the initiative was launched at an international workshop in Geneva, convened by the European Citizen Science Association (ECSA), the Woodrow Wilson International Center for Scholars, and the United Nations Environment Programme (UNEP) in April 2017. The workshop brought together experts and the heads of vector-mosquito citizen science projects from around the world. It quickly became clear that the diversity of approaches should be embraced through the formation of a consortium to serve as a global hub of resources and an engine for mobilising funding for locally customised projects at the country or region level.

The workshop participants agreed on the following vision:

> The Global Mosquito Alert Consortium is a new citizen science initiative that aims to leverage networks of scientists and volunteers for the global surveillance and control of the mosquito species known to carry the following diseases: Zika, yellow fever, chikungunya, dengue, malaria and the West Nile virus. Global Mosquito Alert Consortium will be an open, common set of protocols and a toolkit that is augmented with modular components created to meet both global and local research and management needs.
>
> (Tyson et al. 2018)

The GMAC will start by focusing on four canonical protocols that reflect the goals of existing projects: (1) Real-time surveillance of adult vector mosquitoes; (2) investigation of larvae and breeding sites; (3) tracking of biting and nuisance; and (4) mosquito biodiversity approaches involving specimens and DNA identification techniques. Each protocol is designed with a small set of common, core elements and common metadata documentation and data policies to facilitate interoperability (see also Williams et al. in this volume). These include a common set of data-validation processes and supporting tools, complemented by a directory of experts that can help local projects develop; a common process of data analysis and visualisation; four sets of open, canonical Android and iPhone mobile applications that may be customised for local use (e.g., the Mosquito Alert app has already been translated into Spanish and Cantonese for local pilot deployments in Colombia, Mexico, Puerto Rico and Hong Kong). Common data policies include compatible open source licences for software and open-access licences for data, privacy protections for participants and a set of user agreements.

Looking forward

For GMAC and other citizen science initiatives to realise their promises, policymakers and regulators, in collaboration with technologists, should have an ambitious conversation about global data commons. They need to address the question of how open and resilient big data architectures should be, in particular those used for monitoring vital public health and environmental factors. Experts will also need to consider the challenge and cost of ensuring accuracy when dealing with environmental samples, especially biological and genomics samples. The potential of monitoring for disease-vectors is enormous but methods are needed to validate data and address liability issues.

This is why, throughout this process, GMAC will be working with UNEP to develop a global portal on UNEP's open-access web platform, Environment Live, where both data and techniques can be shared, assessed and improved. In an increasingly interconnected world, GMAC aims to give citizens the tools to make a growing contribution to combatting disease-vector mosquitoes at a scale never previously achieved, regardless of where they live and work.

Part III
Innovation at the
science-policy interface

16
Citizen science for policy formulation and implementation

Susana Nascimento[1], Jose Miguel Rubio Iglesias[2],
Roger Owen[3], Sven Schade[4] and Lea Shanley[5]

[1] European Commission, Joint Research Centre – JRC, Brussels, Belgium
[2] European Environment Agency, Copenhagen, Denmark
[3] Scottish Environment Protection Agency, Aberdeen, Scotland UK
[4] European Commission, Joint Research Centre – JRC, Ispra, Italy
[5] University of North Carolina-Chapel Hill, US

corresponding author email: susana.nascimento@ec.europa.eu

In: Hecker, S., Haklay, M., Bowser, A., Makuch, Z., Vogel, J. & Bonn, A. 2018. *Citizen Science: Innovation in Open Science, Society and Policy*. UCL Press, London. https://doi.org/10.14324/111.9781787352339

Highlights

- Citizen science offers an effective way to connect citizens and policy, bringing societal and economic as well as scientific and political benefits.
- Citizen science has the potential to impact local and national decision-making, empower citizens and lead to better, more transparent government.
- Citizens can get involved by taking part in science-related processes and by understanding and guiding the changes taking place around them.
- Consistent with European Citizen Science Association's Principle 10, current challenges preventing greater take-up of citizen science include diverse legislation, resistance from professional scientists, managing the expectations of participants and data comparability.

Introduction

Citizen science, powered by mobile, online and computing tools, offers an effective way to connect citizens and policymakers. Citizens can get involved by taking part in science-related processes and by understanding and guiding the changes taking place around them. Consistent with the European Citizen Science Association's (ECSA) Ten Principles of Citizen Science, such practices have the potential to impact decision-making at different administrative levels, contributing to monitoring and evaluation, empowering citizens and leading to more effective and transparent government. It can also help raise awareness and, ultimately, foster behavioural change.

Studies in the UK and Germany (Davies et al. 2013; Bramer 2010) have demonstrated a vast potential that remains largely untapped, despite Europe having been at the forefront of citizen science. Recent publications (Haklay et al. 2014) report on established cases of close collaboration between governments and the public, with benefits for both sides, which range from land management to disaster response. However, while citizen science is becoming a valued and useful source of information for governments, adoption is still slow, especially at supranational (e.g., European) level (but see Smallman in this volume).

Successful citizen science experiences at national, regional and local levels, some of which are included in this chapter, can serve as an inspiration for a more integrated approach at the supranational level, as called for in several reports (Serrano et al. 2014; Haklay 2015). These examples cannot only help re-engage citizens, but also empower them in an era when the bond of trust between civil society, science and policy-making needs to be strengthened (see also Mahr et al. in this volume).

More investigation is needed to understand how citizens' knowledge and these novel inflows of data can practically enhance policy-making and implementation processes (see also Shirk & Bonney in this volume). This chapter describes the potential contribution of citizen science to policy formulation and implementation, and the challenges currently preventing its sustained uptake by public authorities in their routine activities. It discusses issues such as how to reconcile bottom-up, grassroots activities with more top-down, policy-driven initiatives. It also presents relevant examples and recommendations that can guide effective partnerships between policymakers and citizen scientists.

Potential citizen science contributions to policy

The potential benefits citizen science can bring to policy formulation and implementation range from providing evidence for assessments through supporting regulatory compliance to community empowerment and awareness raising. Large numbers of volunteers are increasingly willing to take part in these activities, while national, regional and international organisations and initiatives are starting to recognise their role and benefits (ECSA 2016b).

Environmental policies and citizen science

The breadth of citizen science activities in the environment sector is immense, covering an extensive range of policy areas and reaching all corners of the world (Haklay 2015; McKinley et al. 2015; Bowser & Shanley 2013). However, citizen participation in decision-making, especially the role of citizen science in augmenting data collected through official channels, was first recognised in the context of national and international environmental policies.

In 1998 in the Danish city of Aarhus, the United Nations Economic Commission for Europe (UNECE) adopted the Convention on Access to Information, Public Participation in Decision-Making and Access to Justice in Environmental Matters, establishing a number of rights with regard to the environment. The Convention provides, inter alia, for the right of everyone to receive environmental information held by public authorities and to participate in environmental decision-making (UNECE 1998). The EU is party to the Convention since May 2005 (European Council 2005). While the first two pillars of the Aarhus Convention concern two Directives adopted in 2003 (European Parliament and Council 2003), provisions for public participation in environmental decision-making are to be found in a number of subsequent environmental directives, regulations and policy documents, such as the 7th European Union Environment Action Programme[1], the Marine Strategy Framework Directive[2], or the Common Bird Index[3], to name but a few.

In addition to increasing legal provisions, international, European and national policy actors have started to recognise the importance of citizen science activities and the way they support policy (Haklay 2015). This is often linked to understandings of citizen science as a timely, cost-effective source of data, information and knowledge to support evidence-based policy implementation and monitoring, complementing official, authoritative sources.

A growing number of references to the active role of citizen science and crowdsourcing can be found in EU environmental policies and legal documents (e.g., European Commission 2013; European Parliament and Council 2013). However, they are yet to be recognised as effective methods to monitor the implementation of EU Directives, with some authors and organisations calling for a review of existing legislation (ECSA 2016b;

Haklay 2015). Beyond continental Europe, the *Eye on Earth Summit* in Dublin (2013) included citizen science as an important source of knowledge within the diversity of knowledge communities (Haklay 2015). At the technical release of UNEP LIVE in January 2014, UNEP highlighted citizen science and crowdsourcing as the most cutting-edge and exciting tools emerging in the global research arena (ECSA 2016b).

One of the potential benefits of using citizen science to inform environmental policies is to meet the data collection targets of programmes that need to monitor large geographical areas with high frequency, such as in the early detection of invasive alien species (Delaney et al. 2008; also see box 16.1) or monitoring wild birds. In the latter case, networks of observers are using a pan-European Common Bird Monitoring Scheme (University of West of England 2013), contributing to the implementation of the Birds Directive[4] and the generation of the Common Bird Index[5].

Box 16.1. Monitoring invasive alien species of European Union concern

The EU Regulation on Invasive Alien Species (European Parliament and Council 2014) and first list of 37 Invasive Alien Species of EU Concern (European Commission 2016a) establishes a frame that may benefit from biodiversity-related citizen science at the European scale. A mobile application for monitoring alien species has been developed by the MYGEOSS project[6] (see figure 16.1) and investigations have begun into the use of the app in the field and the validation processes required to allow it to feed data into the official European Alien Species Information Network (EASIN[7]). Groundwork has been done to allow in-depth dialogue with relevant stakeholders in the EU, including member state representatives, public servants of the European Commission and scientific networks. This activity is likely to contribute to the process of reporting about invasive alien species to the Commission, which has to be in place by mid-2019. At the national level, in 2012 an initiative launched in the UK to engage citizens in recording data on invasive species so scientists could monitor their spread and effect on the environment (Siegle 2012); and separately an app has been developed to involve citizens in observing alien species, which is proving to be a cost-effective means of gathering data[8].

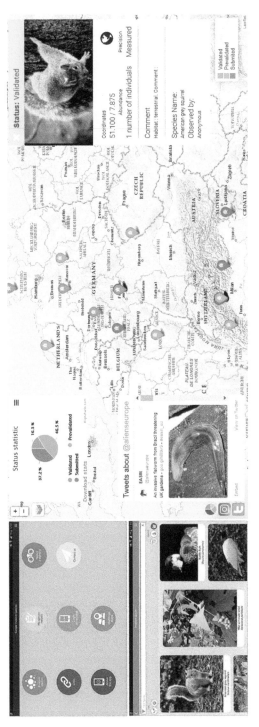

Fig. 16.1 Screenshots of the Invasive Alien Species mobile app (left) and web page (right). Please note this is work in progress; all contained data has been added for testing purposes only.

Some reports confirm the potential of citizen science to serve policymakers by providing evidence to support regulatory compliance, and identify and fill gaps in data and information (University of West of England 2013; Haklay 2015). Therefore, and within the context of the current Fitness Check of EU environmental monitoring and reporting (European Commission 2016b), citizen science has the potential to complement centralised reporting by reducing costs in data collection, validation and verification.

In the UK, there is a long tradition of volunteer naturalists participating in environmental monitoring, with an estimated 100,000 volunteers contributing to recording schemes and societies in 2005 (The Parliamentary Office of Science and Technology 2014). Current UK government action plans and strategies call for volunteers to assist with monitoring in policy-relevant applications, such as the designation of protected areas, ecological impact assessments, the development of environmental biodiversity indicators, and the identification of invasive species and disease outbreaks (see box 16.1). Data collected by volunteers enable the UK to meet its obligations to monitor, report and respond to EU environmental legislation.

The Scottish Environment Protection Agency (SEPA) has also explored the potential for citizen science to support its regulatory and policy efforts (see box 16.2), concluding that it is suitable for assessing impacts of key environmental pressures identified in the Scottish environmental monitoring strategy, such as invasive non-native species, noise and vibration, waste management or greenhouse gas emission monitoring (Pocock et al. 2014a). McKinley et al. (2015) lists a number of citizen science projects and programmes which are already used in environmental science and decision-making in the United States, in particular in the fields of species management, climate change, ecosystem services management, invasive species control, and pollution detection and enforcement. The Irish Environmental Protection Agency (EPA) has developed a mobile application called See it? Say It![9] to help people report environmental pollution (waste, air, noise) – complaints are directed to the local authority, which then has to provide a response within a short time.

Several studies have demonstrated that citizen science projects are cost-effective, especially in the case of large-scale projects. In the UK, a £7 million government investment in volunteer monitoring schemes generated data estimated to contribute £20 million in kind (Makechnie et al. 2011). In France, annual savings of €1–4 million have been estimated as a result of the Citizen Science Biodiversity Monitoring Programme of the French National Museum for Natural History (Levrel et al. 2010; and see Peltola & Arpin; Sforzi et al., both in this volume). In the United States,

Box 16.2. Scottish Environment Protection Agency and citizen science

In recent years, SEPA has recognised the need to take a more strategic approach to its involvement in citizen science, partly in response to the challenges of sustaining and growing an increasing number of projects. The strategic approach comprises several elements, including a high-level strategy outlining how citizen science can help deliver core responsibilities and objectives, published guidance on the types of projects SEPA would support, the coordination of SEPA citizen science activities to ensure alignment with its overall strategy and the provision of relevant IT infrastructure and tools[10]. The Scottish Environment Protection Agency now explicitly states the important role of citizen science in engaging the public in changing attitudes and behaviours towards their environment, improving health and well-being, and developing partnerships, in addition to generating valuable data. The strategy recognises that citizen science is often cheaper, though not zero cost, when it comes to generating data or information, even if it is less precise than SEPA's own professional monitoring. The Scottish Environment Protection Agency has helped communities of interest to maximise the quality of their data through training and access to verification tools. There are, however, a number of issues that SEPA still needs to address, such as capacity building, balancing open communication with SEPA's official policies and messages, improving evaluation, maintaining volunteer motivation and reconciling the cost of citizen science activities with increasing constraints on public resources. The Scottish Environment Protection Agency is working with partners in the UK Environmental Observation Framework Citizen Science Working Group to commission much-needed research on understanding the various motivations for participating in citizen science, and on assessing its costs and benefits for public bodies. There is also a need, as stated elsewhere in this chapter, to provide evidence that engagement in citizen science effects behavioural change by participants and in society more generally.

analysis of 238 citizen science biodiversity projects around the world estimated that the in-kind contributions of 1.3 to 2.3 million volunteers had an economic value up to $2.5 billion per year (Theobald et al. 2015).

Research, science and innovation policies

Speaking more broadly, citizen science is increasingly recognised as instrumental in fostering open and novel science in research, science and innovation strategies, policies and initiatives. For example, in the current EU Research and Innovation policy, it is a key element of one of the five lines of potential policy actions supporting the development of open science in Europe (European Commission 2016c). In the United States, the 2013 Open Government National Action Plan included the initiative to create an Open Innovation Toolkit to promote innovation in federal agencies, including approaches such as crowdsourcing and citizen science. On 30 September 2014, the White House Office of Science and Technology Policy (OSTP) and the Domestic Policy Council (DPC), in collaboration with the Federal Community of Practice on Crowdsourcing and Citizen Science, co-hosted a forum addressing the links between citizen science and crowdsourcing with open science and innovation (see also Robinson et al. in this volume). In this context, recent reports advocate for change to science and research programmes and funding schemes to facilitate the participation of grassroots initiatives driven by citizen scientists and guarantee its sustainability (Serrano et al. 2014; Bonn et al. 2016).

Citizen science is also recognised as engaging the public across the research landscape and guiding research agendas towards issues of concern to citizens. The 'Science with and for Society' programme of Horizon 2020, the current EU Framework Programme for Research and Innovation (see box 16.3), takes an approach called Responsible Research and Innovation (RRI) (see Smallman in this volume for much more on RRI). Responsible Research and Innovation advocates allowing all societal actors (researchers, citizens, policymakers, business, etc.) to collaborate to better align both process and outcomes with the values, needs and expectations of society (European Commission 2015). The involvement of public and civil society stakeholders in processes, outcomes and powerful co-creation is a key component of RRI as a way to build public acceptance of innovation, while making it more effective (Sutcliffe 2011). Crowdsourcing initiatives are mentioned as new ways of involving the public in prioritising innovation and its implementation.

Box 16.3. The EU Research and Innovation policy and citizen science

At a major EU Conference in April 2016 in Amsterdam, the Commissioner for Research, Science and Innovation, Carlos Moedas, outlined his vision for a common EU approach to open science in Europe. He made a call for citizen scientists to contribute to European science as valid knowledge producers by 2020 (Moedas 2016). Open science is one of three goals of the current EU research and innovation policy, first set out in 2015 (European Commission 2016c), together with open innovation and being open to the world. Citizen science is mentioned as a key tool to foster open science in education programmes, promote best practices and increase the input of knowledge in one of five potential policy actions identified in a draft European Open Science Policy Agenda, 'Fostering and creating incentives for Open Science'. It is also one of the eight issues addressed by the recently created Open Science Policy Platform (OSPP), a high-level expert group representing stakeholders, which will propose recommendations for co-designing and co-developing the Open Science Policy Agenda through with relevant actors in science and research in Europe.

Public empowerment and behavioural change

Citizen science is now regarded as a way to empower communities in driving forward policies (Rowland 2012). Some reports confirm that it allows citizens to adopt more active roles in society, protect their environment and drive a more participatory form of democracy (Ala-Mutka 2009; Mueller et al. 2012) and that it provides opportunities for closer interactions, especially with local governments (Irwin & Michael 2003).

The potential of local knowledge and citizen science activities has been demonstrated in several cases of environmental justice including citizen-driven initiatives against water drilling and disposal by an oil company in Peru, noise pollution in a scrapyard in London and hydraulic fracturing ('fracking') in the United States (University of West of England 2013). However, the same report recognises that there are few examples of truly participatory citizen science with evidence that they have influenced decision-making, although this may be related to difficulties in obtaining evidence. Elsewhere, the Environment Agency in England has direct evidence of biological water quality recording by anglers leading to the successful prosecutions of polluters (The Riverfly Partnership 2007).

Overall, though, it appears that many citizen science projects are not benefiting from the more participatory roles of citizens (Mueller et al. 2012). Furthermore, if they are to contribute to more participatory forms of democracy, such activities should be inclusive and accessible to all, not only those who have access to the latest technologies or are well-educated (Haklay 2013; and see Haklay in this volume; Peltola & Arpin in this volume). For example, the participation of local communities in volunteered geographic information initiatives is important in addressing the challenge of building resilient societies (Haklay et al. 2014).

While difficult to measure, evidence suggests that citizen science can positively affect participants' attitudes and behaviours towards the environment (The Parliamentary Office of Science and Technology 2014). Strategic environmental policies, such as the EU Biodiversity Strategy (European Commission 2011), recognise that citizen science is a valuable means of mobilising citizens in biodiversity conservation, while gathering high-quality data. Few studies have analysed changes in attitudes towards the environment and environmental behaviours thanks to public participation in science (University of West of England 2013) but Davies et al. (2013) report that almost half of volunteers recognised a change in the way they thought about the environment and more than one third would change their behaviour towards it. Stepenuck & Green (2015) report some changing attitudes and behaviours, although some appeared to be more superficial than desired. More research is needed to better understand attitudinal and behavioural changes, which could impact attempts to address global challenges such as climate change and biodiversity degradation.

Challenges for citizen science-based policy

The proven and potential benefits of citizen science are offset by challenges ranging from data quality and management, institutional resistance or lack of awareness by public bodies, to persistent social inequalities that limit participation. These obstacles, explored more below, demand further discussion and sustained efforts to co-ordinate responses to them.

Data quality and management

Data quality, comparability and interoperability are considered essential for both evidence-based policy-making and scientific evidence (see Williams et al. in this volume). At the same time, the capacity of citizen scientists to deliver high-quality and reliable data is one of the most debated issues in citizen science. However, studies attest to the accuracy of citizen science models in providing reliable data, including on geographical information (Haklay 2010), bird habitat (Nagy et al. 2012), air pollution (Tregidgo et al. 2013) and ecosystems (Gollan et al. 2012). Instead, the issue of quality in citizen science is related to project design, which demands adequate data validation protocols or mechanisms (Bonter & Cooper 2012). Successful initiatives combine multiple methods to ensure data quality (Wiggins et al. 2011) and operate in different organisational settings beyond more traditionally scientific ones, requiring appropriate quality assurance (Haklay 2017).

Introducing or revising protocols and standards can pose additional challenges for data consistency and its relationship to official and mandatory statistics. This also poses problems for the (re-)use of the data. Consider, for example, air quality monitoring in Europe. Many activities using generally applicable sensor kits are building their own communities, and data is collected and stored in independent information systems. The results neither cover the complete territory of the EU, nor measure in a synchronised or continuous way. Changing this would not only require harmonisation efforts but also sustainability, including long-term storage, curation and archiving of contributions. As a result, citizen science remains largely separate from the knowledge base used to deliver EU policies on environment (but see Volten et al. in this volume).

Data management and interoperability were in fact identified by participants of the Citizen Science and Smart City Summit as critical to long-term benefits from citizen engagement (Craglia & Granell 2014). A later international survey of data management practices revealed that approximately 60 per cent of participating projects followed a dedicated data management plan and many applied standards to the data and metadata generated (Schade & Tsinaraki 2016). Although a majority of projects claimed to provide access to raw or aggregated data, they did not always apply appropriate use conditions or well-defined licences. The detailed underlying issues are also addressed in Williams et al. in this volume and existing solutions are included in the upcoming book from the COST Action (a European framework supporting transnational co-operation in science and technology) on citizen science[11] (Bastin, Schade & Schill 2017). Related work is taken up at the international level by a Data and Metadata Working Group chaired by the Citizen Science Association (2015); and the Open Geospatial Consortium (2016) adopted a Citizen Science Domain Working Group in 2016 regarding geolocation data.

Adoption by public institutions

The current political context increasingly calls for civic involvement, ranging from conventional mechanisms of consultation to direct integration in multiple stages of the policy cycle. However, the actual adoption and impact of citizen science in policy-making is difficult to demonstrate. For instance, although citizen science is positioned as a key tool to foster open science at the European level, mechanisms are still lacking for citizens to impact evidence-based processes for policy-making. Citizen science's weight or importance is not visible at all policy stages, nor is it clearly connected with current public engagement mechanisms such as public consultations or citizen-initiated policy proposals.

Public institutions wanting to engage in citizen science (see box 16.4) also have to consider the resources required to manage expectations from actively engaged citizen scientists and participants. Empowering individuals and communities with information requires constant feedback and dialogue. There is also a perceived danger that alternative messages on environmental issues can develop from public access to raw data, leading to conflicting interpretations. This can be overcome by careful planning of feedback mechanisms and provision of appropriate contextual information.

Box 16.4. Citizen science in the US federal government

While some US federal agencies have supported citizen science projects in the past, a concerted grassroots effort led by the Federal Community of Practice on Crowdsourcing and Citizen Science (CCS)[12] has helped to dramatically increase the visibility, credibility and adoption of citizen science within the US government to address societal and scientific challenges. Through a series of interviews, the CCS identified barriers to adoption by government agencies, including trust, data quality, privacy, cybersecurity and perception of liability risk (Gedney & Shanley 2014). The CCS

---

then developed a set of strategies to address these hurdles, including assembling success stories (Bowser & Shanley 2013), consulting with legal analysts (Gellman 2015; Scassa & Chung 2015a; Scassa & Chung 2015b), streamlining the project approval process (Parker 2016), and providing educational briefings to agency executives and inviting them to speak at citizen science-related events (Shanley et al. 2013). The CCS also collaborated with the White House to build a 'How-To' toolkit[13] and projects database[14] aimed at reducing barriers to entry and increasing government-wide coordination. Lastly, the CCS inspired and informed the White House memo 'Accelerating Citizen Science and Crowdsourcing to Address Societal and Scientific Challenges', issued September 2015, and the Sec. 402 of the American Innovation and Competiveness Act, passed January 2017, as well as other policy directives, providing top-level support for government agency use of citizen science (Holdren 2015; see also GAO 2016; US GEO 2014; USGCRP 2014). The White House memo articulated guiding principles for US federal citizen science and crowdsourcing projects, including (1) applying the principle of 'fitness for use' (i.e., 'ensuring that data have the appropriate level of quality for the purposes of the project'); (2) ensuring that the data, code, applications and technologies generated by federally sponsored citizen science projects are 'transparent, open, and available to the public'; and (3) engaging members of the public in citizen science in meaningful ways such that their contributions are mutually beneficial and publicly acknowledged. The memo also directs each agency to designate a co-ordinator for citizen science and crowdsourcing, and to catalogue federally funded and/or co-ordinated citizen science and crowdsourcing projects, building on the work of SciStarter. The legislation clarifies the authorisation for federal agencies to support citizen science projects, and addresses some administrative and legal issues such as liability.

However, some have questioned the effect of the democratisation of science and technology policy (Wynne 2006; Irwin 2006). Dialogue can be seen as a way of enhancing trust in science, or avoiding public resistance to issues that are economically and politically important. More critical studies argue that engagement tends to be constrained by official perspectives, making participation 'another governance tool among others, e.g., for adjusting, supplementing or enhancing the policy process' (Levidow 2008, 3).

Furthermore, one of the main challenges at the supranational level comes from the diversity of legislation and cultures. European member states, for example, have disparate regulatory frameworks on data management, official measurements and privacy requirements, along with different levels of readiness and previous engagements with citizens and stakeholders. Questions remain about how these issues are addressed across geographic (and thereby administrative) scales; across different institutions (such as the European Commission and the European Environment Agency, see box 16.5); and across supranational agencies, including governmental (e.g., environment protection, mapping and

---

Box 16.5. Environmental knowledge community and citizen science

In order to understand the European challenges and to benefit from the richness and diversity of the European citizen science landscape, five Directorate Generals (DGs) of the European Commission – DG Environment, DG Research and Innovation, DG Joint Research Centre, DG Climate Action and DG Eurostat – together with the European Environment Agency (EEA) agreed to jointly investigate the potentials and limitations when connecting citizen science to European environmental policy-making. A group set up in January 2015 began to consider how data gathered by citizens (using mobile phones, for example) could best be used to complement environmental monitoring and reporting processes in a cost-effective manner, to review the potential of lay, local and traditional knowledge to fill in knowledge gaps and to examine how the involvement of citizens could foster environmental behavioural changes. The participants jointly contributed their experiences and diverse roles in policy-making processes in order to address, among others, the questions outlined in this chapter. This initiative includes direct practical experiences by initiating citizen science demonstrators for European policy-making, which includes the work in support of the European Union Regulation on Invasive Alien Species as reported above. EU-funded research activities such as the Citizens' Observatories[15] are also deemed to contribute to this endeavour.

---

statistical) agencies as well as non-governmental organisations (NGOs) (e.g., the ECSA).

Inequality and power imbalances

Citizen science pushes for more democratic ways of generating, selecting and interpreting high-quality data to inform decision-making. However, citizen science projects are most successful at integrating those citizens who probably already have the most resources to engage in policy in the first place (e.g., time, capital). If one of the main goals of citizen science is to offer more possibilities for citizens to generate knowledge for policy formulation and implementation, underserved communities and unheard voices need to be included in a people-powered science (see Haklay; Novak et al., both in this volume).

Inequalities in the way research findings are taken up by policymakers have also been documented in research close to the field of citizen science. For example, in 'undone science' (Frickel et al. 2010; Hess 2015), community-based participatory research (Bidwell 2009), 'counterpublics' (Hess 2011), or in general community-based science, social movements or civil society organisations have done research that has systematically been unfunded, incomplete or ignored by traditional research bodies. This attests to the asymmetries that citizen science needs to address. Public engagement may therefore require 'control mutuality' between the parties involved, that is, a shared agreement on the influence and control they have over one another (Grunig & Grunig 2001). This would lead to a 'sharing of power' (Seifert 2006, 83), where all parties allow the outcomes of participatory exercises to truly be unpredictable and to have substantial consequences on the processes.

Participatory models to inform policy-making

This section presents an overview of past and present debates around top-down and bottom-up approaches when it comes to the relationship between science and society. From more traditional and one-way connections between experts and non-experts to more recent co-creation models, citizen science remains a contested field of practice, even more so when it moves towards do-it-yourself (DIY) experimentation that places citizens and communities at the centre (Ballard, Phillips & Robinson; Novak et al., both in this volume). These past, present and emerging approaches offer lessons for dialogue, feedback and, ultimately, the co-creation of science and policy.

From top-down to bottom-up science

The interplay between science and society has moved from the top-down approaches of traditional science communication (one-way, from 'experts') towards bottom-up models of public engagement (e.g., two-way dialogue and the co-construction of research agendas and interpretation). Giving a privileged role to the public has been at the core of debate on the relationship between science, technology and democracy (Jasanoff 2004; Latour & Weibel 2005), in which a 'democratic turn' has pushed for a more open agenda (Fischer 2000; Leach, Scoones & Wynne 2005). In practical terms, public engagement has been implemented and tested through consultative and deliberative mechanisms, such as citizen juries, citizen panels, deliberative polls, citizen schools, dialogues, focus groups and consensus conferences (Burri 2009; Coote & Lenaghan 1997; Joss & Durant 1995b), extending to participatory and experimental mechanisms, such as future scenarios, experiential tools, co-design and digital interventions (Chilvers & Kearnes 2016; Nascimento et al. 2016).

In citizen science, it can be argued that more institutionally led (top-down) projects remain abundant (Nascimento, Guimarães Pereira & and Ghezzi 2014). Invitations for collaboration originate from scientific organisations, which largely predetermine the research objectives and citizens' involvement tends to be limited. Even the language can be one-sided, describing citizens as enlisted (Hochachka et al. 2012), recruited (Suomela 2014) or, more typically, as a crowd of data collectors (Devictor, Whittaker & Beltrame 2010) or data processors via their own resources, such as computers and mobile phones (Roy et al. 2012). However, such passive participation is moving towards more active roles, including as interpreters and creators of data in collaboration with scientists and policymakers. Overall, bottom-up contributions should be supported and top-down policy processes engaged to connect the two perspectives so that policymakers are ready to receive data and findings from participants and take action. Increasing numbers of participants are likely to further legitimate official mandates to actually integrate and use data from citizen science.

Citizen science as a contested space

Embracing bottom-up perspectives requires acceptance of a wider range of knowledge co-creation and sharing than traditionally included in evidence-based policy-making. This higher level of participation is focused on citizen empowerment and inclusion in defining the conditions and purposes of evidence. In what some call 'extreme or collaborative science' (Haklay 2013), citizens are mostly seen as equal to scientists when it comes to decisions about research questions, methods or processes (Funtowicz & Ravetz 1992), which can include data analysis by the communities involved in the projects beyond the provision of data or processing resources.

In this way, citizen science can challenge the ways scientists produce knowledge, including their assumptions and standards about what is valid as scientific knowledge. This does not mean a degradation of data quality if data collection is based on systematic, fit-for-purpose observations and protocols agreed from the beginning. Still, citizen science can unsettle traditional beliefs about the uniqueness and complexity of scientific practice if it is then performed by non-professional scientists. In many cases, citizen science participants are experts in their own right as a result of their own professional expertise, or become experts through their voluntary participation (e.g., see Peltola & Arpin in this volume).

Conflicts of interest between parties are a common concern. A recent editorial in *Nature* raised concern that citizen participants might advance their political objectives, such as when 'opponents of fracking, for example, might help track possible pollution because they want to gather evidence of harmful effects' (Editorial 2015). This editorial sparked debate in the citizen science community on social media, mailing lists and blogs, where it was positioned in the wider context: bias as a result of asymmetrical power relations in science and policy; claims of falsification or data fraud outside of citizen science; and the integration of personal motivations, value judgements and social norms in epistemological understandings of objectivity. Open discussion is needed between all parties to transform rigid understandings of what constitutes relevant knowledge for science and policy.

Towards transdisciplinary and DIY trends

It can be argued that the rationale of citizen science in involving diverse groups bring it closer to transdisciplinary frameworks that are visible in different scientific fields. Generally speaking, transdisciplinarity operates both horizontally, to involve and mix different areas of expertise, and vertically, to include stakeholders from civil society and the private and public sectors (Klein 2004). Transdisciplinarity strives to generate comprehensive knowledge through collaborative platforms with both academic and non-academic stakeholders, while also combining frameworks across disciplines. It privileges bringing together all types of knowledge towards a common and practical goal (Nascimento & Pólvora 2015), from a global network of makerspaces publishing their work on Github to localised interventions in an urban neighbourhood monitoring air quality (Balestrini et al. 2016; see box 16.6). It means that inputs for policy formulation and implementation can come from many different places and social groups, as long as they are relevant and can generate high-quality knowledge.

In its more radical forms, transdisciplinarity does not impose a hierarchy of expertise, connecting to emerging citizen science movements that rely on projects initiated and developed by individuals or groups unaffiliated to the scientific establishment. Even where these individuals and groups do have a scientific affiliation or background, their initiatives do not align with conventional or prescribed institutional rules. The DIY movement, or what is sometimes called do-it-together (DIT), has been paving the way for the next steps for citizen science (Nascimento, Guimarães Pereira & Ghezzi 2014; and see Novak et al. in this volume). The 'DIY scientist' is someone who tinkers, hacks, fixes, recreates and assembles objects and systems in creative and unexpected directions, usually using open-source tools and adhering to open paradigms to share knowledge and outputs with others. Do-it-yourself scientists are doing science outside conventional university or lab settings, and instead in makerspaces, Fab Labs, Hackerspaces, techshops, innovation and community-based labs, or even in their homes, garages or schools.

These forms of enquiry recognise different ways of knowing and allow for more out-of-the-box thinking and experimentation. Such emerging practices also bring forward new and valuable sources of data that can contribute to policy-making processes. A DIY environmental science community such as Public Lab[17] uses low-cost techniques to investigate environmental issues, often to improve citizen contributions to decision-making and enable change in the political sphere. Although currently marginal, these practices are likely to grow, along with their challenge to mainstream science and policy-making. Such challenges can be productive and bring about new thinking and practices, not only enriching science and policy but also empowering citizens and communities.

Box 16.6. Citizens and communities building their own sensors

The project Making Sense: Advances and Experiments in Participatory Sensing (H2020 Competitive Project 688620) aims to develop participatory frameworks and tools for citizen-driven innovation[16]. It will show how open source software, open source hardware, digital maker practices and open design can be used by local communities to appropriate their own technological sensing tools and address pressing environmental problems in air, water, soil and sound pollution. The project is developing a Making Sense Toolkit, based on the Smart Citizen platform (see figure 16.2) and in other open source sensors, to be tested in Amsterdam, Barcelona and Pristina. In the pilots, the team is working with communities of interest, including citizens, local associations and civil society organisations, and communities of practice, such as hardware makers and tinkerers (someone who likes to hack, change or repair machines or objects) well-versed in open source technologies and digital fabrication. They meet and collaborate at local Fab Labs and makerspaces to deploy, test and improve readily available open hardware and software tools, and contribute with best practices around community-driven environmental sensing and sense-making. Participants also interact with experts and city officials, collect and share data, visualise and interpret results, and devise responses, either individually or collectively.

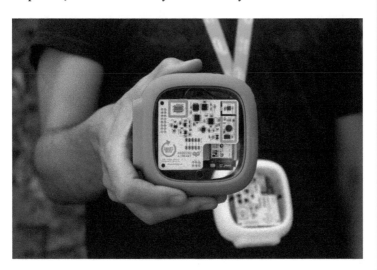

Fig. 16.2   The Smart Citizen Kit, a DIY and open source sensor. (Source: Smart Citizen team; Fab Lab Barcelona | IAAC and MID)

Conclusion: strategies and recommendations

This section offers strategies and recommendations for introducing and coordinating citizen science initiatives within and across different levels of governance. It is not an exhaustive overview but contributes a list of priority areas where action is needed for citizen science to become an integral part of future policy and build effective partnerships between governments and their citizens.

Integrating citizen science data: Citizen science has the potential to complement, validate and enhance data collected through official channels with broader, timely and cost-effective data sources. This has already been showcased in areas such as biodiversity monitoring (e.g., birds and invasive alien species) or compliance assurance (e.g., environmental pollution reporting). Adequate standards and infrastructure are needed to deliver on this potential, including revised data validation protocols, multiple methods for data quality, data interoperability and management, and innovative and robust technologies. To be really effective, this should be complemented by the formulation of more participatory processes, which may imply the review of legislative frameworks.

Developing citizen engagement and empowerment: Citizen science can raise awareness and empower citizens and communities, and potentially improve their relationship with government, official bodies, scientists and other actors. To harness this potential, policymakers and implementers need to embrace more participatory, citizen-centred, inclusive and bottom-up approaches for knowledge and data production, together with formal mechanisms for citizen participation in decision-making and ultimately, the co-creation of policies.

Coordinating across governance levels: Despite the integration of citizen science into national and regional policies and programmes, few networks connect emerging citizen science initiatives with each other or with existing knowledge and policy schemes. Furthermore, existing programmes and policies are mostly linked to environmental monitoring and reporting activities. Co-ordination, with clear definitions of the opportunities, roles and responsibilities at different levels of governance, would strengthen coherence and expand the application of citizen science to policy areas where it has a strong potential, such as climate monitoring, agriculture and food security, urban planning and smart cities, health and medical research, humanitarian support and development aid, science awareness and scientific efforts.

Supporting pilots and practical experimentation: Complex interplay between the many stakeholders complicates the further integration of citizen science and other emerging trends, such as DIY science, in policymaking. Collaboration can be strengthened with empirical studies and practical testing such as demonstration projects (see box 16.6), open and interdisciplinary calls for proposals and projects, or adapted methodologies for community engagement. More needs to be done in terms of mutual learning between such projects and pilots, and pilots can also be appropriate testbeds of community engagement for further integration with policy processes.

Establishing and strengthening communities of practice inside public administration: Previous and ongoing initiatives have proven the effectiveness of sustained mechanisms for civil servants and policymakers to incorporate citizen science in their work (see box 16.4). Examples of mechanisms include creating networks of practitioners and champions across departments, units or agencies; developing adequate communication and capacity-building tools such as roundtables, webcasts, blogs or practical training; identifying obstacles preventing citizen science from being used effectively and widely in specific organisational contexts; and producing best practices which showcase successful projects.

Connecting to current priorities: High-level commitment – from top scientists, management, policymakers and institutions – would promote the use of citizen science in policy formulation and implementation. Understanding of the policy agenda and its pipeline of initiatives should be coupled, when possible, with demonstrations of the potential citizen science impact on constituencies, and this could increase such commitment. A clear policy strategy for citizen science initiatives would help ensure they are perceived as useful for policy, while the wider citizen science community would also benefit from guidance on what policymakers find helpful. The right framework and communication strategy are needed to ensure citizens are heard and feel they are part of the solutions that concern them. Even the careful selection of terminology to describe citizen science in a way that is relevant to policymakers can make a difference.

Acknowledgements

The authors would like to acknowledge the European Environment Agency (EEA) and the Citizen Science Interest Group of the European Protection Agency (EPA) network. They would also like to thank the team and partners of the project Making Sense: Advances and Experiments in Participatory Sensing (H2020 Competitive Project 688620), under the support of the Collective Awareness Platforms for Sustainability and Social Innovation (CAPS) programme. Here a special thank you is due to Fabrizio Sestini (DG CNECT, European Commission). The authors would also like to thank the Alfred P. Sloan Foundation for their support and encouragement of the US Federal Crowdsourcing and Citizen Science initiative through grants to the Commons Lab of the Woodrow Wilson International Center for Scholars. Finally, the authors are grateful for valuable discussions in the European Commission's Environmental Knowledge Community, which helped shape this chapter's recommendations.

Notes

1 Decision No 1386/2013/EU of the European Parliament and of the Council of 20 November 2013 on a General Union Environment Action Programme to 2020 'Living well, within the limits of our planet'.
2 Directive 2008/56/EC of the European Parliament and of the Council of 17 June 2008 establishing a framework for community action in the field of environmental policy (Marine Strategy Framework Directive).
3 http://ec.europa.eu/eurostat/web/products-datasets/product?code=tsdnr100
4 http://ec.europa.eu/environment/nature/legislation/birdsdirective/index_en.htm
5 http://ec.europa.eu/eurostat/web/products-datasets/product?code=tsdnr100
6 http://digitalearthlab.jrc.ec.europa.eu/mygeoss/
7 http://easin.jrc.ec.europa.eu/
8 http://planttracker.org.uk/
9 http://www.epa.ie/enforcement/report/seeit/
10 http://www.environment.scotland.gov.uk/get-involved/
11 https://www.cs-eu.net/
12 https://www.citizenscience.gov/community/
13 https://crowdsourcing-toolkit.sites.usa.gov/
14 https://www.citizenscience.gov/
15 http://www.citizen-obs.eu/
16 http://making-sense.eu/
17 https://publiclab.org/

17

Citizen science and Responsible Research and Innovation

Melanie Smallman[1]

[1] University College London, UK

corresponding author email: m.smallman@ucl.ac.uk

In: Hecker, S., Haklay, M., Bowser, A., Makuch, Z., Vogel, J. & Bonn, A. 2018. *Citizen Science: Innovation in Open Science, Society and Policy*. UCL Press, London. https://doi.org/10.14324 /111.9781787352339

Highlights

- Responsible Research and Innovation (RRI) is emerging as a key approach to mediating the relationship between science and society to tackle social challenges.
- Citizen science has both overlaps with, and divergences from, RRI.
- Citizen science could learn lessons from RRI approaches and processes especially in terms of meaningful citizen participation.
- A more responsible citizen science would need to engage with issues of participation, agenda-setting (including power relations) and acting responsibly – and collectively.

Introduction: Responsible Research and Innovation

Responsible Research and Innovation (RRI), a cross-cutting theme of the European Commission (EC) Horizon 2020 programme, is emerging as a key approach to mediating the relationship between science and society. Bringing together public engagement, open science, gender equality, science education, ethics and governance, and more, RRI aims to align the outcomes of science and innovation with the values of society to address the grand challenges ahead. As the following section will discuss, many of the objectives and outcomes of RRI also have considerable overlaps with the Ten Principles of Citizen Science (ESCA 2015).

Science and innovation are key drivers of developed economies and social change. This is clear from how the car has shaped the structure of cities and transport systems, and how the internet is changing business and social relationships. At a time of increasingly pressing challenges – such as how to feed a growing world population, take care of an ageing population or tackle climate change – many believe that science and innovation will be critical in offering answers.

However, science and innovation brings downsides as well as benefits, and the benefits are not spread evenly geographically or socially. In some instances, science and technology even challenge ways of life: The internet allows companies to produce 'stateless profit' while governments struggle to fund public services; developments in genetics raise questions about the rights of disabled people; and genetically modified (GM) crops threaten non-GM or self-sufficient farmers. It is perhaps understandable that not everyone is as enthusiastic about science as scientists themselves.

Historically, public concerns about science and innovation were seen by scientists and policymakers as 'problems' to be dealt with through more information and education – this is known as the 'deficit model' (Smallman 2014; Stilgoe, Lock & Wilsdon 2014). For example, in 1986 the UK's Royal Society published a report on 'The Public Understanding of Science' (Bodmer 1986), which claimed that improving the general level of public understanding of science was an urgent task, given the importance of science in almost every aspect of life. This soon proved to be an over-simplification of the relationship between knowledge and attitudes, however. For instance, Evans and Durant (1995) found people's attitudes becoming more polarised when they became more informed about a particular area of science or technology. Controversial topics such as embryology research were seen as more controversial by those with higher levels of knowledge (Evans & Durant 1995); while Brian Wynne (1996) highlighted the existence of 'lay expertise', describing how Cumbrian sheep farmers' predictions of how the soil would respond to Chernobyl proved to be more accurate than the 'expert' models.

In the UK, building on this insight and following public controversies around bovine spongiform encephalopathy (BSE) and GM crops, a new approach to science and society was adopted, notably outlined in the UK House of Lords report 'Science and Society' (House of Lords 2000). The report heralded in a new era of 'dialogue', which aimed to involve the public in two-way communication around science so that the public could be assured that their views were taken into account. Various activities followed involving the public in debates about contemporary science and technology, including the UK government–led GM debate (Horlick-Jones et al. 2006; Gaskell 2004).

In Europe and North America, a practice called Participatory Technology Assessment (PTA) arose during the 1980s and 1990s (Griessler, Biegelbauer & Hansen 2011; Joss & Durant 1995). Participatory Technology Assessment is a process (or series of processes) which aimed to broaden the knowledge base of decision-making, in order to make political decisions more informed and rational (Abels, 2007). A number of European countries took up this approach during the 1980s and 1990s, most notably the Danish Board of Technology, which developed and ran a series of 'Consensus Conferences'. The Netherlands also took up the idea, organising a consensus conference on genetic modification of animals in 1993. Such ideas around participation were also taken up more widely by the European Commission's 'Science in Society' Framework 7 Programme (Owen, Macnaghten & Stilgoe 2012). Joss and Durant (1995b) argue that such participatory processes were rooted in this 'dialogue model' of the public understanding of science, in which the key activity is two-way or multi-way communication between scientists and non-scientists, with the aim of creating greater mutual understanding, which may or may not lead to greater accord between scientists and non-scientists (Joss & Durant 1995b).

For many, this move from deficit to dialogue (or public engagement as it became known) remained problematic as the objectives of science – and the assumption that science is an inherent public good – went unchallenged. Dialogue or engagement allowed the public to voice their concerns but this was often in a limited way (Macnaghten, Kearnes & Wynne 2005; Wynne 2006) and appeared to have little impact on policy (Smallman 2017). As Wynne argues, a perceived deficit in knowledge was replaced by a perceived deficit in trust, with two-way communication adopted as a new way for science to win public trust, without putting the objectives and values of the institutions themselves under scrutiny (Wynne 2006).

Drawing on lessons from public engagement, RRI takes up the challenge of listening, taking account of public perspectives and scrutinising the values of science. It aims to build a form of science and innovation that truly reflects wider social needs and values. Indeed, RRI sets out to change the purposes that science is put to – moving away from puzzle-solving and the 'Republic of Science' (Polanyi 1962) view of science as serendipitous, unpredictable and specialist, towards a co-productionist (Jasanoff 2004) perspective. Here, the visions and values of those doing the research and development are understood to be deeply embedded in the knowledge, products and social structures produced. Opening up these visions and values to wider perspectives – and allowing the possibility that non-scientific stakeholders might occasionally take the reins away from the scientists – is key to RRI. This adds new depth to the meaning of the Ten Principles of Citizen Science. Public participation in RRI means interpreting Principle 1: 'actively involving citizens in the scientific endeavour and creating new scientific knowledge' (ECSA 2015) as much more than allowing citizens to taking part and experiencing science from the inside then – it is about citizens working with scientists, policymakers and innovators to set the agenda, anticipate the consequences and work out the best way of making use of, come to terms with or deal with science and its implications (see also Haklay; Novak et al. and Nascimento et al., all in this volume). To give a sport analogy, it is not just about inviting citizens to play in the football team, or helping them understand the rules of the game, but asking them whether they want to play football at all, or whether they would prefer to play hockey or even do some painting instead (see also Ballard, Phillips & Robinson and Gold & Ochu in this volume).

A variety of definitions of RRI have emerged (see for instance Owen, Bessant, & Heintz 2013; Owen, Macnaghten, & Stilgoe 2012; RRI Tools 2016; Sutcliffe 2011; von Schomberg 2013). Although each has a slightly different focus, they share common features: Firstly, RRI is seen as a way to focus research and innovation on societal challenges. Secondly, there is agreement that RRI will achieve this goal by:

(a) ensuring that wider perspectives shape research and innovation by involving all relevant stakeholders throughout the research and innovation process;
(b) opening up the values and visions within science and innovation to wider debate and influence;
(c) making sure that research is able to anticipate and respond to risks; and
(d) framing responsibility as a collective rather than individual activity.

RRI advocates believe that the mistakes of the past can be reduced by following these principles to ensure that technologies are 'ethically acceptable; socially desirable and sustainable'. (von Schomberg 2013, 64).

The recent EU-funded RRI tools project (www.rri-tools.eu) set out to develop this framework beyond a theory and to operationalise RRI. This involved identifying and describing case studies to bring the concept to life, and developing a set of processes and outcomes to help researchers implement this approach.

The project described RRI as 'Involving society in science and innovation "very upstream" in the processes of R&I [Research & Innovation] to align its outcomes with the values of society'. It has identified three outcomes that RRI projects should be aiming for (see box 17.1 for more detail):

1. Learning outcomes (engaged publics, responsible actors, responsible institutions);
2. Research and innovation outcomes (ethically acceptable, sustainable and socially desirable research outputs); and
3. Solutions to societal challenges.

A series of process requirements have also been developed to help researchers understand how to implement RRI and how to measure their progress in this implementation (see box 17.2). Significantly, these outcomes, processes and principles are seen to apply to across the spectrum of research – from basic to applied research. Some activities might want to emphasise some aspects more than others, but RRI is seen as a useful tool and necessary approach for all areas of research.

While these ideas might appear to be challenging, there is growing evidence that this approach offers opportunities – not just in minimising the risk of future controversies, but in opening up new business models, as the case study in box 17.3 illustrates.

Further to the outcome and process dimensions of RRI, RRI and the Ten Principles are mutually reinforcing in guiding citizen science engagement, processes and outcomes.

Overlaps with citizen science

Responsible Research and Innovation's commitment to openness and desire to involve stakeholders in the whole of the research and innovation process demonstrates clear overlaps with the practice of citizen science. It is important to highlight, however, that there are also clear divergences between the two.

Box 17.1. RRI outcomes

1. Learning outcomes
   - Engaged publics
   - Responsible actors
   - Responsible institutions

   RRI leads to empowered, responsible actors across R&I systems (researchers, policymakers, businesses and innovators, CSOs, educators). Structures and organisations should create opportunities and provide support to actors to be responsible, ensuring that RRI becomes – and remains – a solid and continuous reality.

2. R&I outcomes
   - Ethically acceptable
   - Sustainable
   - Socially desirable

   Responsible Research and Innovation practices strive for ethically acceptable, sustainable and socially desirable outcomes. Solutions are found in opening up science through continuous, meaningful deliberation to incorporate societal voices in R&I, which leads to relevant applications of science.

3. Solutions to societal challenges

   Focus on seven grand challenges:
   - Health, demographic change and well-being;
   - Food security, sustainable agriculture and forestry, marine, maritime and inland water research, and the bioeconomy;
   - Secure, clean and efficient energy;
   - Smart, green and integrated transport;
   - Climate action, environment, resource efficiency and raw materials;
   - Europe in a changing world – inclusive, innovative and reflective societies;
   - Secure societies – protecting freedom and security of Europe and its citizens.

   Our societies face several challenges, which the EU has formulated as the seven 'Grand Challenges' – one of the three main pillars of the Horizon 2020 programme. To support European policy, the EU requires R&I endeavours to contribute to finding solutions for these Grand Challenges.

   Source: https://www.rri-tools.eu/about-rri

Firstly, citizen science encompasses a range of different levels of engagement, from encouraging citizens to participate in the scientific process by observing and gathering data, up to involving them in the design and implementation of scientific projects (Silvertown 2009; and see Novak et al. in this volume). Some approaches to simply involve citizen scientists in roles such as data collection – for example, classifying galaxies in the 'Galaxy Zoo' project – have been criticised for leaving citizens in passive research roles (Mroz 2011) and treating them as free labour rather than genuine partners. Questions have also been raised about the quality of their input and motivations for being involved (Editorial 2015). Moves have therefore been made to improve the support and training of citizen scientists and to encourage them to take on more active and in-depth roles (see also Nascimento et al. in this volume).

Such a participatory approach appears to be a cross-over with the ethos of RRI, but even with meaningful public participation, vital

---

Box 17.2. Process dimensions of RRI

To reach the RRI outcomes, practising a more Responsible Research and Innovation requires that processes are:

Diverse and inclusive: Involve early a wide range of actors and publics in R&I practice, deliberation and decision-making to yield more useful and higher quality knowledge. This strengthens democracy and broadens sources of expertise, disciplines and perspectives.

Anticipative and reflective: Envision impacts and reflect on the underlying assumptions, values and purposes to better understand how R&I shapes the future. This yields to valuable insights and increases capacity to act on knowledge.

Open and transparent: Communicate in a balanced, meaningful way methods, results, conclusions and implications to enable public scrutiny and dialogue. This benefits the visibility and understanding of R&I.

Responsive and adaptive to change: Be able to modify modes of thought and behaviour, and overarching organisational structures, in response to changing circumstances, knowledge and perspectives. This aligns action with the needs expressed by stakeholders and publics.

Source: https://www.rri-tools.eu/about-rri;
visit site for summary and more details

Box 17.3. RRI in practice: HAO2 – involving citizens in technology design

Hao2 (Hao means 'good' in Chinese) is a company that develops and sells 3-D virtual environments. As well as the RRI focus on outcomes that address the needs of society, its principles of diversity, inclusion and engagement, as well as responsiveness and adaptive change, form the backbone of the company.

For example, many people working in the software industry have autistic spectrum disorder, but are often expected to work in ways and environments that are challenging and uncomfortable to them. Nikki Herbertson, founder and CEO of Hao2, noticed how staff with autism working in her software company became much more sociable in online environments such as the virtual world game *Minecraft*. She therefore investigated the potential of 3-D virtual world applications to enable staff to communicate with each other. The company involved people with autistic spectrum disorder – people who are rarely involved in such a process – in developing this new product. They were so successful in their approach that that since 2010 the company has entirely focused on promoting 3-D virtual world products and services to help organisations improve services, especially for people with disabilities.

Hao2's products are now used in a range of settings, from businesses to education, and Hao2 has won numerous awards. Hao2 has built a successful company by involving more diverse groups than simply the product developers in the process of innovation, and building RRI into their DNA. Using RRI has also allowed Hao2 to build products strongly focused on solving societal problems, increasing opportunities for those with autism and other complex needs.

> It was quite clear from the outset that the only people that could really deliver the insight that we needed from a research and development point of view would be people with autism. And it was absolutely critical that they were not just a focus group, but actually that they were the citizen researchers alongside me looking at the options and then designing the solutions in a sustainable way.
>
> Hao2 Founder and CEO Nikki Herbertson

questions about power and agenda-setting can remain unanswered (see also Novak et al.; Gold & Ochu, both in this volume). Opening up such questions to wider scrutiny, debate and participation is key to the RRI agenda and is an approach that is being taken on by 'Extreme Citizen Science' (ExCiteS) (http://www.ucl.ac.uk/excites).

Unlike 'contributory' citizen science, which typically asks citizens to participate in scientific data collection and often appeals to those who have an interest in, or enthusiasm for, science, Extreme Citizen Science opens up participation in all aspects of research – including data collection, analysis and agenda-setting – to people from a wide range of backgrounds (Haklay 2013). Involving those who are not usually able to participate in such activities means that Extreme Citizen Science has the potential to open up the range of voices, values and visions directing and shaping the scientific 'project' and to include wider societal perspectives (Stevens et al. 2014). This latter point, particularly if engagement is also aimed at encouraging reflection, sharing purpose and anticipating uses and risks, offers a key way for RRI and citizen science to work together, to develop more responsible and socially relevant science and innovation.

Developing responsible citizen science – and responsible science

Building on the foundations of Extreme Citizen Science and taking account of the ECSA Ten Principles of Citizen Science could bring RRI and citizen science closer together to develop a notion of responsible citizen science and see its realisation. Wider lessons can also be drawn from the RRI and citizen science communities. The projects in boxes 17.4 and 17.5 illustrate some of this learning.

Box 17.4. RRI and citizen science in action 1 – the Swedish Challenge Driven Innovation programme

Challenge Driven Innovation is a research and innovation funding programme developed by Sweden's innovation agency, Vinnova, and launched in 2011. It aims to fund collaborations in research and innovation that address societal challenges and involve partners from different parts of society.

(continued)

To make sure the programme focused on the issues society wanted to address, stakeholders were involved from the start of the project through consultations and workshops. In this way, participants developed the three principles upon which the funding model would be based.

- Policy issues must be prioritised, and a challenge-oriented approach adopted;
- Subject areas and sectors should be intermixed, so a multidisciplinary approach rather than a traditional focus on separate disciplines was adopted;
- The user perspective must be the starting point for innovation, thus building an Extreme Citizen Science approach from the start.

With these citizen-developed principles in mind, a series of funding calls were launched. All problem-oriented, they placed no restrictions on which stakeholders, sectors, research topics or disciplines could apply. Instead, they asked for all necessary stakeholder groups to be involved – including citizens and end users – to allow the projects to address the selected challenges. Examples of funded projects include those focused on urban farming, getting more people into the labour market, making socially deprived areas more attractive and creating meeting places.

As well as funding projects that focus on real social problems, the programme appears to have had other significant impacts. Firstly, it has generated a shift in the funding organisation, away from an unspoken focus on technical innovations to a much broader concept of innovation. This led to the launch of a social innovation programme in 2015, which set out to involve civil society members to a greater extent than previous projects. Secondly, working practices at Vinnova have also changed as a result of the programme. The range of stakeholders who receive funding from Vinnova has widened, dialogue and collaboration between officers in various departments has increased and the organisation has taken up the important focus on societal challenges.

Source: www.rri-tools.eu; visit site for more details

Box 17.5. RRI and citizen science in action 2 – Xplore Health

Xplore Health (https://www.xplorehealth.eu) is a European educational programme aiming to bridge the gap between research and secondary STEM (Science, Technology, Engineering and Mathematics) education.

Originally the project focused on building pupils' understanding of the research process through a series of online tools. The project has, however, evolved over time. Inspired by RRI and citizen science, it now focuses on empowering secondary school students to participate in R&I processes and in R&I decision-making, with a focus on making it more ethically acceptable, socially desirable and sustainable. It aims to train students to become active citizens of the knowledge society, to be able to make informed decisions and to contribute to addressing societal challenges.

With this in mind, Xplore Health combined their online activities with an innovative participatory research project, Ment Sana (Healthy Mind). Ment Sana is a Community Based Participatory Research (CBPR) project, in which educators, learners, researchers and policymakers work together to design and implement health interventions for students and with students.

The project started in 2015 with a needs assessment, where students chose the topic of stress and depression from a list of health issues and built a collective agenda of interests. Next, a number of research projects were designed and implemented in collaborations between researchers, higher education students and secondary school students. These projects culminated in a catalogue of recommendations for policymakers, which were presented in May 2016 at a final congress with more than 350 students and high-level policymakers from the Catalan Government and the NGO Federació de Salut Mental de Catalunya.

This participatory process gave students the opportunity to learn science through science, to develop scientific inquiry, critical thinking and engagement skills, but also to consider what important questions should be addressed with science – and to help address these questions (see also Edwards et al. and Harlin et al. in this volume). Participants agreed that the process strengthened both the research process and its outcomes, helping to do excellent research and find solutions adapted to the needs and expectations of end users. Most importantly, the research focus and approach was dramatically transformed by the involvement of citizen researchers.

Source: www.rri-tools.eu; visit site for more details

Conclusions

Responsible Research and Innovation and citizen science are both emerging and developing, meaning that it is perhaps too early to set out a concrete path ahead. It is, however, clear that RRI has the potential to deepen interpretations of and contributions to the Ten Principles of Citizen Science. Over the next few years, the following issues are likely to demand attention.

1. Participation

How does citizen science involve citizens and reflect their contributions in all aspects of research? As well as involving participants at an early stage in establishing what science should be done and which questions it should tackle, citizens also need to be involved in anticipating possible future uses and misuses. Mechanisms exist for doing this – for example, the UK's ScienceWise programme (www.sciencewise-erc.org.uk) has developed strong methodologies for involving citizens in discussions about new and emerging science. Questions remain about how these approaches are incorporated in research.

2. Agenda-setting

How does citizen science involve citizens in meaningful discussions of current and future research, without their expectations being shaped by the values of scientists themselves? Public dialogue activities, for deliberate or accidental reasons, are often shaped by the aspirations and values of the scientific community such that public participation in science sees citizens co-opted into the 'world view' of science and scientists (Smallman 2016; Thorpe & Gregory 2010). For instance, the need to bring emerging technologies to life for citizens to form meaningful opinions about them means that scientists' understandings of these technologies become embedded in the minds of the participants, restricting possible futures (Smallman 2016). Questions remain about how to meaningfully engage citizens in abstract scientific ideas without limiting their thinking, in both public and private sector research.

3. Acting responsibly – and collectively

How does citizen science develop an idea of shared responsibility that takes account of all of the actors and implications of scientific developments? Responsibility has traditionally focused on the roles of the scientist, their responsibility for their research and the tensions between academic freedom and responsibility (see for example Douglas 2003). Questions remain about how to promote and enact shared responsibility as part of the move to involve wider voices in scientific research.

Science and innovation are arguably the biggest drivers of change in the early twenty-first century (both positive and negative). Such significant levers of power are too important to be left to a small group of researchers. For science to reach its full potential, it must be set free of its laboratories and take its rightful place – at centre stage in everyone's lives. That means developing a truly responsible approach to science, with citizens at its heart.

Acknowledgements

I would like to thank Olivia Brown for her research support and my RRI Tools Colleagues Steve Miller, Rosina Malagrida and Karin Larsdotter for their work developing the case studies.

18
Conservation outcomes of citizen science

Heidi L. Ballard[1], Tina B. Phillips[2] and Lucy Robinson[3]

[1] *University of California, Davis, CA, US*
[2] *Cornell Lab of Ornithology, Ithaca, NY, US*
[3] *The Natural History Museum, London, UK*

corresponding author email: Hballard@ucdavis.edu

In: Hecker, S., Haklay, M., Bowser, A., Makuch, Z., Vogel, J. & Bonn, A. 2018. *Citizen Science: Innovation in Open Science, Society and Policy*. UCL Press, London. https://doi.org/10.14324/111.9781787352339

Highlights

- Different models of citizen science (contributory, collaborative and co-created) can contribute to different types of conservation outcomes.
- Contributory projects, often with large spatial and temporal-scale datasets, may be most likely to contribute to conservation indirectly via research.
- Collaborative and co-created projects, which often include intensive involvement of participants in local conservation issues, may be more likely to contribute directly to site and species management, as well as indirectly via education and capacity building.
- Citizen science project leaders can employ a theory of change approach to design and execute citizen science programmes to achieve conservation outcomes.

Introduction

As environmental problems mount and funding for environmental agencies continues to decline (James, Gaston & Balmford 2001), citizen science is often seen as a cost-effective alternative for agencies that need to routinely gather large amounts of data from diverse locations (e.g.,

Frost-Nerbonne & Nelson 2004). Citizen science can also have many broader conservation outcomes, including social as well as environmental benefits. Like conservation biology, environment-based citizen science projects have the ultimate goal of advancing understanding of natural systems and protecting biological diversity (Dickinson et al. 2012). A key difference between traditional conservation biology and citizen science is the inclusion of members of the public in collaborative research with professional scientists (see Danielsen et al. 2009 re. indigenous knowledge). The inclusion of the public and the data generated from citizen science can be used by decision-makers to impact policy and natural resource management (McKinley et al. 2015) and thereby impact conservation outcomes. We further argue that this is most effective when citizen science research is closely paired with, and used to inform, environmental stewardship.

In recent decades, there has been a proliferation in the number and variety of citizen science projects with targeted scientific goals aimed at gathering large amounts of data to answer questions at scales unattainable through traditional methods (Bonney et al. 2014). Other projects may also emphasise the impact on volunteers themselves, through explicit educational outcomes that may be cognitive, affective and/or behavioural in nature (Jordan, Ballard & Phillips 2012; Phillips, Bonney & Shirk 2012). The recent dramatic increase in conservation programmes that include citizen scientist-collected data (Theobald et al. 2015) suggests that involving the public in scientific research may also contribute to conservation outcomes.

Although several typologies have been proposed to capture the variety of citizen science projects (e.g., Bonney et al., 'Public Participation', 2009; Danielsen et al. 2009; Shirk et al. 2012; Wiggins et al. 2011), this chapter uses the three-model typology based on participants' level of involvement in the scientific process, first introduced by Bonney et al. ('Public Participation', 2009) and then refined by Shirk et al. (2012). The *contributory* model of citizen science is researcher-driven and focused mostly on large-scale data collection by volunteer participants. It has its roots in disciplines that have historically embraced volunteer involvement such as ornithology (Greenwood 2007), palaeontology (Harnik & Ross 2003) and astronomy (Barstow & Diarra 1997). *Collaborative* projects typically originate with researchers but may include input from participants in multiple phases of the scientific process, such as designing data collection methods and analysing data. This model has its roots in volunteer monitoring, particularly water quality projects in which sharing data with the wider community has the potential to affect local issues (Whitelaw et al. 2003). *Co-created* projects involve participants in all aspects of the scientific process including defining research questions, interpreting data and disseminating findings (see also Haklay; Novak et al., both in this volume). These projects have their origin in participatory action research or community science initiatives, often aimed at addressing public health or environmental justice issues (Fernandez-Gimenez, Ballard & Sturtevant 2008). Broadly speaking, none of these three models is better or worse than the others, but they may vary in the ways in which they contribute to conservation because they differ in numbers of participants, intensity of time and commitment required by participants, and locus of control in terms of who is setting the research agenda.

Defining conservation outcomes for citizen science

Despite the recent surge in citizen science projects globally, the contributions that all three models of citizen science projects can make to conservation have only recently begun to be examined (Conrad & Hilchey 2011; Ballard et al. 2017; Sullivan et al. 2017). Conservation biology as a field also suffers from a relative lack of such evidence of impacts. According to Margoluis et al. (2013), one reason for the lack of evidence is that conservation initiatives are often chosen based on assumptions of what might work rather than on proven success in similar contexts. Further, the efficacy of conservation biology initiatives is not often measured, and when it is, the processes for documenting and measuring impact are seldom shared with other conservation organisations (Margoluis et al. 2013). As such, there is significant scope for the field of citizen science to add to the evidence base for successful and unsuccessful approaches in conservation, and for conservation research to inform citizen science practice (see Kieslinger et al. in this volume for more on evaluation). In response to the need for conservation organisations to better evaluate the conservation impacts of their work (Miller et al. 2004; Spooner et al. 2015), in 2008 the Cambridge Conservation Forum (CCF) developed a conceptual framework to enable organisations to systematically evaluate the effectiveness of their conservation activities (Kapos et al. 2008). This framework was based on an extensive review of current conservation research and the input of 36 conservation organisations. The CCF identified seven categories of activity that lead to targeted improvements in the status of species, ecosystems or landscapes. Two categories of activity have a direct impact on the conservation target – species management and site management – while five influence conservation indirectly – research, education, policy, livelihood and capacity building.

Ballard et al. (2017) adapted the CCF framework to examine natural history museum (NHM)–led citizen science programmes at three NHMs, and found that 59 per cent of programmes contributed towards at least one of the conservation outcomes identified by the CCF (see also Sforzi et al. in this volume on museums and citizen science). In that study, long-term monitoring programmes and those focused on a single site or small geographic area contributed most frequently to conservation outcomes. Sullivan et al. (2017) also modified the CCF framework to document the ways in which eBird data, a project in which users record their own bird observations, were being used in support of conservation science and action. This chapter similarly applies the CCF framework to citizen science programmes that represent the three models described above, to examine whether, and how, each model may be more or less likely to lead to conservation outcomes. This strengths analysis helps to identify the most effective features of each model with regards to conservation outcomes, which could potentially be applied to the others. In line with Ballard et al. (2017), the analysis combines species management and site management into a single category for the purposes of this discussion. Importantly, the programmes analysed here have a variety of goals in addition to conservation; conversely, not all conservation activities can or should be expected of them.

This chapter examines three case studies, one for each of the project models, looking first at the evidence of the conservation outcomes as defined in table 18.1, which has been adapted from the CCF framework (Kapos et al. 2008). It then examines these outcomes for each of the three models to consider how citizen science can leverage the strengths of different types of projects to influence conservation outcomes. This is achieved by looking specifically at the relative extent of a project's outreach; spatial and temporal data coverage; useful data and peer-review publications; contributions to knowledge of global systems; leveraging of, and contributions to, local ecological knowledge; adaptive management and social capital; and contributions to conflict resolution and policy and advocacy.

Table 18.1 Definitions of conservation activities (adapted from Kapos et al. [2008] and Ballard et al. [2017]).

Conservation activity type	Definition and examples
Direct contributions to conservation outcomes	
Species and site management	Managing species and populations (e.g., captive breeding); and managing sites, habitats, landscapes and ecosystems.
Indirect contributions to conservation outcomes	
Research	Research aimed at improving the information base on which conservation decisions are made (e.g., surveys, inventories, monitoring and mapping).
Education	Education and awareness-raising to improve understanding and influence people's behaviour (e.g., campaigns, lobbying and educational programmes).
Policy	Developing, adopting or implementing policy or legislation (e.g., management plans, trade regulations and actions to enforce conservation goals).
Livelihoods	Enhancing and/or providing alternative livelihoods to improve the well-being of people impacting species/habitats of conservation interest, (e.g., through sustainable resource management, income-generating activities, etc.).
Capacity building	Actions to enhance specific skills among those directly involved in conservation.

Comparative citizen science contributions to conservation

This section presents three examples of citizen science projects (table 18.2) selected because (1) they serve as representative examples from around the world of the three models of citizen science defined above, and (2) they are long-standing programmes so evidence of their contributions to conservation are readily available on the internet and in peer-reviewed literature. It is important to note, however, that these projects are just one example of each of the three models, and that the structures, goals and topical foci of other projects in each model can vary widely. For example, contributory projects are typically focused on a specific taxonomic group

Table 18.2 Summary of the three examples

Project title	Model	Location and scope	Conservation activity type (from Kapos et al. 2008)
GBIF: Global Biodiversity Information Facility (see box 18.1) https://www.gbif.org	Contributory	Global	Research
Hudson River Eel Project (EELS) (see box 18.2) http://www.dec.ny.gov/lands/49580.html	Collaborative	Regional (New York, US)	Site and species management, education, capacity building
Community group–led ecological restoration (see box 18.3) http://www.landcare.org.nz/Regional-Focus/Manawatu-Whanganui-Office/Citizen-Science-Meets-Environmental-Restoration	Co-created	National (New Zealand)	Site and species management, education, capacity building

---

Box 18.1. GBIF: Global Biodiversity Information Facility – contributory citizen science

Citizen science contributions to GBIF-mediated data

Kyle Copas, GBIF Secretariat, Denmark

GBIF, the Global Biodiversity Information Facility (https://www.gbif.org), is an open-data research infrastructure for biodiversity information funded by the world's governments. The GBIF network supports and enhances capacity for providing free and open access biodiversity data by sharing common standards and data formats, open-source software and peer-to-peer professional development. As such, it fits the contributory model of citizen science.

Establishing direct connections between GBIF and the CCF framework can prove difficult, not least because 'raw' species data mediated by GBIF are rarely cited explicitly in policy and on-the-ground conservation management and protection, even if noteworthy exceptions do occur (e.g., Secretariat of the Convention on

(continued)

Biological Diversity 2014; US Fish and Wildlife Service 2014; US National Oceanic and Atmospheric Administration 2014). However, substantive uses of GBIF-mediated data appear in peer-reviewed papers at a rate of more than one a day, signalling that GBIF produces clear indirect conservation outcomes through facilitating research.

An example of GBIF contributions to research is its role in GEOBON (Group on Earth Observations Biodiversity Observations Network). GEOBON has developed its concept of Essential Biodiversity Variables (EBVs), a minimum set of measurements needed to capture and track the major dimensions of biodiversity change over time (Pereira et al. 2013). In late 2015, the GBIF Secretariat sought to understand how and where citizen science already contributes to EBVs, and the global agendas they support, by reviewing citizen science contributions to species occurrence datasets available through GBIF.org.

The results (Chandler et al. 2017) showed that species occurrence datasets gathered largely or entirely by citizen scientists contributed up to 349 million of the 640 million species occurrence records available through GBIF.org, as of 1 March 2016. The contributions are uneven across taxa, although citizen science programmes account for 70 per cent of all GBIF-mediated records for animals and 87 per cent for birds (largely due to eBird data). Citizen science contributions also show biases at regional and national scales (table 18.3). However, placed in the context of the research team's broader finding that fewer than 10 per cent of all relevant citizen science programmes contribute data to GBIF, improving publishing tools and incentives for citizen science programmes could do much to close the large worldwide gap in data sharing.

Table 18.3 Geographical distribution of occurrences contributed to GBIF by regional location of occurrence

Continent	Number of occurrence records	Per cent of total citizen science contributions
North America	202,269,978	58.0 per cent
Europe	119,671,494	34.2 per cent
Oceania	17,987,545	5.2 per cent
Central and South America	4,327,079	1.2 per cent
Asia	2,727,302	0.8 per cent
Africa	1,785,960	0.5 per cent

Box 18.2. Hudson River Estuary Eel Project – collaborative citizen science

Collaboration between a state agency, local residents and schools

Chris Bowser, New York State Dept. of Environmental Conservation, US

The Hudson River Estuary Eel Project (EELS, http://www.dec.ny .gov/lands/49580.html) began in 2008 at two sites on the Hudson River, and as of 2017 had expanded to a dozen sites with over 750 volunteers. American eels hatch in the Atlantic Ocean and drift/ swim to the North American East Coast. Many continue their journey upstream to fresh water to grow into adults before returning to the ocean years later to reproduce. This species is in decline in much of its range, and this project provides crucial baseline data about the young eel population in the Hudson River. Volunteers coordinated by the New York State Department of Environmental Conservation (NYDEC) catch and count thousands of juvenile American eels (*Anguilla rostrata*), known as 'glass eels' for their transparent appearance at this lifecycle stage, each year and release them above dams or other barriers to their migration.

As a catch and release programme, the project also restores the migration patterns of thousands of eels by moving eels upriver from a dam/obstruction. The EELS primarily involves teachers and river-based organisations who use the experience of wading through streams with nets and other equipment to provide local high school students with authentic science field skills, often over many weeks. This is a collaborative citizen science project because, at some sites, participants have taken on leadership roles to collaborate with the project coordinator from NYDEC over the course of the project's evolution and expansion. In some cases, this involved participants modifying aspects of the protocol that were then adopted as new methods across sites, and in other cases community-based organisations and teachers approached the project coordinator to develop and implement a site in their own stretch of the river.

The contributions to conservation education are documented by the teachers integrating the content into their curriculum to help students learn about the biology and ecology of this unique species and the Hudson River ecosystem. The contributions to site and

(continued)

species management, then, come from the integrated nature of the project, where both monitoring and stewardship takes place in tandem throughout the project. The involvement of the local community and young people, who adopt their own EELS sites and some of whom participate for multiple years, also indicates a contribution to conservation capacity building as defined by Kapos et al. (2008). The project has documented an overall increase in the number of eels caught over the monitoring period, increasing from an average of 17.5 eels caught per day across all sites in 2008 to 215 eels in 2016; this may indicate increasing populations, though more information is needed (Bowser 2016; see table 18.4). In addition to using nets for catch and release each spring, volunteers and project co-ordinators have collaborated to develop low-cost eel ladders at several sites, which are made from large plastic tubing and netting that allow eels to climb the ladder into buckets where they are counted and released up stream during summer months.

Table 18.4 Total eels caught and eels caught per day as a catch per unit effort (CPUE) combined for all sampling sites in that year

Year	Total YOY glass eels	CPUE YOY glass eels	Total elvers	CPUE elvers	Total eels caught	CPUE Total eels caught
2008	2,388	16.6	181	1.8	2,569	17.5
2009	7,740	34.8	430	1.7	8,170	36.5
2010	10,603	21.6	1,411	3.2	12,014	24.8
2011	6,964	16.1	1301	3.4	8,265	19.5
2012	85,166	128.9	1,432	1.9	86,598	130.8
2013	103,123	188.3	1,647	2.3	104,770	190.6
2014	49,760	124.9	683	1.5	50,443	126.5
2015	48,158	114.6	1,298	3.3	49,456	117.8
2016	142,770	221.5	2,383	3.6	145,153	215.1
Total	**456,672**		**10,766**		**467,438**	
Average		**95.3**		**2.5**		**97.7**

Source: Bowser 2016
Note: In this study, eels are separated into two age classes: young of year (YOY) glass eels and elvers. 'Glass eels' are just entering the Hudson River system in the spring of the sampling year (which includes recently pigmented eels in late spring), and 'elvers' are fully pigmented eels that have been in the Hudson River system for at least a year.

Fig. 18.1 Local students checking eel nets for a daily survey of glass eels in a local stream. (Source: Hudson River Eel Project)

Box 18.3. Community group–led ecological restoration network – co-created citizen science

Grassroots citizen science in New Zealand: Quantifying community-led conservation gains

Monica A. Peters (Hamilton, NZ) and Ngaire Tyson (New Zealand Landcare Trust)

Prior to the thirteenth century, New Zealand's unique suite of flora, fauna and fungi had evolved in isolation with no land mammals, other than two species of diminutive bat. A history of land use change and the introduction of new biota have had disastrous effects on native ecosystems. In response to ongoing threats to indigenous biodiversity and continued habitat decline, a recent study investigated community group–led monitoring and ecological restoration in New Zealand (Peters, Eames & Hamilton 2015). Some 540 self-mobilising groups operate largely independently of one another, but identify as a part of a large, loosely defined network of community-based restoration practitioners, that contribute both to active restoration and monitoring through citizen science approaches (see https://www.naturespace.org.nz/groups).

(continued)

Contributions to site and species management for conservation have been documented at the community group level: Major biodiversity gains have been achieved through sustained invasive species control or eradication; revegetating cleared land and riparian margins with native species; restoring wetland hydrology; and translocating threatened species to their former habitats. A recent study identified that nearly half of the groups (49 per cent, $n = 282$) carried out their own monitoring or grassroots citizen science, primarily to determine their restoration management outputs (e.g., number of rodents trapped), rather than the conservation outcomes of their activities (e.g., increases in desirable avifauna species resulting from predator control) (Peters et al. 2016). Contributions to conservation research cannot be substantiated currently because monitoring results are not widely used beyond the scope of the groups' own projects, owing to differences in data formatting, monitoring methods and objectives, and questions around data quality (Peters, Eames & Hamilton 2015). For this reason, quantifying community conservation efforts nationally through groups' own data is challenging and needs to be addressed. The government's ambitious 'Predator Free 2050' plan to rid New Zealand of key introduced species may support greater co-ordination between groups and promote more strategic data collection in the future.

Based on this co-created model, key recommendations for countries with dispersed community-led restoration initiatives include the following:

1) Providing greater support from agencies/NGOs and funders to promote and support strategic intra-group co-ordination;
2) Co-funding contractors to work across groups to enable consistent data collection; and
3) Using a partnership approach from the outset to design monitoring programmes that meet the information needs of both groups and partners (e.g., a guide is currently being produced for the Auckland Council to ensure consistency when council staff work with community-based organisations).

or geographic region, (i.e., eBird [ebird.org], the Coastal Observation and Seabird Survey Team [COASST, https://depts.washington.edu/coasst/], and the Monarch Larvae Monitoring Project [https://mlmp.org/]), in which participants affiliate with, and contribute to, a specific research or monitoring question. In the example in box 18.1, however, the Global Biodiversity Information Facility (GBIF) is a global and taxonomically inclusive platform to which many citizen science projects contribute. Therefore, these three projects simply serve as illustrative examples. Each project is described in a separate box as listed in table 18.2.

These examples highlight several important points for citizen science projects that wish to contribute to conservation.

1. It is possible to evaluate and document the ways a citizen science project contributes to the key conservation activities outlined by Kapos et al. (2008), but evidence must be deliberately collected. This evidence is often difficult to collect and often requires additional funding beyond project implementation alone, which is also a challenge for the field of conservation more broadly, as noted above. The lack of evidence of conservation impacts in some of these citizen science examples may not indicate a lack of impact but that projects must devote greater resources to evaluating their own activities and outcomes.

2. Citizen science projects may not only indirectly impact conservation through research and education, but also directly through site and species management. Both the EELS and the New Zealand community-based restoration projects closely integrate stewardship with citizen science activities, through the catch and release of juvenile eels, or invasive species controls and revegetation, respectively. Specifically, volunteers in both projects are trained and then implement the scientific monitoring as essential and complementary to the direct stewardship activities that impact habitats and species. Other citizen science projects are finding success in this approach, for example in coastal eelgrass systems where volunteers plant eelgrass and monitor it repeatedly in Maine, US. (Disney et al. 2017), or when volunteers assess and weigh marine debris on beaches and then dispose of it (Thiel et al. 2017). With respect to education outcomes, volunteers can gain awareness of the need for both scientific monitoring and stewardship actions for enhancing long-term conservation of species and habitats and making evidence-based management decisions. Importantly, scientists and land managers not only benefit from the restoration work on the ground

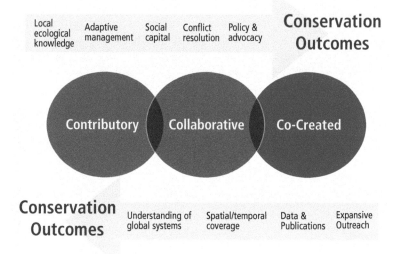

Conservation Outcomes

Fig. 18.2 Relative strengths of three models

but also the regularly collected data they need to manage it effectively. Combining activities in this way means that citizen science projects can achieve both short- and longer-term impacts for conservation while fostering volunteers' passion and commitment for conservation.

3. Some models may be better suited for particular activities that contribute to conservation (see figure 18.2). For example, *collaborative and co-created* projects that facilitate in-depth interaction and shared practice among participants and between scientists and participants tend to allow participants to gain a deeper awareness of environmental and community-based advocacy issues, and often increase trust between scientists and the public (see also Fernandez-Gimenez, Ballard & Sturtevant 2008). The local and regional scale many of these projects operate in also facilitates the inclusion of local expertise and may promote enhanced social capital, adaptive management opportunities, improved conflict resolution and policy and advocacy initiatives (Fernandez-Gimenez, Ballard & Sturtevant 2008). At the other end of the spectrum, contributory projects with larger participant numbers and large spatial and temporal coverage tend to produce data that is highly utilised and disseminated in peer-reviewed publications (see also Sullivan et al. 2017).

Large databases can accommodate a proportion of error while remaining high quality, which in turn improves knowledge of global systems.

Improving citizen science contributions to conservation

This chapter provides examples of how engaging the public in conservation research can contribute to desired outcomes but questions remain about (1) the specific pathways by which conservation goals can be reached in citizen science and (2) which models of citizen science best support or facilitate each of these pathways. Margoluis et al. (2013) suggest the use of 'results chains' to describe how the implementation of project activities and assumptions about how projects operate link to relevant short- and long-term impacts. Citizen science projects could apply this tool alongside the notion of 'theory of change' (Weiss 1995), a planning and evaluation tool increasingly used in conservation biology, to articulate conservation pathways in citizen science. Theory of change has its origins in the field of evaluation, and is a graphical representation of the process by which clearly identified goals are reached (Weiss 1995). Theory of change provides explanatory linkages between project activities and outcomes, usually with 'if . . . then' statements, and seeks to explain how, and why, the desired change is expected. Developing a theory of change requires the articulation of assumptions about why certain activities will lead to intermediate outcomes as well as the identification of indicators of success for measuring whether intermediate outcomes were achieved (see the Center for Theory of Change: http://www .theoryofchange.org). Results chains then include evidence of results added to the theory of change such that evidence of the specific pathways by which a citizen science project leads to one or more conservation outcomes can be properly examined. This would allow the field to identify successful strategies for documenting and even measuring intermediate, but necessary, steps or outcomes that are important for achieving ultimate conservation impacts. In fact, systematically and rigorously analysing the evidence of intermediary results from citizen science projects following Margoluis et al. (2013) could also provide cautionary scenarios for the potential misuse of, or over-emphasis on, citizen science in achieving conservation outcomes. While documenting the results chains for the specific citizen science projects in this chapter is beyond the scope of the chapter, this is a way forward for the field to become more critical of the way citizen science may, or may not, be contributing to conservation outcomes.

Conclusions

Conservation biology at its core seeks to directly impact biodiversity through site and species restoration and preservation (Kapos et al. 2008). One of the main lessons from examining a spectrum of citizen science programmes is that citizen science, conversely, tends to affect conservation *indirectly* through the application of research findings, education of stakeholders, policy changes and individual and community-level actions. Direct contributions may primarily occur when citizen science is coupled with related restoration and stewardship activities. Although the mechanisms for how these ultimate conservation outcomes are reached have not been well-studied, these indirect pathways may have a significant impact on conservation goals. These case studies demonstrate the need for better tracking of the onward use of citizen science data (and indeed any research data) for environmental conservation purposes, to ensure that citizen science can be targeted where it is most effective or most needed, and that its contribution to conservation is recognised. While evidence for the conservation outcomes of citizen science is still lacking in many cases, more projects are beginning to evaluate conservation outcomes, which will help build a better understanding of what structures and approaches produce specific intended conservation outcomes. Most importantly, this chapter has highlighted the ways each model of citizen science may support different types of conservation outcomes. Project designers can therefore take into account the strengths and structures from each model to design for the conservation outcomes they seek, as well as explicitly state their theory of change and document evidence for the intermediary results throughout their projects.

19
Capacity building in citizen science

Anett Richter[1,2], Daniel Dörler[3], Susanne Hecker[1,2],
Florian Heigl[3], Lisa Pettibone[4,5], Fermin Serrano Sanz[6],
Katrin Vohland[4] and Aletta Bonn[1,2,7]

[1] *Helmholtz Centre for Environmental Research – UFZ, Leipzig, Germany*
[2] *German Centre for Integrative Biodiversity Research (iDiv) Halle-Jena-Leipzig, Germany*
[3] *University of Natural Resources and Life Sciences, Vienna, Austria*
[4] *Museum für Naturkunde Berlin, Germany*
[5] *Ludwig-Maximilian-Universität München, Munich, Germany*
[6] *Fundación Ibercivis, Zaragoza, Spain*
[7] *Friedrich Schiller University Jena, Germany*

corresponding author email: anett.richter@idiv.de

In: Hecker, S., Haklay, M., Bowser, A., Makuch, Z., Vogel, J. & Bonn, A. 2018. *Citizen Science: Innovation in Open Science, Society and Policy*. UCL Press, London. https://doi.org/10.14324/111.9781787352339

Highlights

- Strategic capacity-building programmes have been initiated at the European and national scale leading to the development of the Socientize Green and White Paper for Citizen Science in Europe and the *Greenpaper Citizen Science Strategy 2020 for Germany*.
- These programmes have broader relevance in informing national and supranational programmes elsewhere in the world.
- Capacity building involves five main steps: (1) identifying and engaging different actors, (2) assessing capacities and needs for citizen science in the setting under focus, (3) developing a vision, missions and action plans, (4) developing resources such as websites and guidance, as well as (5) implementation and evaluation of citizen science programmes.
- Capacity building is an iterative and adaptive process that needs a sound engagement of all involved actors from society, science and policy.

Introduction

Citizen science builds on long traditions as well as on new developments. Collaboration between professional scientists and volunteers committed to research is not new and has been practised in various forms for centuries (Silvertown 2009). Clubs, expert associations, museums and universities have always played a pivotal role in this collaboration (Miller-Rushing, Primack & Bonney 2012; Ballard et al. 2017). In recent decades, however, the increased specialisation of science and reduced social recognition of expertise can be understood as having led to a gap between the ideas and activities of citizens and the practice of scientific research (Gibbons 1999). As a result, the ambition to open science to citizens (again) has developed both in civil society and among scientists, fuelled by technical advances (e.g., Mazumdar et al., in this volume) and calls for stronger participation in research itself (e.g., Danielsen et al.; Haklay, both in this volume; Silvertown 2009; Bonney et al. 2014).

Today, voluntary participation in science is undergoing a revival and the field of citizen science is rapidly growing (Kullenberg & Kasperowski 2016). This has been aided by new means of technology, online communication tools, social media and accessible databases and repositories (Williams et al., in this volume). The recent trend towards professionalisation of citizen science has resulted in the almost simultaneous establishment of national and international citizen science associations in the United States, Australia and Europe (Göbel et al. 2016) to support and advance citizen science through communication, co-ordination, knowledge sharing and education (Haklay 2015). Aligned with this, the American government has established infrastructures to monitor and collect inputs from citizen science for environmental policies (www.citizenscience.gov). The European Commission has also reinforced citizen science by promoting it through their research and innovation programme (Horizon 2020; and see Nascimento et al.; Smallman, both in this volume) and developing and supporting targeted citizen science activities and capacity-building programmes (e.g., citizens' observatories, Socientize). In addition, several landmark reviews and guides aiding practitioners in the establishment of citizen science have been released, often led by UK scientists (Roy et al. 2012, Unit 2013, Pocock et al. 2014b). As citizen science is becoming more formalised and widely accepted in both science and society, capacity building paired with political commitment is now required to support its potential (Newman et al. 2012). Capacity building refers to a framework for individuals and organisations, which focuses on process-orientated goals to strengthen and maintain the capabilities to set and achieve their own development objectives over time (Eade 1997; UNDP 2009).

This chapter presents key findings from capacity-building programmes at the European level and in Austria and Germany. It showcases capacity development for citizen science in various settings and synthesises key experiences and outputs of strategic citizen science development. The chapter therefore distils the principles of citizen science capacity building to inform capacity building elsewhere, illustrating the current political dimensions of citizen science in Europe to draw out key lessons that could also be applied in other contexts beyond Europe.

Socientize: White Paper on Citizen Science for Europe

Socientize (2012–2014) was a consortium project initiated by the European Commission under the Directorate General for Communications Networks, Content & Technology (DG CONNECT) and co-ordinated by the University of Zaragoza with other institutions from Spain, Portugal, Austria and Brazil. It was influential in increasing recognition and appreciation of citizen science research experiments. The main aim of Socientize was to co-ordinate actors involved in citizen science to set the basis for a new open science paradigm in the framework of current citizen science development in Europe. Socientize presented the added value of collaboration and knowledge sharing through digital tools by involving some 12,000 citizens in a range of science projects from mapping flu outbreaks and labelling images of cancer cells to collective music creation (Lanza et al. 2014). Socientize created a multi-channel platform for discussion and developed Green and White Papers on Citizen Science for Europe, applying an open, iterative and inclusive approach. Within its first year, the Green Paper (Socientize 2013) presented an analysis and mapping of citizen science projects, and identified ongoing programmes and initiatives paying special attention to researchers outside academia. These trends were analysed, best case studies were promoted, and cross-cutting concerns and draft policy options addressed key areas in need of change. A wide audience was reached through common digital technologies such as YouTube, Google Hangouts and WordPress forums as well as more specific collective intelligence tools such as allourideas, Thinkhub or LimeSurvey. Socientize also ran a number of virtual workshops, moderated open consultations with questionnaires and facilitated online discussions.

The later White Paper (Socientize 2015) included proposed actions and measures to address the key challenges of science-society-policy interactions. Arranged at the macro-, meso- and micro-levels, these correspond to strategies for policymakers and science funders, and recommend plans for citizen science mediators and facilitators as well as actions for citizen science practitioners (table 19.1).

Table 19.1 Actions and measures of the *White Paper on Citizen Science for Europe*

Actions	Measures
Targeted funding	Designing funding schemes and launching programmes specific to citizen science. Targeted calls will achieve a broader uptake and will keep established networks and systems going. Programmes should contribute to a deeper analysis of citizen science practices and outcomes.
Mainstreaming citizen science	Embedding citizen science into existing funding schemes. Like science communication, citizen science should become an integral part of ongoing scientific activities. Research should be given greater credit for the inclusion of citizen science strands covering multiple disciplines, addressing the public's needs and concerns.
Education	Updating educational programmes to promote and recognise new forms of community engagement and digital skills in the curriculum.
Evaluation and assessment	Expanding current academic reputation systems and evaluation criteria to account for social impact and engagement.
Access to technology	Broadening access to technology and improving the systems required to make the most of the power of networked communities, paying special attention to the digital divide in Europe.
Data policy	Clear ethical guidelines are needed for EU-wide data policy. Stakeholders are asked to share public datasets and research data infrastructures (to promote quality, reliability, interoperability of data) as well as data handling tools and methods (such as algorithms, descriptive, predictive, visualisation, decision-making). This requires attention to intellectual property rights,

Table 19.1 (continued)

Actions	Measures
	fundamental personal data protection rights, ethical standards, legal requirements and scientific data quality.
Dissemination and support	All strategies and policy actions must be communicated by providing appropriate knowledge-based guidance.

Source: Socientize 2015

The White Paper recommendations led to the establishment of the European Citizen Science Association (ECSA, www.ecsa.citizen -science.net) and were endorsed and embedded in ECSA policy development. High-level guidance and support on road mapping was presented to the European Commission, member states, local and regional authorities, and private actors. The White Paper provided the basis for several actions and policies related to public engagement in science directed by the European Commission (e.g., in the Science with and for Society programme 2018–2020). Currently, Socientize is run and supported by the Ibercivis Foundation, which is coordinating the Spanish national citizen science platform (www.ciencia-ciudadana.es) and developing the co-creation of citizen science road mapping in Spain.

Österreich forscht: Development of citizen science in Austria

In Austria, capacity building for citizen science is closely connected with two comprehensive citizen science initiatives: the establishment of the online platform *Österreich forscht* (www.citizen-science.at) and the Center for Citizen Science (www.zentrumfuercitizenscience.at/en/citizen-science). Two national citizen science funding schemes were initiated, namely Sparkling Science by the Federal Ministry of Science, Research and Economy (BMWFW) and Top Citizen Science by the BMWFW together with the Austrian Science Fund. The platform *Österreich forscht* is an independent, bottom-up initiative managed by early career researchers at the University of Natural Resources and Life Sciences (BOKU), Vienna. It was launched in March 2014 to connect Austrian citizen science actors from different disciplines and institutions. Since 2016 the platform has been financially supported by BOKU, which has led to the formation of the Citizen Science Network Austria (CSNA, www.citizen-science.at /netzwerk). Since summer 2017, Austrian institutions (NGOs, universities and companies) have now formally expressed their commitment to citizen science in a joint Memorandum of Understanding of the CSNA. The CSNA aims to further develop citizen science in Austria, secure and foster quality and method development, publicise projects to interested publics and enhance dialogue between different actors. The first Austrian citizen science conference was held in Vienna in 2015, organised by the team of the platform initiators. The conference is now an annual event and connects scientists working in, or about, citizen science and enables citizens to connect with project managers. Until 2016, most curation of the platform has been done on a volunteering basis by two people committed to spend a great amount of time into the curation. In 2016, it had 38 different citizen science project partners from more than 30 different institutions (Pettibone, Vohland & Ziegler 2017). As a bottom-up initiative, the partners meet annually to decide on tasks for the upcoming year in a democratic way to integrate all views. To do so, they use the following guidelines:

- Decisions on management and development of the platform are taken by all platform partners.
- Platform partners can participate in the creation of content on the platform, such as writing and publishing articles and news about their projects. This fosters cohesion and ownership of the platform.
- The annual Austrian citizen science conference is hosted by different platform partners each year. This helps to integrate and establish citizen science in the different host institutions by (a) raising awareness about participation in science with new stakeholders and (b) allowing the host institution to choose a special focus for the conference through which it becomes more involved in the platform development.

The Austrian Center for Citizen Science was initiated by the BMWFW as a top-down approach and was established at the Austrian Agency for International Cooperation in Education and Research (OEAD) in June 2015. It aims to be an information and service centre for researchers, citizens and experts from different disciplines, and to establish links with interested communities beyond Austria. Six Austrian universities now refer to citizen science as an important part in their service agreement with the BMWFW.

Thanks to these two initiatives, the term 'citizen science' is now established not only in Austrian science communities, but also in the media and research policy. Future steps for capacity building in citizen science in Austria could include funding long-term citizen science projects and networks, as well as fostering science awareness in the general public.

GEWISS: Developing a citizen science strategy for Germany

The 'Citizens Create Knowledge' (GEWISS) capacity-building programme was funded by the German Federal Ministry of Education and Research (BMBF) aiming to strengthen citizen science capacity in Germany through a series of capacity-building activities. The objectives were to build a strong German citizen science community network, to assess the current state and needs of citizen science in Germany and – building on this – to develop the *Citizen Science Strategy 2020 for Germany* (Richter & Pettibone 2014; Bonn et al. 2016). GEWISS employed a modular programme with the following steps:

- Workshops for networking and capacity needs assessment: Organisation of more than 10 national dialogue forums hosted by different partners, including a high-profile think-tank, to identify the needs of citizens and researchers, connect diverse actors working or interested in citizen science and engage decision-makers and funders (Richter et al. 2017).
- Resource development: Development of technical and organisational resources to develop guidelines for citizen science (Schierenberg et al. 2016), three film clips (e.g., www.youtube.com/watch?v=cE1kpXLkGbo) and three training workshops (for reports, see www.buergerschaffenwissen.de).

- Online platform: Collaboration with projects to develop an online platform to increase their visibility, enhance public awareness and allow citizens to link with each other (www.buergerschaffenwissen .de).
- Strategy development: Development of the *Greenpaper Citizen Science Strategy 2020 for Germany* through the GEWISS workshops, moderated online consultation with 1,000 online visits and over 50 formal position papers from civic society and science organisations (Bonn et al. 2016).
- International conference: Hosting the first European citizen science conference in Germany in collaboration with ECSA (www .ecsa2016.eu).

The development of the Citizen Science Strategy 2020 for Germany was the central strategic policy instrument of the GEWISS programme (Bonn et al. 2016). The core of the strategy development was the combined dialogue forum input from 10 workshops with over 700 participants from 350 organisations, including scientific institutions, environmental groups, informal science clubs, science shops, funding organisations, state and federal agencies, and local interest groups (Bonn et al. 2016; Richter et al. 2017). This was followed by a six-week online consultation with over 1,000 website visitors and over 400 comments submitted on the draft of the Green Paper as well as over 50 formal position papers from civic society and science organisations. The open, iterative and transparent consultation process was important to facilitate ownership of various stakeholder groups by including their perspectives directly both in the strategy development and implementation. The Green Paper presents aims, potentials and challenges of citizen science in Germany with five visions of citizen science (table 19.2). The strengthening, establishment and integration of citizen science into science, society and policy were identified as core fields, alongside potential actions for each. Next steps ideally include translation into action plans by the different interest groups (Bonn et al. 2016).

Table 19.2 Priorities and visions for citizen science in Germany (Bonn et al. 2016)

Priorities for citizen science in Germany	In the year 2020, citizen science in Germany is . . .
Integration	. . . an integral part of social and scientific debates, and an approach that brings benefits for science, politics and society. At the same time, the various forms of participation – from co-operation to the active co-design and active co-production of research – are valued, recognised and lived in science, society and politics.
Empowerment	. . . an important part of citizens' lives which enables individual, formal and informal learning, empowers citizens to participate in research processes and allows them to engage with science.
Recognition	. . . a scientifically accepted, established and practised research approach which puts both participatory and transdisciplinary research into practice and unleashes innovative potential in research processes by including a wide range of knowledge sources and extensive participation.
Participation	. . . a politically accepted process of citizen participation for the generation, quality assurance and dissemination of knowledge and an expression of participation and encounter between science and society that is supported and sponsored by policy.
Innovation	. . . a participation format characterised by the use of web-based infrastructures which – as trustworthy environments that are in compliance with data protection regulations – promote knowledge exchange and co-operation in the context of citizen science projects.

The Green Paper was launched and presented to the German Ministry of Research and Education (BMBF) during a high-profile event in Berlin in March 2016. Based in part on its recommendations, the German Ministry launched a new funding scheme for citizen science projects in summer 2016. The call resulted in over 300 project proposals (BMBF 2017), which is one indicator of the growing capacity for citizen science in Germany as a result of the above initiatives.

From experiences to principles of capacity building in citizen science

The examples of capacity-building processes for citizen science presented above demonstrate capacity building at both individual and organisational levels. Building on UNDP recommendations (UNDP 2009), five critical steps to capacity development in citizen science can be identified (figure 19.1), which are ideally consecutive and iterative.

1) Actor engagement

First, actors are identified to include diverse interests related to citizen science: scientific enquiry, education, public engagement and more. Actors are approached in different ways, for example, through online platforms, face-to-face meetings and networking opportunities. The goal is to build a setting where individuals and key groups can share their experiences, learn from each other and build a shared identity. The strength and breadth of actor engagement drives the quality of the subsequent discussions and outcomes.

2) Capacity and needs assessment

The actors assess each other's capacities, based on current activities, shared goals (see Robinson et al., this volume) and community needs.

3) Visions, missions and action plan(s)

Actors formulate visions for the development of citizen science. This can include defining the goals of citizen science within the community, establishing guidelines and developing a strategy or action plan, for example, in a Green Paper followed by a White Paper (Socientize 2015; Bonn et al. 2016; Pettibone et al. 2016; Richter et al. 2017).

4) Resource development

A response plan is implemented building on the action plan, including the creation of funding opportunities, the development of tools and guidelines, and provision of training and educational material to publicise and improve the quality of citizen science.

5) Evaluation

Finally, the implementation is evaluated and reflected on to foster further development.

The nature of citizen science activities means that capacity building takes place at all geographical levels, from local to regional to national and international scales. Crucially, strategic capacity building comprises the development of policy instruments to frame citizen science in science and policy, and to further develop citizen science as an integral component of the science-society interface. In each setting, forums for discussions and other opportunities for participation are needed to bring perspectives from science, society and policy together. These discussions help to improve visibility of existing activities and assess the state and needs of citizen science actors corresponding to steps 1 and 2 of the

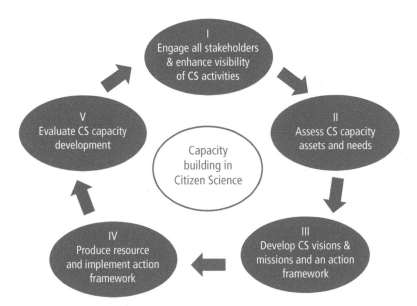

Fig. 19.1 Steps towards capacity building in citizen science. (Source: Adapted from UNDP 2009)

Table 19.3 Overview and links to selected outcomes of capacity-building programmes

Step	Capacity development	Country/region	Measure	Outcomes	Resources
I.	Engage stakeholders and enhance visibility	Europe – Austria Germany	Platforms and community building	> 100 citizen science (CS) projects now visible	www.socientize.eu, ecsa.citizen-science.net www.citizen-science.at www.zentrumfuercitizenscience.at www.buergerschaffenwissen.de
II.	Assess CS capacity assets and needs	Europe Germany	Conferences and accompanying research	Workshop, conferences and public consultation	www.ecsa2016.eu www.citizen-science.at/konferenz
III.	Develop citizen science	Europe Germany	Development of science policy documents and strategies; Develop CS visions and missions and develop action framework	Green and White Papers	www.socientize.eu/sites/default/files/SOCIENTIZE_D5.3.pdf www.ecsa.citizen-science.net/documents www.socientize.eu/sites/default/files/white-paper_0.pdf www.buergerschaffenwissen.de/sites/default/files/assets/dokumente/gewiss_cs_strategy_englisch_0.pdf.
IV.	Produce resources and implement action framework	Europe – Austria Germany	Development of practical resources and establishment of new funding programmes	Guides and video clips	www.buergerschaffenwissen.de/citizen-science/ressourcen Citizens create knowledge www.youtube.com/watch?v=md2Lgg5D62E www.youtube.com/watch?v=Z_fwsMAtM64 www.ec.europa.eu/programmes/horizon2020/en/h2020-section/science-and-society
		Austria Germany		Funding programmes	www.zentrumfuercitizenscience.at/en/ausschreibungen www.sparklingscience.at www.bmbf.de/foerderungen/bekanntmachung-1224.html
V.	Evaluate CS capacity development	Europe	Scientific integrative network	Paper, workshops, guidelines, recommendations	www.cost.eu/COST_Actions/ca/CA15212 www.cs-eu.net

capacity development cycle (figure 19.1). The developments of Green and White Papers in the policy area are instrumental in helping enhance governance structures and developing organisational structures and frameworks, corresponding to step 3. The implementation and development of practical resources such as guidelines was accomplished by supporting the development of new citizen science–funding schemes, both nationally and internationally, corresponding to steps 4 and 5 (figure 19.1). The first criteria for assessing citizen science projects and aid funding scheme development have recently been suggested (Kieslinger et al. 2017). However, the true evaluation of capacity-building programmes can probably only take place in a few years so that it can assess the sustainability of measures undertaken and the long-term implementation of action frameworks as outlined (table 19.3). The facilitators of capacity-building programmes either obtain their mandates from the 'top' (programmes which are therefore attached to overarching policy goals) or they are developed as a result of vision from the 'bottom' via volunteer teams, existing project members or a consortium of institutions (reflecting the shared needs of a community). Both approaches are appropriate depending on the context, and the case studies presented show that either approach can engage citizen science projects, non-governmental organisations and governmental authorities, local, regional and international organisations, as well as interested individuals in developing and advancing citizen science. This in itself can be considered an important milestone for achieving steps 1 and 2 of the capacity development cycle (table 19.3).

Lessons learned from capacity building in citizen science in Europe

The most important lesson from capacity-building programmes is the need for in-depth understanding of prospective stakeholders and important actors in citizen science. Citizen science is a broad field and actors share different goals and approaches. This heterogeneity is seen by some as incompatible with understandings of citizen science; by others, this is welcomed as a rich diversity of citizen science. Sensitivity is therefore required within emerging communities as they develop more concrete plans and strategies, to ensure that the diversity of approaches is accurately represented and supported. Scientific actors are often the easiest to reach, as they can participate in capacity-building efforts as part of their paid work. Other stakeholders might be reached consistently, but to a lesser extent. For example, educational actors, who engage in citizen science projects in curricular or extracurricular activities (Richter et al. 2016; Makuch et al. this volume) or policy actors whose statutory monitoring requirements are fulfilled by citizen science data (Nascimento et al; Parker et al., both in this volume), have only recently coalesced into identifiable stakeholder groups.

Other groups and entities that bridge science and society, such as science communicators, environmental groups or science jobs, that is, independent and participatory research support taking up concerns of the society (www.livingknowledge.org), might be reached through attendance at conferences. Participants can rarely be engaged in centrally organised daytime meetings, but possibilities to engage them can be harnessed through online tools, consultations and open discussions at science shops, museum events or via exhibition stands at public events.

Individuals who live in deprived communities or people with a different cultural background as well as unemployed individuals and/or individuals with formal education have yet to be included more specifically as stakeholders in most capacity-building programmes, and steps are needed in this direction to foster even broader societal integration. Citizen science should also harness the opportunity to learn from, and build on, experiences in other networks, such as participatory health research or science shops, which have gone through the same capacity-building processes. Capacity building for citizen science is a highly political process enabled by participants from science, society and policy communities. Therefore, strong, integrative capacity-building processes can foster the development of appropriate policies and support schemes for citizen science in the future.

Final remarks

The chapter presented the functionality and mode of operations of capacity building for citizen science and provided insights into the practice of the processes involved in capacity building in Europe and at the national level (Austria and Germany). Steps and activities established and implemented showcased how people, governments, international organisations and non-governmental organisations are needed as partners and actors for the development of capacities for citizen science. The capacity building as presented in the case studies improved governance structures and developed organisational structures and frameworks to strengthen and enhance citizen science, and hopefully encourages other initiatives at local, regional or national level to foster engagement in policy-making related to citizen science.

Acknowledgements

The 'Citizens Create Knowledge' GEWISS project received funding through the German ministry of education and research (BMBF grant agreement No 01508444). The Citizen Science Network Austria received funding through the University of Natural Resources and Life Sciences Vienna from 2016–2018.

20
Citizen science in environmental protection agencies

Roger P. Owen[1] and Alison J. Parker

[1] *Scottish Environment Protection Agency*

corresponding author email: roger.owen@sepa.org.uk

In: Hecker, S., Haklay, M., Bowser, A., Makuch, Z., Vogel, J. & Bonn, A. 2018. *Citizen Science: Innovation in Open Science, Society and Policy*. UCL Press, London. https://doi.org/10.14324/111.9781787352339

Highlights

- Environmental protection agencies (EPAs) in Europe and the United States are increasingly making use of citizen science for environmental protection, including engaging the public and awareness-raising, empowering action by communities, monitoring and data collection, and providing sound evidence on which to make decisions.
- To increase the impact of citizen science for environmental protection, improvements are needed in data management and infrastructure support, communication of data quality, sensor development and communication with citizen science data providers.
- Innovations in technology and organisational practices are enabling a citizen-agency dialogue based on good data and feedback on use of evidence. Citizen science has the potential to transform environmental protection by inviting the public to work with agencies to generate knowledge and find solutions.
- Effective case studies include citizen scientists acting as an agency's 'eyes and ears' (see box 20.2, Improving the effectiveness of sentinel systems: Irish Environmental Protection Agency) in addressing local environmental matters, including identifying environmental concerns for additional research and action (see box 20.3, Regulatory action spurred by citizen science: The Clean Air Coalition of Western New York and Tonawanda Coke Corporation); Promoting environmental education in schools (see box 20.4, The Enviróza school programme); and further legitimising the high-level strate-

gic delivery of a responsible authority's environmental protection mandate (see box 20.1, From Opportunism to a Strategic Approach: Scottish Environment Protection Agency [SEPA]).

Introduction

Citizen science has the potential to transform environmental protection by involving the public in work with agencies to generate knowledge and find solutions (Shirk & Bonney; Ballard, Phillips & Robinson, both in this volume). Environmental protection agencies are increasingly turning to citizen science to assist in the achievement of environmental protection (Hindin 2016; NACEPT 2016); and innovative technology is enabling more citizen-agency dialogue and feedback on use of evidence (Wiggins & Crowston 2015; Novak et al. in this volume). Environmental protection agencies in Europe and the United States have protection of the environment as their primary responsibility. In addition to that overarching mission, key policy drivers often include encouraging the wise use of resources, protecting public health and enabling sustainable economic development. This policy balance is critical, particularly as defining sustainable economic growth in an environmental context can be challenging, especially in respect of resource use and disposal of waste. Environmental protection agencies rely heavily on good-quality evidence about the health of ecosystems, pressures on these natural resources and the effectiveness of regulatory and other interventions. This chapter demonstrates that citizen science has an increasingly important role in providing evidence, raising environmental awareness and empowering the public (Nascimento et al.; Smallman, both in this volume). The case studies as presented reflect examples of how EPAs are increasingly making use of citizen science to engage the public in environmental issues and provide sound evidence on which to make decisions.

Citizen science and community science

Citizen science is incredibly diverse (Wiggins & Crowston 2015), and the diversity of citizen science approaches is evident when looking at the types of projects that EPAs are involved in; these include air and water quality monitoring, collection of baseline data and identification of hotspots, networks of sensors, environmental justice efforts, educational projects, public engagement and more (boxes 20.1 to 20.4; NACEPT 2016).

The diversity of environmental citizen science is reflected in how – and by whom – projects are initiated and implemented. Often, citizen science projects are contributory in nature in that they are initiated and defined by an institution or researcher within an institution (such as an EPA), who then solicit input and participation from members of the public (boxes 20.1, 20.2 and 20.4; see also Novak et al.; Ballard, Phillips & Robinson, both in this volume). For example, the volunteer rainfall observation network in Scotland is driven by the objectives of the hydrometric network in the Scottish EPA and UK Meteorological Office. Although EPAs often initiate and implement projects that directly support the priorities of the agency, they also often encounter projects that are initiated and defined by the goals of community members, who may work independently or in collaboration with scientists but maintain ownership over the entire scientific process (Dosemagen & Gehrke 2016). Although this chapter includes these projects in the definition of citizen science, the terms 'community science' (e.g., Dosemagen & Gehrke 2016) or 'community citizen science' (e.g., NACEPT 2016; Chari et al. 2017) are also used to describe this type of project. The Angler's Riverfly Monitoring Initiative in the UK (The Riverfly Partnership 2017) is one example where an interested and concerned community group has acted as the main driver for monitoring environmental quality. Community goals are usually rooted in a specific environmental issue and often related to environmental justice; most often, communities identify a local environmental issue, work to understand that issue, and then use that understanding to advocate for improvements to local environmental and human health. The goals of these projects can include increased regulation and enforcement of violations, and communities often approach EPAs or other government organisations for help in achieving their goals (box 20.3).

Uses of citizen science by EPAs

Analysis of citizen science initiatives among EPAs in the European EPA Network and in the United States reveals several principal goals.

Education and awareness-raising

Many projects emphasise the value of environmental education and scientific literacy for both children and adults, especially in providing participants with an understanding of the places where they live and work

(Haywood 2013; Newman et al. 2017). Citizen science also supports a shared understanding of the importance of environmental protection. There are many current examples of EPAs encouraging citizen science activities to raise awareness of environmental issues such as the identification of contaminated land sites in Slovakia (see box 20.4: Enviróza school programme), the recording of invasive riparian plants in Scotland (Plant Tracker) and the mapping of toxic algal blooms by the South East Alaskan Tribal Toxin partnership in the United States (NACEPT 2016). Participants of citizen science projects often also educate their local communities. For example, both students and teachers that participate in the Enviróza school programme are able to share their knowledge of contaminated sites with their families and communities (see also Peltola & Arpin in this volume).

Empowering action by communities

Knowledge and evidence of environmental issues can be powerful motivators for community action to directly address an issue or exert pressure on governments to take action or reprioritise resources (box 20.3). For example, litter data from organised litter picking campaigns such as 20 years of data from the UK Marine Conservation Society's Beach Clean (Marine Conservation Society 2018) empower communities to conduct dialogue with sewage treatment companies and manufacturers of litter-related products. Similarly, the Great Nurdle Hunt is a citizen science project organised by an environmental charity in North East Scotland which collects data on the prevalence of nurdles, pre-cursor plastic pellets, in the strandline of beaches (FIDRA 2018). This simple monitoring scheme has now extended across the UK and parts of Europe. The evidence of nurdle pollution has led to Operation Clean Sweep (OCS), an international programme originally designed by the plastics industry and supported by The British Plastics Federation and PlasticsEurope. The OCS manual provides practical solutions to prevent loss for those who make, ship and use nurdles with the key message that good handling practice can easily reduce pellet loss.

Citizen science can also contribute to positive interactions between EPAs and the public (see also Sforzi et al. in this volume). Beyond education and data collection, citizen science can allow for the development of a shared agenda with members of the public for environmental protection. Encouraging public input and open collaboration and responding to community concerns means that EPAs demonstrate a commitment to serving the public and investigating their concerns and priorities. Citizen science can thereby provide opportunities for local, regional and national action that will improve environmental protection – from individual action to national policy. This facilitating role for citizen science can be beneficial for both members of the public and EPAs, which gain recognition of their environmental protection mandate through public participation.

Monitoring and data collection

Generally, data generated through citizen science has been met with scepticism by professional scientists and policymakers (Kosmala et al. 2016; Bonney et al. 2014), but recently more EPAs have come to regard such data as potentially contributing significantly to evidence needs. Volunteer monitoring has a long history in EPAs – in the case of rainfall measurement, this extends as far back as the early twentieth century. Other examples include discerning biological indicators of water quality and pollution incident recording. In some cases, as in rainfall observing, the quality of data recording has consistently met recognised quality standards over many decades; in other cases, the quality of data has improved markedly over the past two decades (Crall et al. 2011). Furthermore, modern sensor technology has enabled more technically demanding measurements by interested members of the public, as in recent air quality monitoring surveys in The Netherlands (see also Volten et al. in this volume). This data is proving to be genuinely useful to the relevant Dutch environmental authority (Snik et al. 2014).

Citizen science can invigorate environmental research by generating data that allows for a deep understanding of environmental quality and environmental issues (Hampton et al. 2013). Citizen science is uniquely suited to some areas of research and in some cases it is the only option. For example, by harnessing the contributions of people all over the world over long time periods, citizen science can provide datasets with rich complexity over space and time (McKinley et al. 2015). Citizen science can also fill in gaps in environmental information, including issues that are currently not regulated, such as the emergence and spread of invasive non-native species (see also Nascimento et al. in this volume). Finally, citizen science can allow EPAs to better understand priority issues like climate change. For example, Evolution MegaLab invites the public to record phenotypic variation in garden snails hypothesised to be driven by temperature changes and regional climate change (Evolution MegaLab 2018).

Influencing decision-making and policy

Although the link between public participation in environmental activities and policy-making is not well established (see also Nascimento et al. in this volume), the connection to decision-making in EPAs has become stronger in recent years (McKinley et al. 2015; Haklay 2015; McElfish, Pendergrass & Fox 2016). Public campaigns backed by evidence generated through citizen science can strengthen their influence on government, national agency and local authority policy-making and policy implementation. Examples from the European EPA Network include air quality strategy in The Netherlands (see Volten et al. in this volume), controls on the fly-tipping of waste (for example ZeroWaste Scotland) and many more (see boxes 20.1 to 20.4). Citizen science approaches can be a powerful tool for identifying emerging issues and developing solutions through new knowledge generation, for example, in water resource management and the assessment of human impacts in remote regions (Buytaert et al. 2014). The approach can be especially useful in crisis situations when community or advocacy groups work to initiate government action on a specific issue (Conrad & Hilchey 2011; box 20.3).

Advantages and challenges

Citizen science is cost-effective, but it is not free (Shirk & Bonney 2015; see also Danielsen et al. in this volume). However, investment in citizen science can have significant impact and progress can support citizen science contributions to environmental protection (Geoghegan et al. 2016; NACEPT 2016), especially when that investment helps maximise the advantages of citizen science and overcome the following challenges. For many of these advantages and challenges, the European Citizen Science Association's Ten Principles of Citizen Science are a useful guide (ECSA 2015).

Data management and infrastructure support

Many citizen science initiatives suffer from a lack of infrastructure to efficiently hold, manage, analyse and interpret citizen science data. This is true especially where resources are limited, which particularly includes community initiatives. This has hampered the usefulness of many citizen science projects where the aggregation of data is of key importance (see Williams et al. in this volume). Environmental protection agencies can help by providing resources to enable the construction of data management tools and facilitating their use, or by providing easy access to tools that are already available. The Scottish EPA is addressing this by developing a web-based platform for handling citizen science initiatives (box 20.1). Investment in data management tools will greatly improve the potential for data sharing between projects and between projects and agencies (see also Sforzi et al. in this volume).

Communication of data quality

EPAs and other institutions are often reluctant to accept and use citizen science data due to an implicit assumption that citizen science data is inherently of lower quality than traditional science data (Kosmala et al. 2016; Bonney et al. 2014). There are several strategies to ensure appropriate data quality to enable the use citizen science data for environmental protection (see also Williams et al. in this volume):

1. Citizen science projects can integrate multiple mechanisms to ensure high-quality data from participants (Wiggins et al. 2011; Freitag, Meyer & Whiteman 2016).
2. Providing extensive metadata allows citizen science projects to communicate the 'known quality' of the data so that it can be used appropriately by EPAs (Bowser, McMonagle & Tyson 2015).
3. EPAs can create standards for data quality and communicate the quality of data needed for different purposes. Citizen scientists can produce 'data fit for purpose' through careful project design in communication with representatives from EPAs (Wiggins et al. 2013; Roy et al. 2012; NACEPT 2016).
4. EPAs and organisations involved in citizen science can engage in partnership strategies and approach a problem with a combination of professional data collection and citizen science (NACEPT 2016).

Sensor development

Recent development of advanced monitoring techniques has opened up possibilities for citizen science to include measurements from low-cost, portable, small sensors to monitor air and water quality. However, the individual performance of these advanced monitoring technologies is often not considered sufficient to be used directly by EPAs, especially when it comes to regulation and enforcement. Environmental protection agencies can help by focusing on networks of these sensors to obtain a synoptic picture of environmental quality (see Volten et al. and Ceccaroni & Piera in this volume), supporting a system for evaluation or certification of sensor technologies, providing information on appropriate technologies and developing guidance on messaging and interpretation of data from advanced monitoring technologies (Hindin et al 2016).

Communication with citizen science data providers

Encouraging the active participation of volunteer environmental observers in organised projects brings benefits for EPAs and for the individuals and communities involved but, not surprisingly, also generates expectations on government agencies. If agencies wish to empower and engage the public on environmental issues by providing them with the means to gather relevant data, good feedback on the value of the data is important to keep volunteers motivated (Geoghegan et al. 2016). This feedback may, for example, relate to data quality, consistency of protocols or the utility of the dataset for agency work. Significant resources are required to provide effective and continuing feedback on any citizen science initiative. Environmental protection agencies can set up systems to provide this information to providers through appropriate data portals and other dialogue mechanisms.

Timely feedback on the usefulness of the data in effecting environmental improvement is also needed. Many engaged participants want to know if their efforts to measure or report on environmental issues have made a difference and contributed to positive change (Trumbull et al. 2000) or become more engaged in environmental issues. Timely feedback can be successful where there are direct and reasonably rapid results, as in the pollution alert system set up by the Irish EPA (see box 20.2). Where volunteer data has to be aggregated over time or across a large number of contributors, or where resulting interventions take years to have a measurable effect, managing expectations of volunteers can be much more difficult. It is important that EPAs address this problem. For example, the Anglers' Monitoring Initiative in the UK, jointly organised with the Environment Agency, reports back regularly on the outcomes of volunteer ecological quality assessments (including some successful prosecutions) to maintain motivation and participation (The Riverfly Partnership 2017).

Integrating citizen science into the work of the agency

Many EPAs have begun citizen science work and are engaged in collecting data through volunteer participation. In a more limited set of examples, EPAs and other government agencies have responded to citizen science data collected by community groups or incorporated these data into EPA work (box 20.3; NACEPT 2016). The impact of citizen science can be increased when EPAs look to fill their data needs through citizen science projects initiated and conducted both in and outside of government. Boxes 20.1 to 20.4 illustrate the breadth of EPA engagement with citizen science through initiating projects (boxes 20.1, 20.2 and 20.4) and by responding to community-initiated data (box 20.3).

Box 20.1. From opportunism to a strategic approach: Scottish Environment Protection Agency (SEPA)

The Scottish Environment Protection Agency, and its predecessors, has worked with members of the public to monitor the environment since before the term 'citizen science' came into widespread use. For example, 130 volunteer 'Rainfall Observers' have been providing daily records of rainfall from locations throughout Scotland for up to 40 years. These records support weather and flood forecasting, improve models to predict climate change and support the management of water resources (Scottish Environment Protection Agency 2017a). More recently, SEPA and partners have helped initiate a growing number of citizen science projects, including Plant Tracker (Plant Tracker 2017), River Obstacles (River Obstacles 2017), and Anglers Riverfly Monitoring Initiative (The Riverfly Partnership 2017).

SEPA has recently recognised the need for a more strategic approach to citizen science, partly in response to the challenges of sustaining an increasing number of projects. This strategic approach has comprised:

- A high-level strategy, signed-off by SEPA senior management, outlining when and how citizen science would help deliver core responsibilities and objectives.
- Published guidance on the types of citizen science the agency would support.
- Co-ordination of citizen science activities to ensure alignment with the overall strategy.
- Provision of relevant infrastructure support such as IT capacity and training.
- Development of a web-based portal (Scottish Environment Protection Agency 2017) to provide access to tools and resources, data input and feedback mechanisms.

To support this, SEPA commissioned 'Choosing and Using Citizen Science' (Pocock et al. 2017). This 'Blue Guide' offers a step-by-step approach to assessing whether the use of a citizen science approach is appropriate, through project design, initiation and promotion, to project reporting and evaluation (see figure 20.1).

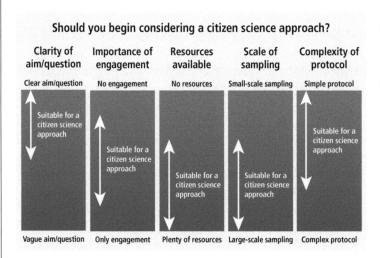

Should you begin considering a citizen science approach?

Fig. 20.1   Key design elements for citizen science projects

Through the EU LIFE–funded 'Scotland's Environment Web', SEPA has also partnered on the development of a mobile app and web-based infrastructure to support citizen science projects. Examples include the 'Learn About Air' citizen science teaching pack for schools (Scottish Environment Protection Agency 2017b) and Scotland's Environment Citizen Science Portal for data entry (Scottish Environment Protection Agency 2017c).

SEPA's involvement in citizen science has taken a contributory approach until recently. It still needs to address building capacity (see also Richter et al. in this volume) in 'co-production' projects involving volunteers in project management, balancing open communication with SEPA policies, improving evaluation and maintaining participant motivation in the context of continuing constraints on public resources.

Box 20.2. Improving the effectiveness of sentinel systems: Irish Environmental Protection Agency

The advent of smartphones has provided opportunities to engage the public in innovative ways that include the use of people as a network of sentinels of environmental problems. The Irish EPA developed a smart phone application called See it? Say it! to help people to report pollution in their towns and villages, including the illegal dumping of waste (Irish Environmental Protection Agency 2017). Users can take a photograph of the incident and add a description and contact details (see example photographs in figure 20.2). The app adds GPS location coordinates and the report is automatically sent to the relevant enforcement agencies or local authority. This complements a 24-hour complaints phone line. Reports are also delivered to FixYourStreet.ie which all Irish Councils monitor.

The uptake of the application has been good; 29 per cent of all environmental complaints on Fix Your Street now arrive via See it? Say it! and 1,500 complaints were received in 2015 when the app was launched. Figure 20.3 provides a snapshot of reported incidents throughout Ireland on specific days in 2015.

Fig. 20.2 Examples of postings received from all over Ireland in July 2016. (Photos: Elena Bradiaková)

Fig. 20.3 Distribution of reported pollution incidents, summer 2015

(continued)

This method of reporting has proved useful to regulators and local authorities given its locational accuracy and the detail provided by photographic evidence. The main types of incidents reported include the dumping of waste, backyard burning, noise from commercial sites and pollution incidents such as fish kills. At the same time, public interest and engagement on local environmental issues is increasing, so careful resource planning is needed to meet demand.

**Box 20.3. Regulatory action spurred by citizen science:
The Clean Air Coalition of Western New York
and Tonawanda Coke Corporation**

The residents of Tonawanda, New York, became concerned about chronic illness in the local community and the odours and smoke from 53 industrial plants nearby. Residents formed a 'bucket brigade', or community group that uses a low-cost canister to independently conduct air quality testing. The group tested for Volatile Organic Compounds (VOCs) using grab samples collected in buckets purchased at a hardware store. Initial sampling showed extremely high levels of benzene and high levels of formaldehyde. The Coalition approached the New York Department of Environmental Conservation and the US EPA with the data collected, who then conducted a comprehensive, year-long air quality study including four permanent air monitors. This study confirmed that benzene levels were 75 times higher than the EPA guideline, and that there were high levels of five additional air pollutants, and identified Tonawanda Coke Corporation as the predominant source. As a result, Tonawanda Coke Corporation installed new air controls, resulting in a 92 per cent decrease in benzene emissions. A representative from the US Department of Justice became aware of the issue, which resulted in an individual conviction against the Tonawanda Coke environmental manager for misleading information about chemical emissions (James-Creedon 2016).

Box 20.4. The Enviróza school programme

Enviróza is a citizen science game for primary and secondary schools, created by the Slovak Environment Agency. It was financed by the EU Cohesion Fund as part of the Operational Programme Environment (2007–2013) and launched at the start of the 2013/2014 school year, under the auspices of the Ministry of Environment of the Slovak Republic.

Using Enviróza, participants (teachers and pupils) seek out and identify contaminated sites, publish their data online and score points for doing so. Through accompanying competitions, they also inform the public about this issue, contributing not only to environmental conservation but possibly also to participants' health.

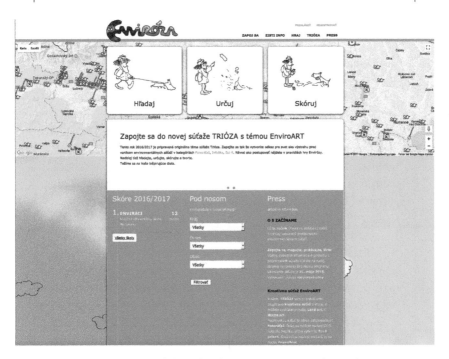

Fig. 20.4 Homepage of the school programme Enviróza website. (Source: http://www.enviroza.sk/)

(continued)

Fig. 20.5 A total of 25 new contamination/contaminated sites were added by schools. Field inspection and evaluation was then carried out by a SEA expert to allow their classification in the ISCS. (Photo: Irish Environment Protection Agency)

Enviróza is intended to update information about selected contaminated sites registered in the Information System of Contaminated Sites (ISCS) and to identify new sites (known as school-identified sites) that display signs of serious contamination. Data gathered by participants is processed by the Slovak Environment Agency (SEA), integrated into the ISCS and thus made available to state authorities as well as professionals and the public.

Enviróza's educational goal is for participants to gain information about existing contaminated sites and the state of their environment based on first-hand observations in the field. In the process, pupils develop skills with maps and navigational tools; work with data and use ICT; and gain experience working as a team as well as thinking critically and expressing their opinions. The programme provides teachers with an experiential learning tool for environmental education and incorporating the issue of contaminated sites into other school subjects, including mathematics, information technology, biology, chemistry, art and civics.

Conclusion

Environmental protection agencies can support and encourage citizen science by providing access to innovative mobile sensors and applications; data capture infrastructure and communication; advice on funding; and demonstrating the practical uses of evidence generated through citizen science. Most importantly, EPAs can support citizen science by validating its use and integrating it into agency work. Investment in a few key areas can support the contribution that citizen science makes to environmental protection (Blaney et al. 2016; NACEPT 2016). Small investments can support big progress, for example, building in feedback mechanisms between projects and agencies; supporting data management and standards; and creating technology standards and support.

Recent advice and recommendations directed at EPAs and citizen science projects have provided guidance to support this transformative approach through key contributions. In December 2016, the National Advisory Council for Environmental Policy and Technology recommended that the US EPA invest in citizen science, improve technology and tools, enable the use of citizen science data, and adopt a positive, co-operative agenda that increases the use of citizen science data. The Council also recommended that the US EPA provide guidance and communicate data quality needs for different data uses. The UK Environmental Observation Framework provided recommendations including that citizen science activities be rigorously evaluated and that research should be encouraged on the 'difficult-to-quantify' benefits of citizen science (Blaney et al. 2016).

Citizen science projects and participants can also work to maximise the value of their efforts for environmental protection. Citizen science groups can (1) engage early with EPA staff to increase the relevance of citizen science project objectives and outcomes for the responsible agencies, set expectations and guide projects for maximum impact; (2) ensure project planning includes careful communication and feedback loops to ensure equitable collaboration, motivation and learning; and (3) seek to co-design projects with staff at EPAs and other organisations to ensure that the intended use, accessibility, quality and constraints of the data are factored into project planning and implementation.

Citizen science can have a significant impact in achieving environmental protection by generating new knowledge, raising awareness and empowering community members. Over the next decade, EPAs will continue to turn to citizen science to work towards environmental protection in collaboration with the public, and additional case studies will support the potential uses of this powerful approach. Citizen science will enhance science and open opportunities for EPAs, community groups and other organisations towards the shared goal of environmental improvement.

Acknowledgements

Box 20.1 was provided by Dr. Paul Griffiths, Environmental Quality State of Environment Unit, Scottish Environment Protection Agency. Box 20.4 was provided by Elena Bradiaková, Contaminated Sites Expert, Slovak Environment Agency. This chapter was supported in part by an appointment to the Internship/Research Participation Program at the US Environmental Protection Agency (EPA), administered by the Oak Ridge Institute for Science and Education through an interagency agreement between the US Department of Energy and the EPA.

Part IV
Innovation in technology and environmental monitoring

21
Citizen science technologies and new opportunities for participation

Suvodeep Mazumdar[1], Luigi Ceccaroni[2], Jaume Piera[3],
Franz Hölker[4], Arne J. Berre[5], Robert Arlinghaus[4] and Anne Bowser[6]

[1] *Sheffield Hallam University, UK*
[2] *Earthwatch, UK*
[3] *Institut de Ciències del Mar, Barcelona, Spain*
[4] *Leibniz Institute of Freshwater Ecology and Inland Fisheries (IGB), Germany*
[5] *SINTEF, Norway*
[6] *Woodrow Wilson International Center for Scholars, Washington DC, US*

corresponding author email: s.mazumdar@shu.ac.uk

In: Hecker, S., Haklay, M., Bowser, A., Makuch, Z., Vogel, J. & Bonn, A. 2018. *Citizen Science: Innovation in Open Science, Society and Policy*. UCL Press, London. https://doi.org/10.14324 /111.9781787352339

Highlights

- New technologies supporting data collection, data processing and visualisation, and the communication of ideas and results create a wide range of opportunities for participation in citizen science.
- Technologies are especially beneficial for opening additional channels for public involvement in research, allowing participants to contribute through a range of activities and engaging newer audiences.
- There is a range of existing resources to help project co-ordinators develop and maintain citizen science technologies.
- It is important to consider issues such as participant demographics, affordability and access, and fitness for purpose when selecting technologies.

Introduction

In the latter part of the nineteenth century, there was a paradigm shift with the institutionalisation of scientific activities through the establishment of research institutions and a growing emphasis on rigour, processes and protocols (see also Mahr et al. in this volume). Members of the public remained contributors to scientific research throughout this process, albeit in selected areas of study including astronomy, archaeology, ecology and the natural sciences. During this time, researchers primarily involved citizen science volunteers in data collection initiatives, with observations interpreted and analysed by professional scientists (e.g., the Audubon Christmas Bird Count). Such data collection generally followed a paper-based approach, with volunteers either systematically recording observations or individually sending evidence such as photographs or specimens to professional scientists, along with key metadata such as observation time and location (Miller-Rushing, Primack & Bonney 2012).

The recent proliferation of Information and Communication Technologies (ICT) such as mobile technology, the rise of Web 2.0 (e.g., moving beyond static web pages towards user-generated content and social media) and the ubiquity of high-speed internet has resulted in a further paradigm shift, this time in citizen science (Silvertown 2009). The rising interest in, and popularisation of, science and technology, as well as the push by governments and institutions for Science, Technology, Engineering and Math (STEM) education, have further created an excellent environment for individuals and communities to participate in scientific research (see Haklay in this volume). Participation itself now takes numerous forms extending far beyond data collection, such that the very conceptualisation of a citizen science project can now be initiated by individuals and their communities rather than scientists (see Ballard et al.; Novak et al., both in this volume).

This chapter discusses the new tools and technologies that have influenced citizen science and, as a result, revolutionised how citizens and communities can participate and engage in research. The following section presents a high-level overview of the various tools and technologies used in citizen science as well as resources to allow projects to develop similar tools and technologies. This is followed by a discussion of how key technological developments have created and expanded opportunities for citizen participation. The chapter concludes with key policy implications, as well as a brief discussion of how the future of citizen science may be shaped by, and benefit from, emerging technologies and online services.

Overview of citizen science technologies

New technologies facilitate scientific research by supporting the collaborative collection of data and dissemination of information in real-time (Mooney, Corcoran & Ciepluch 2013). These platforms also support social interactions and organisation between public participants and scientific researchers, and among public participants and their communities. As such, citizen participation in democracy is now transitioning from one-way broadcasts to two-way dialogues, empowering more people to express their voices and drive change. This is also true in the context of scientific research.

Citizen science participation in data collection can be explicit (when citizens collect the data themselves) or implicit (when contributors share geolocated photographs, videos or messages on social media). Explicit data collection can now be carried out through a wide range of new instruments, devices, tools (including do-it-yourself, or DIY, technologies) and mobile apps that can be easily built, bought or borrowed by citizens, communities and enthusiasts. However, the use of ICT does not always guarantee high data quality and participant engagement.

On the contrary, adopting suboptimal ICT can hurt projects through hidden costs including poor usability and lack of appropriate functionality (Wiggins 2013). Different mechanisms for data collection should usually be considered, based on user preferences, demographics and constraints (see box 21.1). For example, participants less familiar with technologies like mobile apps may prefer to provide data via more traditional forms such as pen-and-paper-based data sheets. Facilitating participation through a range of channels can help avoid age-dependent bias, as well as biases that may exclude low resource communities.

Researchers have identified several technologies that are promising for the field of citizen science, including wireless sensor networks, online gaming (Magnussen 2017) and, perhaps most importantly, the development and adoption of smartphones and mobile applications (Newman et al. 2012). Technology development has steered the direction of citizen science and offered new mechanisms for engaging volunteers. While some projects build their own tools and technologies, there are a number of resources to help projects recruit and communicate with volunteers, collect, share, store and manage data, and enhance participation (table 21.1).

Project websites

Most citizen science projects have a presence on the web to (1) provide information, (2) recruit and (3) manage volunteers, and (4) allow citizens to contribute to research by collecting or analysing data. Initiatives like Project BudBurst (Johnson 2016), where volunteers provide information on plant phenology cycles, employ websites with information and basic

Table 21.1 Different types of technologies and supporting resources used in citizen science

	Supporting resources	Purpose
General purpose technologies		
Project websites	Development frameworks.	Make it easier for users to build websites.
Project catalogues	Existing catalogues and directories of citizen science projects.	Allow users to list projects and/or conduct research.
Web 2.0 and social media	Most social media platforms use application programming interfaces (APIs) to make it easier to create posts and access data. Third-party tools like TweetDeck and Hootsuite allow posts on multiple accounts/platforms.	Help users collect data from, or through, social media sites and communicate with volunteers.
Technologies to support data collection and analysis		
Mobile websites and apps	Tools to support responsive design and hybrid apps.	Make it easier for projects to develop websites that are accessible on mobile devices or tablets.
Smartwatches and wearables	Development kits.	Help users automatically collect data as they go about their everyday activities.
DIY sensors and the Internet of Things (IoT)	DIY sensor kits.	Help users build sensors for large-scale, ongoing data collection.
Drones	Drone kits.	Help users collect data in difficult to reach environments.
Data analysis tools	Platforms that process, visualise and export data.	Help users answer research questions by analysing data and detecting trends.
Mapping technologies	Mapping platforms.	Allow projects to publish data on maps and integrate various data layers to support analysis.

Table 21.1 (continued)

	Supporting resources	Purpose
Improving the citizen science experience		
Virtual reality and augmented reality	Virtual reality headsets.	Create an immersive experience to augment or replace real world environments.
Open data and supporting resources	Data standards; data storage and management platforms.	Collect, store, and manage open and interoperable data in a publicly accessible repository, enabling access and use beyond the lifetime of a particular project.

forms for data collection. Test My Brain, for example, provides more visual approaches to data analysis, such as by allowing volunteers to sift through images to perform tasks such as counting craters or matching or classifying images. The websites of virtual citizen science projects like EyeWire (Kim et al. 2014) also employ real-time communication such as chat systems or forums to help participants and create a more supportive community.

While websites broadly facilitate participation for users, the increasing availability of *development frameworks* support and empower project leaders. Development frameworks help project owners create websites and other tools to support citizen science projects without the need to write complex software from scratch. At the most basic level, WordPress, Django, Wix and Weebly are examples of frameworks that provide means for interacting with participants through features like content management, authoring (a content authoring feature is used to create multimedia content typically for delivery on the World Wide Web), authentication, blogging and basic input via forms. Such frameworks also support responsive design to deliver content appropriate for display on mobiles, desktops and tablets. For more advanced users, frameworks such as PhoneGap and Ionic help developers write websites in HTML and JavaScript, which can be easily packaged as mobile applications. Ushahidi, Inc. and Open Data Kit (ODK) provide a way to easily develop customised surveys and set up websites and mobile applications that can be distributed to crowdsource information. These frameworks also allow project owners to aggregate, visualise and analyse the data collected.

Development frameworks simplify the web development process. However, it is important to ensure that project websites are appropriately designed for their target users. Although customising templates supported by web hosting platforms is an apparently inexpensive solution, it often comes with hidden costs such as poor usability and awkward workflows (Wiggins 2013). Newman and colleagues (2010) explored the various factors that should be considered when developing websites for citizen science that are particularly relevant to websites that involve interactive maps.

Project catalogues

Websites like SciStarter, The Federal Catalogue of Crowdsourcing and Citizen Science, iNaturalist, Natusfera, Citsci.org and Zooniverse serve as project catalogues, or online directories that benefit citizen science by helping participants find projects to contribute to and collecting information for researchers to analyse. Many of these platforms also support participation directly. For example, iNaturalist and Natusfera allow citizen science volunteers to find biodiversity monitoring projects and directly upload biodiversity data. Some platforms, like Citsci.org, allow participants to create their own citizen science projects to initiate data collection and analysis via websites and/or mobile applications. Other platforms, most notably Zooniverse, provide cyberinfrastructure supporting data analysis via tasks such as classification, annotation and tagging (in a variety of fields such as arts, biology, literature and planetary science). Unlike development frameworks, which were designed for use in any context, these project catalogues are designed specifically to support citizen science.

Web 2.0 and social media

Web 2.0 and social media offer new means for citizens to express themselves and connect with others via open and free platforms. Citizen science has benefitted from social media platforms like Twitter, Facebook and Instagram that help project co-ordinators recruit and communicate with participants. In addition, data generated from online platforms such as Twitter can be automatically processed and analysed to provide citizen-generated data on critical events and emergencies (Gao, Barbier & Goolsby 2011; Shaw, Surry & Green 2015). The very nature of social media has also paved the way for global communities to self-organise, develop and become more sustainable, helping promote grassroots or bottom-up citizen science activities (see also Hecker et al. in this volume).

Mobile websites and apps

Technological developments in smartphones are revolutionising citizen science: Web-based data capture, analysis and presentation tools and apps are in common use, and a wide range of next-generation environmental sensors to be coupled to smartphones are under development. From online recording and real-time mapping to digital photography, there are tools for most tasks (Tweddle et al. 2012). In terms of actually making the record, many field recorders still use pencil and notebook or record cards (although increasingly relying on GPS handsets for geolocation) and this may be the most efficient method for capturing data in the field for many experts. However, communications technology has facilitated the ability to make records, especially incidental records, through smartphone apps. Currently, there are apps linking directly to iRecord (for efficient data flow) for recording ladybirds, butterflies, orthopterans, mammals and invasive non-native species. These provide the ability to take a photograph (or potentially, for species such as orthopterans to make a sound recording), capture location via GPS and store the record for later upload to iRecord. These apps are an ideal tool for widening participation, especially when observing species that are relatively large or immobile, conspicuous and easy to identify. Records still need to be verified for them to become scientifically useful though, and one important advantage of interoperable data systems is that there is the potential to bring together records from many different websites and smartphone apps to facilitate efficient verification (Pocock et al. 2015).

A collection of recommendations specific for citizen science that provides support and advice for planning, design and data management of mobile apps and platforms that will assist learning from best practice and successful implementations can be found in Sturm et al. (2017). Smartphones support many of the same data collection functions as desktop computers, allowing volunteers to provide observations and opinions through web forms and supporting simple data analysis tasks. More complex tasks are harder to support through mobile apps or mobile websites so some projects are not accessible via mobile devices (e.g., EyeWire). However, the ability to deliver content via mobile phones and tablets provides an excellent opportunity for citizen science projects to involve participants at all times, even while they travel. Further, mobile devices may facilitate

Fig. 21.1 The Project BudBurst website is designed to recruit and train participants, collect and publish data and provide education materials. The project also supports a mobile application mainly designed to facilitate data collection. The app is coded in HTML5, which is easier to develop and maintain but has less functionality than a native app available for Android or iPhone.

© 2017 NEON, Inc. All rights reserved.

Fig. 21.1 (continued)

access by larger, more diverse populations – access to the internet via mobile and tablet exceeded desktop for the first time in November 2016 (Gibbs 2016). Many citizen science projects therefore support both web-based and mobile participation. Project BudBurst hosts a website for desktop users, as well as an HTML5 website for mobile users (figure 21.1) (HTML5 is a markup language used for structuring and presenting content on the World Wide Web), and has explored an additional gamified app (Bowser et al. 2013). The iRecord Dragonflies mobile web application is another example of this approach.

Responsive design enables websites to be viewed according to the device being used to access them, by adapting layouts, media items and other content to different resolutions and screen-sizes. For projects that seek to host a website and a mobile site or app, styling tools employing responsive design, including Bootstrap and Boilerplate, can greatly simplify this process. Alternately, *hybrid apps* are web pages packaged into mobile apps that can run on multiple operating systems without the need for a web browser. The process of developing hybrid apps too can be greatly simplified by using frameworks such as Ionic, PhoneGap and Cordova. Finally, *native apps* are apps that are developed individually for different mobile operating systems using different programming languages. Native apps require greater investment and development effort but support a more interactive experience, and enable developers to use the phone's hardware to a greater extent.

Smartwatches and wearables

The increasing development of wearables and smartwatches offers the opportunity to explore new forms of engagement and data collection (e.g., Tse & Pau 2016; Nieuwenhuijsen et al. 2015). Smartwatches or wearables can provide information on the environment, human health and mobility using a wide range of sensors such as accelerometers, GPS, cameras, microphones, heart rate sensors, barometers, compasses and air quality sensors. Smartphones, smartwatches and wearables also facilitate lifelogging, recording activities throughout the day to help people understand how their habits and routines relate to external variables such as environmental conditions. For example, AirBeam provides wearable sensors and the AirCasting Android app for collecting air quality information as citizens travel around cities.

Do-it-yourself sensors and the Internet of Things

Do-it-yourself technologies have recently become popular, mainly due to the development of makerspaces or hackerspaces (see box 21.2; see also Novak et al. in this volume). These collaborative spaces offer different tools and facilities, including equipment such as 3-D printers, laser cutters and computer-controlled machines, for making, learning, exploring and sharing technologies. Open to a diverse community (from kids to budding entrepreneurs), makerspaces seek to provide hands-on learning, support community interests and creative expression, and foster critical thinking, particularly linked to STEM education. Makerspaces are also used as incubators and accelerators for business start-ups. In addition to persistent spaces like Fab Labs, participatory technology development is also supported through events like hackathons (see box 21.2; see also Gold & Ochu in this volume).

While traditional sensors are developed by engineers and experts, citizens and enthusiasts can now make use of DIY devices such as Arduino and Raspberry Pi. These are essentially basic computers to which different sensor modules can be attached. A large variety of modules can be used, including GPS sensors, accelerometers and cameras. Projects such as Smart Citizen (Diez & Posada 2013) use DIY sensors to help participants upload environmental data for analysis. Another example is the Cosmic Pi project, which aims to use low-cost, pocket-size detectors to detect cosmic rays.

Drones

Unmanned aerial vehicles (UAVs), or 'drones', are powerful platforms for monitoring and reporting, especially in terrains that are difficult to access on foot. In some areas, drones have a bad reputation because of their role in military missions (for surveillance or bombing) and due to privacy concerns. However, drones and other DIY aerial platforms can be used for social good (Choi-Fitzpatrick 2014). For example, members of Digital Democracy worked in Guyana with the local Wapishana people to build DIY drones to monitor and map deforestation (MacLennan 2014).

The 2010 Deepwater Horizon oil spill in the Gulf of Mexico illustrates the complex nature of aerial data collection by citizen science communities, since BP (the company responsible of the spill) and the US government explicitly denied monitoring access to journalists, citizen groups and scientists. While the word 'drones' typically evokes a high-technology approach, this is not always the case. In the Public Labs model, aerial mapping based on DIY balloon and kite systems serves as a powerful alternative strategy to monitor the environment (Dosemagen, Warren & Wylie 2011).

Data visualisation tools

Data visualisation is helpful for feeding analysed data back to participants and for presenting results to policymakers. There is a range of tools available to support project co-ordinators and volunteers in processing, analysing and visualising data; and many also come with plug-ins or modules to provide further analytic capabilities (see also Williams et al. in this volume). Earthwatch's Freshwater Links and UCL's Extreme Citizen Science: Analysis and Visualisation (ECSAnVis) are examples where users can visualise data coming from a variety of remote databases. Simple tools like Google Charts (an interactive web service that creates graphical charts from user-supplied information) also provide a quick means of visualising data online as configurable charts and graphs. As mentioned earlier, more complex frameworks such as Ushahidi, Inc. and ODK support both data collection and data analysis/visualisation.

Mapping technologies

Spatial data analysis is often critical to understanding variables ranging from biodiversity presence and distribution, to local environmental conditions, human population and transportation patterns. Many websites and most citizen science apps provide feedback to participants through maps, using map layers to add collected information as point data (e.g., iNaturalist displays points for observations of different species) and to overlay information such as heat maps (e.g., the Environment Hamilton's INHALE Hamilton project presents air quality information in this way) or geometries (e.g., Safecast presents levels of radiation and air quality data in this way, among others).

GIS tools have a long history of expert use, but mapping technologies have only recently been made easily available to non-expert users. Tools like Google Maps and OpenStreetMap paved the way for location services such as routing, searching, trip planning, traffic estimation and other routine tasks now used on a daily basis. Overlays can be created fairly easily using the methods made available in standard mapping tools such as OpenStreetMap, Google Maps, OpenLayers and Mapbox. Currently, the largest citizen science mapping initiative is Google Local Guides, with 50 million volunteers in October 2017.

Virtual reality and augmented reality

Virtual reality and augmented reality may be viable and cost-effective ways to improve data collection – for example, measuring the colour of the sea in the Citclops project (Wernand et al. 2012), train citizen science volunteers with personalised and immediate feedback, track individual data quality and improve retention and motivation, for example, increasing patient engagement in rehabilitation exercises using computer-based citizen science (Laut et al. 2015).

The availability of smartphones has resulted in investigations into how augmented reality can be embedded into standard interfaces – for example, overlaying objects on top of on-screen displays of camera views or base map layers. In addition to employing gamification approaches, which use the motivational elements of games to engage users, virtual or augmented reality can provide engaging applications to support citizen science through the increased recruitment of volunteers. And virtual reality can improve data quality and participant engagement by allowing users to dynamically interact in immersive environments (Klemmer, Hartmann & Takayama 2006).

Open data and supporting resources

Open data are both a resource for citizen science and an output of most citizen science initiatives. Open data policies implemented by governments, businesses and universities have begun to make large volumes of data available, which is openly accessible for the public to query, process and analyse. Many citizen science projects also make their data available as downloadable raw-data files, queryable databases or processed visualisations. Technical developments supporting open data include data standards, which promote interoperable data collection and sharing (Williams et al. in this volume) and are developed and maintained by organisations like the Open Geospatial Consortium (OGC). Expanded data storage, such as scalable databases and cloud storage, also supports open data, with computational, storage and hosting resources available from providers either for free (e.g., WordPress and Google Sites as general technologies; CitSci.org for citizen science) or on-demand (e.g., Amazon Web Services), offering much needed help for citizen science projects.

Box 21.1. Collaborative research on sustainable fish stocking in Germany

Angling clubs are fishing rights holders in Germany, and any changes to the governance and management of fisheries depends in part on decisions made by these clubs. Fish stocking is the practice of raising fish in a hatchery and releasing them into a river, lake or the ocean to supplement existing populations, or to create a population where none exists. Stocking may be done for the benefit of commercial, recreational or tribal fishing, but may also be done to restore or increase a population of threatened or endangered fish in a body of water closed to fishing. Stocking is a contested issue, whose success or failure depends on a range of social, ecological and evolutionary factors (Arlinghaus et al. 2014). To learn about successful and unsuccessful stocking practices, as well as associated genetic and other ecological risks, researchers partnered with 18 angling clubs in Lower Saxony on a transdisciplinary research project called Besatzfisch (which translates as 'stocked fish').

Working in close collaboration with the angling clubs, the research team developed an experiment involving radical stocking density treatments of northern pike (*Esox lucius* L.) and common carp (*Cyprinus carpio* L.) in angler-managed flooded gravel pits. Workshops were used to develop specific goals, objectives and hypotheses and to allocate treatment to 24 angler-managed flooded gravel pits. Outcomes were monitored jointly through a series of workshops, creating opportunities for reflexive learning.

Anglers participated in fish surveys and completed angling diaries to monitor carp. The research team chose paper-and-pencil-based diaries to allow anglers of all age groups to participate. Surveys of club anglers were also used to understand attitudes, norms and other human dimensions related to stocking and to behaviours (Arlinghaus et al. 2014; Gray et al. 2015; Fujitani et al. 2016). Results showed that the integration of anglers into the experiments was instrumental in improving ecological knowledge.

This project shows how citizen science using paper-and-pencil-based diaries, workshops and flooded gravel pits can support the co-production of knowledge. Given the age of many club anglers, it is likely that an app would reduce participation and bias the study towards the younger demographic segment. The benefits of ICT-enabled versus non-ICT-enabled citizen science approaches should therefore be carefully weighed depending on the target audience and project goals.

Box 21.2. Participatory technology development

Making and hacking democratise the creation of the hardware and software that aid in research, just as citizen science democratises the scientific research process itself. Fab Lab and TechShop are names used for two particular types of makerspaces:

> The *fabrication laboratories* (Fab Labs)[1] programme was initiated in 2001 by Professor Neil Gershenfeld of the Massachusetts Institute of Technology (MIT) and it has since become a collaborative and global network. Fab Labs are currently governed by the Fab Foundation, which lists more than 1,000 Fab Labs from all over the world (including 700 in Eurasia, 300 in America, 40 in Africa and 8 in Oceania; Gershenfeld 2008).
>
> *TechShop* was a chain of makerspaces started in 2006 in California. It was supported by monthly fees from members, which supported access to machines and tools. TechShop defined makerspaces as part prototyping and fabrication studios and part learning centres. As of 2017 there were 10 locations in the United States: three in California, one in Arizona, one in Arlington, Virginia (near DC), one in Michigan, one in Texas, one in Pittsburgh, Pennsylvania, and one in Brooklyn, New York, as well as four international locations. On November 15, 2017, with no formal warning, the company closed and announced they would declare bankruptcy under Chapter 7 of the United States bankruptcy code (immediate liquidation).

Hackathons also have promoted the development of new technological products to facilitate citizen participation. Hackathons are short-term, collaborative design events where volunteers, often including computer programmers, engineers and designers, create new technologies for a prize or other reward. These new technologies are usually software projects and applications, but they can include hardware products as well. Hackathons may be sponsored and organised by companies, educational institutes, non-profit organisations or government agencies. The US National Aeronautics and Space Agency (NASA), for example, routinely hosts the International Space Apps Challenge, a 48-hour event where teams use public data to solve challenges in hardware, software, citizen science and information visualisation (Bowser & Shanley 2013).

Practical and policy considerations

This chapter has explored a wide range of technologies used in citizen science, and offered examples of existing resources available to researchers and project co-ordinators, ranging from web development frameworks to virtual reality headsets. The majority of these resources were not developed specifically for use in citizen science. Therefore, it is important to understand how these resources are being used in a citizen science context, as well as to assess their strengths and limitations for different citizen science contexts.

To complement their existing database of citizen science projects, SciStarter is compiling a database of tools and technologies that citizen science volunteers can build, borrow or buy. This database will help project co-ordinators and volunteers to:

- Find information about different tools and technologies, and determine which are suited to their needs;
- Access new tools and technologies by linking to blueprints, lending libraries and online marketplaces; and
- Identify gaps in existing hardware and infrastructure, which could be filled by new collaborations between the citizen science and maker movements, bringing two participatory paradigms into closer alignment.

Another opportunity lies in the collection and development of relevant data and metadata standards to promote the collection, sharing and use of interoperable citizen science data. This could include standards for citizen science observations that follow the structure of a common model, such as the ISO 19156 model for Observations and Measurement. A citizen science profile for this has been suggested in the Sensor Web Enablement for Citizen Science work within OGC (Williams et al. in this volume).

Citizen science tools and technologies also need to be maintained as well as developed. On the one hand, building new technology for use in a citizen science project offers extensive customisation and opportunities for collaborative or participatory design. However, on the other hand, these technologies must then be maintained by the core project team, rather than relying on external developers. It is also important to consider how and where technologies will be deployed. For example, sensors used in the WeSenseIt project[2] were installed in river banks, which are often difficult for citizens to access so professional help was required to maintain and service sensors (Mazumdar et al. 2016).

There are numerous policy considerations to the development or procurement and use of citizen science technologies. Data quality is a critical issue in citizen science, especially in policy contexts such as monitoring and regulation (see also Brenton in this volume; Williams et al. in this volume). It is important to consider fitness for purpose in all aspects of project design, including when designing or selecting citizen science technologies. For example, while some environmental monitoring sensors may align with regulatory standards, others may not (Volten et al. in this volume).

Funders and policymakers have both made it clear that citizen science activities should produce open and interoperable data. For example, recent guidance on crowdsourcing and citizen science issued by the Director of the US Office of Science and Technology Policy suggests that, 'federal agencies should design projects that generate datasets, code, applications and technologies that are transparent, open and available to the public, consistent with applicable intellectual property, security, and privacy protections' (Holdren 2015). Guidance in the EU tends to recommend a balanced approach to openness and emphasise interoperability. For example, in 2015 the European Commission's Horizon 2020 Framework Programme issued a call for the 'Coordination of Citizens' Observatories and Initiatives' (SC5-19-2017) seeking a team of researchers to help 'promote standards' and 'ensure interoperability'.

Some citizen science projects, particularly those run by government agencies, may be limited in the types of technologies they can use. The US Crowdsourcing and Citizen Science Act (15 U.S.C. § 3724 [2017]) tasks one government agency, the General Services Administration (GSA), with specifying the appropriate technologies and platforms to support citizen science activities. While these guidelines would strictly apply to all federal employees, citizen science projects hoping to influence government decision-making would be wise to consult any published list of GSA guidelines for citizen science technologies and tools. Additional policy guidance in the United States, the EU and elsewhere is likely to be issued as citizen science continues to grow.

Conclusions

As citizen science has evolved, new technologies have emerged to enable citizens and communities to contribute to citizen science in a variety of ways. The relevance, contribution and importance of technology in citizen science therefore demands much more attention from practitioners and communities. Mobile technologies will continue to revolutionise the field, an innovation particularly valuable for engaging new communities and stakeholders in citizen science, including younger populations and participants in developing countries. Technologies also support a wide range of project governance models. Many future citizen science endeavours will harness the power of social networking to larger effect in all aspects of research, with members of the public collaboratively conducting research, validating and publishing results. Resources that support technology development by making it easier to build websites, apps, sensors and maps similarly lower the barrier to entry for top-down and bottom-up models of citizen science alike. At the same time, the use of various technologies should be carefully considered, taking into account participant demographics, affordability and access, and fitness for purpose.

Acknowledgements

The authors would like to acknowledge the support of the following projects: EU FP7 WeSenseIt and Citclops; Horizon 2020 Seta, STARS4ALL and CAPSSI; Crowd4Sat; EyeOnWater @ Vendée Globe; and ATiCO; as well as the Alfred P. Sloan Foundation.

Notes

1 https://www.fablabs.io/labs
2 http://www.wesenseit.com/

22
Maximising the impact and reuse of citizen science data

Jamie Williams[1], Colin Chapman[2], Didier Guy Leibovici[3], Grégoire Loïs[4], Andreas Matheus[5], Alessandro Oggioni[6], Sven Schade[7], Linda See[8] and Paul Pieter Lodewijk van Genuchten[9]

[1] *Environment Systems Ltd, Aberystwyth, Wales UK*
[2] *Welsh Government, Aberystwyth, Wales UK*
[3] *University of Nottingham, UK*
[4] *CESCO, MNHN, Paris, France*
[5] *Secure Dimensions GmbH, Munich, Germany*
[6] *Institute for Electromagnetic Sensing of the Environment, CNR, Milan, Italy*
[7] *European Commission – Joint Research Centre, Ispra, Italy*
[8] *International Institute for Applied Systems Analysis, Laxenburg, Austria*
[9] *GeoCat BV, Bennekom, The Netherlands*

corresponding author email: jamie.williams@envsys.co.uk

In: Hecker, S., Haklay, M., Bowser, A., Makuch, Z., Vogel, J. & Bonn, A. 2018. *Citizen Science: Innovation in Open Science, Society and Policy*. UCL Press, London. https://doi.org/10.14324/111.9781787352339

Highlights

- Open data and open standards promote interoperability, which in turn allows citizen science data to be more widely discovered and used.
- Data reliability is essential for citizen science data to be trusted and align with environmental regulation and monitoring requirements from governments.
- Contextualising data with metadata, including descriptions of their purpose and methods of dataset creation, allows users to evaluate their possible reuse.
- Reuse of project results is ensured through the use of open data, open standards and by having good data reliability, metadata and documentation.

Introduction

There is an increasing number and diversity of citizen science projects, which can potentially generate new data at a lower cost than professional data collection (De Longueville et al. 2010; Antoniou, Morley & Haklay 2010; Friedland & Choi 2011) and arguably with greater value than those generated by expert knowledge alone (Fischer 2000; and see Danielsen et al. in this volume). When considering Ten Principles of Citizen Science (ECSA 2015), in particular openness and accessibility, these citizen science data have the potential to be a valuable source of information for decision-making and policy formation on local, regional and national scales. However, for the data to realise their full potential, a number of factors have to be considered.

This chapter identifies the factors that affect citizen science data using examples from environmental monitoring and geographic information. These factors include open data standards and interoperability; data reliability and alignment with government environmental regulation and monitoring requirements; the contextualisation of data to enable users to evaluate its possible reuse; and the reuse of project results (see box 22.1). This chapter addresses each of these factors in turn to help specialists and non-specialists alike to better plan citizen science projects.

Data contextualisation

Data are only meaningful if they can be interpreted. Therefore, it is vital to know the context within which a particular dataset has been created,

---

Box 22.1. Key factors when initiating a citizen science project

- **Data contextualisation** – communicating the context in which a particular dataset has been created.
- **Data interoperability** – enabling seamless reuse of resources (in this case, data and processing) across different systems.
- **Data quality** – data quality has long been identified as the crucial challenge for the use of citizen science data.
- **Data reuse** – data ownership and future accessibility.

---

including, for example, units of measure, measurement devices, pre-processing procedures, quality assurance (QA) mechanisms, uncertainties and intended use. As soon as a dataset needs to be understood by anybody other than by its creator (for example, by a customer, reviewer or peer), then this contextual information should be explicitly provided, ideally with the dataset itself.

Some of the contextual description of why, how, when and by whom a dataset was created can be unambiguously provided using standard vocabularies and code lists (for example, for units of measure or particular statistical processing algorithms). However, other information may bring uncertainty when interpretation takes place. The description of data provenance (i.e., the processing steps applied) is a typical case in which it remains difficult to keep track of a complete, up-to-date and reproducible instruction. This holds particularly for datasets subject to intense experimentation. In citizen science, these issues are complicated by the training required to provide detailed contextual information about data, which may not be readily available to participants.

Context again becomes relevant if a given dataset is applied to another, initially unintended, purpose. In such cases, the effects of the contextual change on possible interpretations of the data have to be carefully examined. For example, in establishing if occurrences of fly fishing derived from social media can be used for an indication of river health.

The creation of metadata is key to capturing contextual information. It would normally consist of the title, description, number of participants/observations, contact details, and temporal and geographical extent of the data. If the data are combined with other data, legal constraints and data quality aspects such as lineage information should also be included. For the purpose of assessing data quality, the identity of the observer, location accuracy and potentially the device accuracy would also be required.

Citizen science projects are constantly looking for new participants, and ways to make their project's data available through as many means as possible, therefore making the efforts of the voluntary work as effective as possible. Wide discovery of citizen science resources is important for maximising impact, creating additional value and encouraging reuse, beyond the scope of the original project.

In support of this, scientists, journalists and citizens are continuously looking for datasets relevant to their field of study, typically querying a search engine or open data catalogue for datasets using keywords and a location/timeframe of interest. The use of common vocabularies makes metadata meaningful for different usages, for example, DCAT enables interoperability with the open (government) data, schema.org enables discoverability and ISO19139 connects to the spatial data infrastructure (SDI) community.

Data interoperability

As this book illustrates, citizen science can be extremely varied, with diverging objectives and research questions from within and across different subject areas and geographic scales. The level and type of citizen participation also varies greatly (see Haklay; Ballard, Phillips & Robinson, both in this volume). Consequently, data are generated, analysed and presented in a variety of ways. The case-by-case, tailor-made management and handling of datasets might serve their original intended purpose (for example, assessing water quality or monitoring birds) but is likely to reduce interoperability, in other words, the communication, exchange and use of data. This not only includes the interaction between machines, but also between machines and humans (users), and between communities of people themselves.

Interoperability, which enables the seamless reuse of resources (in this case, data and processing) across different systems, can be reached by applying community-wide agreements. For example, agreeing on the use of software tools, data standards and best practices, or by improving data accessibility and exchange through ad hoc tools or community practices. It can address the data themselves, but also the processes and services that generate and exchange data between any two parties (for example, between citizens and academics working on a national-funded research project, and between citizens and local decision-makers collaborating for the benefit of the local community). Such processes help to i) ease the integration of data from different sources; ii) improve the reuse of data in other contexts; and iii) save resources in the development of data management and handling tools. Semantic interoperability provides interoperability at the highest level to exchange data with unambiguous, shared meaning and improve quality, efficiency and efficacy. Semantic interoperability adds to the possibilities of data sharing as well as the meaning of the data, linking any data elements with metadata and terms of vocabulary.

There is a high diversity in the details of modelling, encoding and describing datasets, as well as in the communication protocols for data storage, processing and access. Museum collections, institutional biodiversity datasets and international projects have adopted extensions of the Darwin Core in multiple encodings, while the SDI favour other mark-up languages for sharing geographical data and metadata. Initiatives such as the PPSR_CORE Program Data Model Metadata Standard/data-sharing protocol are progressing this field. However, each initiative has a form of standardisation in mind, which prevents each new citizen science project from designing their own ontology relating to their domain. Once a known subset of standards for specific domains has been developed, it will be much easier to write connectors (mappings) to interact between those standards, therefore greatly improving the data-sharing potential.

Data standards (i.e., user-accepted norms on data models, formats and exchange protocols) are the key to achieving interoperability. The main challenges include the agreement of standards within a user group and across communities. The Open Geospatial Consortium (OGC), for example, has a long history in the technical specification of geographic

---

Box 22.2. EC INSPIRE Directive

'The INSPIRE Directive aims to create a European Union spatial data infrastructure for the purposes of EU environmental policies and policies or activities which may have an impact on the environment' (European Commission 2017). It is based upon these common principles:

- "Data should be collected only once and kept where it can be maintained most effectively.
- It should be possible to combine seamless spatial information from different sources across Europe and share it with many users and applications.
- It should be possible for information collected at one level/ scale to be shared with all levels/scales; detailed for thorough investigations, general for strategic purposes.
- Geographic information needed for good governance at all levels should be readily and transparently available.
- Easy to find what geographic information is available, how it can be used to meet a particular need, and under which conditions it can be acquired and used".

---

data and metadata (i.e., data documentation) models and encodings. The Sensor Web Enablement for Citizen Science (SWE4CS) proposal for a standard in the OGC exemplifies the need for flexibility when modelling and exchanging citizen science data (OGC/CS DWG 2016). In some regions, standards are complemented by more conceptual frameworks, such as the legally binding European Directive 2007/2/EC to establish an Infrastructure for Spatial Information in the European Community (INSPIRE) (European Commission 2017; see box 22.2), and the US Crowdsourcing and Citizen Science Act of 2016 (Congress 2016).

Experts are now widening the extension of these standards into the citizen science community. Currently, SWE4CS (Citizen Observatories 2015), exists in parallel to a Volunteered Geographic Information (VGI) extension to the INSPIRE standards (Reznik et al. 2016). SWE4CS also enables the inclusion of concepts from other standards that are established in other areas. See box 22.3 for examples of common metadata standards for dataset discovery.

As previously mentioned, metadata plays an important role in the data-sharing process. Beyond just providing an understanding of what is published, machine-to-machine understanding needs standardised interfaces with common exchange formats. This is not a requirement for citizen science data but would maximise their use and reuse. Few citizen science projects currently adopt standards for web services or data encodings, as most projects have yet to realise the benefits of sharing their data or are unaware of how best to do so. Agreements and technology implementations are needed, first to adapt established practices to new interoperable systems, and second to stimulate new projects to adopt standards and tools.

Ongoing global initiatives such as the Group on Earth Observations Biodiversity Observation Network (GEO BON), which 'aims to improve the acquisition, co-ordination and delivery of biodiversity observations and related services to users including decision-makers and the scientific community' (GEO BON 2017), and Global Biodiversity Information Facility (GBIF) 'an open-data research infrastructure funded by the world's governments and aimed at providing anyone, anywhere access to data about all types of life on Earth' both provide guidance on aspects of interoperability within environmental monitoring (GBIF 2017). However, more needs to be done to get novice, local-scale citizen science projects to adopt these standards.

Box 22.3. Examples of common metadata standards for dataset discovery

Dublin Core	Vocabulary for resource description, used as a base vocabulary in other vocabularies (http://dublincore.org/documents/dces/). DC is the default schema in Catalogue Service for the web, a metadata transfer protocol standard by Open Geospatial Consortium (http://www.opengeospatial.org/standards/cat)
ISO19139	XML/XSD-based vocabulary to describe spatial datasets (https://www.iso.org/standard/32557.html), commonly used in the GIS domain (INSPIRE).
DCAT	Data Catalog Vocabulary is a Resource Description Framework vocabulary to describe datasets maintained by W3C. Used in open data portals (http://www.w3.org/TR/vocab-dcat/).
VOID	Vocabulary of Interlinked Datasets is a vocabulary to describe linked datasets maintained by W3C (https://www.w3.org/TR/void/).
schema.org	Initiative of the main search engines to enable crawling web content as structured data. Contains a concept for dataset (http://schema.org/Dataset).
SDMX	Vocabulary to describe datasets in the statistical domain (https://sdmx.org/).
Datapackage	Vocabulary to describe (and embed) datasets, maintained by Open Knowledge Foundation (https://specs.frictionlessdata.io/data-package/).

Data quality

It is commonly agreed that the lack of knowledge about data quality limits the use of citizen science data (Flanagin & Metzger 2008; Haklay 2010; Goodchild & Li 2012; Fowler et al. 2013; Hunter, Alabri & Ingen 2013). Furthermore, citizen science projects are designed to be carried out by non-experts, with controlled data collection methods to support scientific integrity (Craglia & Shanley 2015). With project goals of i) enlarging participation and consequently data collection over space and time, and ii) ensuring that information embedded in the dataset varies according to the field of interest, context such as location, date and rules for data standardisation should be specified as part of the project, but parameters for participation, such as skill level, interpretation and observation intensity, should remain flexible. However, where data collection protocols are not respected by participants or simply implemented incorrectly, the resulting data could be of lower quality and potentially include misleading information. To further confound this, a learning effect has been reported in many citizen science studies, for example, participants get better at identifying different species (see box 22.4; Peltola & Arpin in this volume), though their ability may differ between activities. To balance this, the iSpot crowdsourcing qualifying system, for example, uses a reputation score for participants over eight groups of species. The contributor's reputation per species group acts as a quality measure of trust and can be used to evaluate their identifications over alternatives. Using this system, Silvertown et al. (2015) reported improvements in accuracy when multiple identifications were recorded, as well as the ability to quantify the level of confidence in observations.

Some data quality issues can be addressed by using a QA process, either human or automatic, to produce metadata on data quality. This quality information establishes trust in an observation and the volunteer who produced it, in a similar way to the trust traditionally placed in experts (Alabri & Hunter 2010; Hunter et al. 2013; Bishr & Kuhn 2013; Zhao et al., 'A Spatio–Temporal VGI Model', 2016). This trust is then transferable to the data themselves (Leibovici et al. 2017a).

A human-based QA process, such as peer verification, allows project participants to help identify and validate the observations provided by new users (Warncke-Wang et al. 2015; Antoniou & Skopeliti 2015). This peer verification, crowdsourcing the quality assessment or 'wisdom of the crowd' (Surowiecki 2005), enables some control in the same way that Wikipedia allows editing of an article to support convergence towards shared narratives. In Wikipedia, the data themselves are subject to peer verification quality improvements and the edits are logged (Warncke-Wang et al. 2015; Mobasheri et al. 2015). For citizen science, most of the time the data will not be as modifiable as in Wikipedia. However, they will have a quality that may be identifiable, and according to the level of quality and reliability that was attributed to the data, it may be reused regardless of whether it has been validated or not. Nonetheless, multiple citizen science observations in the same location or made at the same time can allow for application of a similar process as in Wikipedia editing. Volunteered geographic information (VGI) such as OpenStreetMap (OSM) data follows this principle (Haklay 2010).

Peer verification, such as expert verification, is not without issues (Wiggins et al. 2011; See et al. 2013). The volume of data to be checked and verified can be overwhelming, and errors made in human verifying may still have implications. The development of geo-computational QA offers greater scalability with constant reliability of assessment, ensuring better comparability (Kelling et al. 2011; August et al. 2015; Meek, Jackson & Leibovici 2016; Leibovici et al. 2017a). With this, stakeholders setting up a citizen science project can define the QA with its requirements based on rules defining the levels of quality, which are then transformed into a workflow of quality controls, generating the metadata on data quality.

It is possible to use automatic QA to complement peer assessment, accumulating trustworthiness in the volunteers (Leibovici et al. 2017b) (see box 22.4). The impact of these different QA methods can vary, so as part of the whole data curation process the QA has to be designed, agreed to for the usage of the citizen science data, and published as part of the metadata (Higgins et al. 2016).

Assessment standards are still to be finalised to communicate metadata on data quality, and will be an addition to the interoperability discussed above. Currently, the ISO19157 metadata standard for geographical data is applicable to citizen science as it produces geolocated data. This includes 'usability' but omits quality dimensions, such as trust and number of participants. Therefore, on top of the 'producer' model represented by ISO19157, citizen science demands a 'stakeholder' model to assess the participant (as a sensor) and a 'consumer' model through which peers give feedback on an observation (Meek, Jackson & Leibovici 2014; Leibovici et al. 2017a).

Data reuse

Supporting and planning for the reuse of data collected through citizen science activities is key for realising their long-term value. There are several aspects that need to be considered when planning for sustainable data management, such as the intellectual property rights (IPR) associated with contributions from citizens with respect to patents and copyrights (Scassa & Chung 2015a). The raw data contributed to a citizen science

Box 22.4. Quality assurance in an invasive species survey of *Fallopia japonica* (Japanese knotweed) in Wales (Leibovici et al. 2017b).

This example is typical of a plant identification survey and illustrates the different dimensions of quality that are important in citizen science. Participants were trained to identify Japanese knotweed and were sent to the Snowdonia National Park in Wales to locate, capture geolocated pictures and answer questions on invasive species. Reliability in the location and identification of the plant were the most important quality assessment criteria.

When it is not possible to manually assess each observation for accuracy, rules can be established to help assess the observation based upon factors such as proximity to cultivated land and forest, rivers or paths, but these can in turn be compromised with poor positioning (propagation of error). Modelled Earth observation data can also be used to assess the likelihood of species presence although this includes problems such as satellite imagery from different dates to that of volunteer data capture. Confirmation from multiple observations or closeness of observations can also be used.

Combining these imperfect rules and quality controls could lead to an improved QA. Furthermore, adding bespoke rules concerning the interaction of the factors may also improve the final assessment, and therefore their reusability.

project have no copyright, but the form in which the data are presented may qualify for copyright, for example, with photographs or written text. Therefore, it is important for citizen science projects to consider what contributions might be subject to IPR and the form in which the contributions are made.

To help developers of citizen science projects consider these issues, Scassa and Chung (2015a) provide a typology that categorises citizen science projects into four types: i) classification or transcription of data; ii) data collection; iii) participation as a research subject; and (iv) problem-solving, data analysis or development of ideas. They argue that there will be minimal IPR issues related to the first three categories based on examinations of the form of participation in different citizen science projects, for example, those found on Zooniverse. However, they also identified examples of projects that collect photographs or written text and therefore may be subject to copyright issues. The fourth type of citizen science project has potential patent issues since citizens may engage in inventive activities that could lead to patent rights, of which developers of citizen science projects should be aware. Within the EU there are additional database rights provided through the EU Database Directive. Licensing is one way to handle these IPR issues. For example, OSM has an Open Database License that specifies use of the data by anyone for any purpose provided attribution is given to the project and its contributors as a whole.

When initiating a citizen science project, the two primary high-level considerations regarding IPR and citizen science are (1) what background IPR will be used (for example, knowledge and data) and what restrictions is it subject to; and (2) if the project wishes to allow access to the knowledge and data (and to what level – see box 22.5) generated by the project (foreground IPR). Guidance on IPR is available from multiple governing bodies, organisations and institutions. Ultimately, how the IPR for any citizen science project is handled should be set out in the terms of participation in a project (the 'terms of use') (Scassa & Chung 2015a) so that participants are clear on these conditions and can agree to them during registration, prior to data collection.

A further consideration for data reusability is personal privacy. Protecting participating citizens' privacy is a key priority in a citizen science project (Bowser et al. 2014). When individuals provide data as part of a citizen science project, data may be stored with the individual's personal details. The contribution may sometimes require that the citizen's identity is known or can be known if necessary, or the citizen might wish to restrict, or actively promote, the attribution of their contribution with their personal details. Location-based information, recorded using mobile devices, can further reveal the position of individuals as well as their movements. This information could inadvertently be used to locate individuals in space and time, and in some cases, identify their home address or workplace. Regardless of the intended use of the personal information collected during and after the project, it must be stated in projects' 'terms of use' at the point of registration or, as a minimum, prior to commencement of data collection.

Many countries have data protection laws that protect individuals; however, these vary from country to country (Dyson et al. 2014). For Europe, the EU Regulation 2016/679 will protect EU citizens in terms of the processing and free movement of personal data. This regulation comes into force in 2018. Some principles are that users must be able to control their personal data at any time, including the inspection and deletion of

Box 22.5. Open data

The proliferation of open data can bring new opportunities in environmental monitoring (and other areas), by allowing cross-validation, data conflation, or increased temporal or spatial coverage. Citizen science data have a role to play in this open data movement, where open data is defined as the following:

> Open data is publicly available data that can be universally and readily accessed, used, and redistributed free of charge. Open data is released in ways that protect private, personal, or proprietary information. It is structured for usability and computability. (Verhulst & Young 2016)

One way in which citizen science data can be released as open data is using the Creative Commons (CC) open data licensing framework. Creative Commons encourages sharing under any of its licences as a way to create a more open data culture. Examples of CC licences that conform to the open definition and which could be suitable for use with citizen science applications (providing this intent was stated in the project's terms of participation, prior to citizen participation) are:

- Creative Commons (CC0),
- Creative Commons Attribution 4.0 (CC-BY-4.0),
- Creative Commons Attribution Share-Alike 4.0 (CC-BY-SA-4.0).

By releasing project data under one of the established CC licences, thus allowing any restrictions on their use to be fully understood, the likelihood of data reuse greatly increases.

their personal record. Personal data can only be collected for a particular purpose and the user must agree to this (i.e., give prior consent) before the data are exchanged. The personal data collected should also be limited to what is absolutely necessary, which requires knowing this information for a given project in advance. This regulation means that implementation of pan-European citizen science platforms can be challenging. Moreover, this becomes problematic if there is personal data exchange to countries outside of the EU, where there is none, or a variation in personal data protection legislation. One example would be the

United States, which has passed the Crowdsourcing and Citizen Science Act of 2016. This act endeavours, 'While not neglecting security and privacy protections', to make data collected through a citizen science project open and available, in machine-readable formats, to the public. As part of this process, federal agencies are required to inform participants on the expected uses of a project's data and if project results will be made available to the public. Furthermore, federal agencies would retain ownership of such data.

Data may also be collected about people, species or other entities that exist in real life. In this context, privacy and security not only play a major role to ensure the protection of personal data from citizens but also the well-being of the objects observed. Two types of 'objects' can be identified: i) primary objects about which the citizen is collecting information, for example, the ancient tree in a photograph; and ii) secondary objects that are recorded with the primary object, either by accident or because it was not possible to record just the primary object, for example, if the ancient tree was located alongside a school with children playing outside, who are also captured in the photograph. The spectrum for protection of these objects is manifold but can be condensed to conditions that apply when making observations electronically available. For example, the observations in time and location for endangered species are one type of information that may not be made available to the general public (primary

---

Box 22.6. Data contextualisation, interoperability, quality and reuse, in practice

An example of the key considerations for citizen science projects in practice is Geo-Wiki, which is an online platform and set of mobile tools for improving global land cover datasets (Fritz et al. 2012). Geo-Wiki was designed to address the problem of the high spatial disagreement that can be observed when different global land cover products are compared (Fritz et al. 2011) and to use data collected by citizens to create improved hybrid land-cover maps (See Fritz et al. 2015). In the online tool, citizens are asked to interpret land cover using medium-resolution satellite imagery from Google Earth and Bing; and more recently images from the Sentinel-2 satellite have been added. In the mobile apps, citizens are guided to locations and asked to classify the surrounding land cover and land use, supplementing their observations with geo-tagged photographs. More

(continued)

opportunistic tools are also available for recording land cover and land use at any location.

To involve citizens in the data collection process, the Geo-Wiki team have run a number of citizen science campaigns that have lasted a few weeks to six months. Various incentives are used to encourage participation, from prizes to co-authorship on scientific papers. More details of various campaigns can be found in See et al. (2015); Sturn et al. (2015); and Laso Bayas et al. (2017).

The four key issues discussed in this chapter have been tackled by the Geo-Wiki project. The raw data collected during the first set of citizen science campaigns have now been published in an open-access repository, PANGAEA (Fritz et al. 2017). Data from a more recent campaign focusing on classification of imagery for crop-land has also been published in PANGAEA (Laso Bayas et al. 2017). This publication of the data supports both the **interoperability** and **reuse** of the project by encouraging reuse of the data for applications such as land cover map development (as training data) and for the evaluation of land cover maps (i.e., validation). Although the data do not follow a specific metadata standard, they are supplied with accompanying metadata that explains each of the data fields. The **contextualisation** is provided through the narrative that accompanies the publication of the data (Fritz et al. 2017; Laso Bayas et al. 2017) and the land-cover definitions used are generic enough that they can be applied to many other land-cover products.

The data have been published open access in raw form so that users can apply their own data quality measures to the observations and filter them based upon the needs of their own applications. The **data quality** has been analysed and reported in a number of different papers using a variety of methods, ranging from comparison with authoritative or expert data sources to different conflation methods such as majority voting when multiple observations are available for a single location (See et al. 2013; Laso Bayas et al. 2017; Salk et al. 'Local Knowledge', 2016; Salk et al., 'Assessing Quality', 2016; Salk et al. 2017; Zhao et al. 2017). Overall the reliability has been good and lessons have also been turned into recommendations to further improve it, such as methods to enhance the training of the citizens, more effective use of real-time feedback, and so forth.

objects). In general, observations that contain children or sexual content as well as violent language must be redacted according to the Western law. Furthermore, personal identifiers such as car license plates, doorbell signs and people who did not agree to be visible (secondary objects) must also be removed. This ensures the privacy and well-being of the secondary objects. Automatic or semi-automatic processing of this typically involves functions that can be identified as a privacy/security extension to QA, where the objective is to be compliant with the governing law.

Conclusions

Citizen science data can act as timely evidence for various decision-making processes that impact on citizens' lives and surroundings, including environmental policy. However, it is only with good management of data and metadata, particularly when it comes to data reliability, that citizen science data can fulfil their role of empowering citizens.

Establishing the evidence that citizen science can be used effectively for policy will take time (see Nascimento et al.; Shirk & Bonney, both in this volume). However, in order for policy to realise its full benefit, the ability to share and use data across platforms and stakeholder groups is essential (Higgins et al. 2016). To maximise the impact and reusability of citizen science data, citizen science projects should therefore adopt standards for web services or data encodings and, where possible, adapt previously collected observations to these standards. This allows other citizen science initiatives on the same or complementary topic to reuse the data generated.

Being able to ingest, conflate and disseminate citizen data across systems not only supports the generation, assessment and sharing of citizen science data as evidence suitable for decision-making, but also improves its impact and reusability. This is achieved through the transmission of data to other interoperable systems, used by other projects and purposes, to stimulate more targeted research, societal benefits and potentially commercial revenue.

The field of citizen science is both long-established and continually evolving. New technologies and understandings provide the potential to increase the impact of citizen science projects. The Ten Principles of Citizen Science offer some guidance in the area of maximising the impact and reuse of citizen science data (ECSA 2015). However, further technical work is needed in the areas of domain-specific citizen science metadata and data quality (among others). Additionally, specific guidance on the areas of IPR and privacy would contribute towards citizen science data reaching their full potential. In support of this, dedicated organisations such as the European Citizen Science Association (ECSA), the Australian Citizen Science Association (ACSA) and the Citizen Science Association (CSA), and other organisations with domain-specific working groups, such as the OGC and the Committee on Data for Science and Technology (CODATA), among many others, continually work towards these common goals.

23
Enhancing national environmental monitoring through local citizen science

Hester Volten[1], Jeroen Devilee[1], Arnoud Apituley[2],
Linda Carton[3], Michel Grothe[4], Christoph Keller[5], Frank Kresin[6,7],
Anne Land-Zandstra[5], Erik Noordijk[1], Edith van Putten[1],
Jeroen Rietjens[8], Frans Snik[5], Erik Tielemans[1], Jan Vonk[1], Marita
Voogt[1] and Joost Wesseling[1]

[1] *Dutch National Institute for Public Health and the Environment, Bilthoven, The Netherlands*
[2] *Royal Netherlands Meteorological Institute, De Bilt, The Netherlands*
[3] *Radboud University Nijmegen, The Netherlands*
[4] *Geonovum, Amersfoort, The Netherlands*
[5] *Leiden University, The Netherlands*
[6] *Waag Society, Amsterdam, The Netherlands*
[7] *University of Twente, Enschede, The Netherlands*
[8] *SRON Netherlands Institute for Space Research, Utrecht, The Netherlands*

corresponding author email: hester.volten@rivm.nl

In: Hecker, S., Haklay, M., Bowser, A., Makuch, Z., Vogel, J. & Bonn, A. 2018. *Citizen Science: Innovation in Open Science, Society and Policy.* UCL Press, London. https://doi.org/10.14324/111.9781787352339

Highlights

- Citizens are highly motivated to contribute to air quality measurements that complement existing measurement networks because of their high spatio-temporal resolution;
- Data needs to be assimilated, for example, using models;
- Low-cost sensors need to be developed further, and their application calibrated and validated;
- Easily accessible expert information and feedback is needed to support participants;
- Environment protection agencies (EPAs) can both support and benefit from citizen science using small sensor networks.

Introduction

A key motivation for national environmental protection agencies (EPAs) to support and participate in citizen science is to allow these knowledge institutes to get out of the well-known scientific 'ivory tower'. Citizen science is one way to shape science-society relationships in a more interactive and reflexive way. Reflexivity means scientists being aware of the potential societal effects of their research and taking these into account in their choice of research objects, methods and approaches. It is assumed that the reorganisation of governmental scientific advice along the lines proposed by reflexive scholars will increase the accountability, quality, effectiveness and legitimacy of scientific expertise in society (Funtowicz & Ravetz 1993; Nowotny, Scott & Gibbons 2001; Jasanoff 2003; and see also Smallman; Mahr et al., both in this volume).

At the same time, EPAs can be useful partners to more local citizen science projects because this relationship facilitates better data collection (see also Owen & Parker in this volume). Often, the initiative for an air quality citizen science project is (at least partly) taken by a municipality. For municipalities, citizen science provides the opportunity to improve the connection between citizens and their living environments by studying the environmental conditions of their direct local vicinity. For EPAs to support this, however, a lot of learning is needed, including about the governance of long-term data collection, the dissemination of results and the use of platform technology with open data (see Williams et al. in this volume).

Citizen science is not only beneficial for its organisers, but also provides participants with the opportunity to democratise science, and to learn about the scientific topic of focus. In a relatively recent literature review, Haywood (2014) collects claims about the benefits of citizen science for its participants (table 23.1). These benefits are dependent upon interactions at the local level and the way collected data is made available to participants (e.g., in maps and with adequate explanation).

This chapter summarises the involvement of the Dutch National Institute for Public Health and the Environment (RIVM) in a series of citizen science projects, to draw out some of the potential benefits and challenges of EPA involvement. It then describes the RIVM's roadmap to further develop the use of citizen science in its national monitoring programme, to provide an example for other official, national-level institutions that may seek to benefit from citizen science.

From ad hoc citizen science to national measurement networks

The RIVM began its involvement in citizen science around 2012 with its ad hoc participation in several air quality citizen science projects. The projects were varied, and it took a while before the significance of the citizen science movement was fully recognised. In 2016, the RIVM and the Dutch Ministry for Infrastructure and the Environment (responsible for air quality in the Netherlands) agreed to start a programme to innovate its national air quality measurement network (LML). The project should enable small sensors and citizen science to become an integral part of the monitoring procedures. The innovation programme has a timeframe of five years (2016–2020).

In short, the ambition is to make citizen science data an integral part of standard procedures and models for determining air quality. This is a way to not only motivate participation in citizen science, since it is more rewarding when the measurements are actually used, but also to make citizen science sustainable, which is often lacking. The final goal is to have a hybrid, flexible network using data from different types of sensors, including reference instruments, intermediate- and low-cost, and satellite observations. The data can be contributed by different parties, including citizen groups, cities, non-governmental organisations (NGOs) and official measurement institutes.

There are several reasons why the RIVM decided to participate in citizen science. Innovation of the air quality monitoring network is in part affected by the fact that such a network is expensive, making better and more cost-efficient solutions welcome. In addition, advances in micro technology (especially sensor technology) and the spread of smartphones have democratised the ability to perform air quality measurements. This means that practically all stakeholders and citizens can do air quality measurements if they want to, perhaps because they do not trust the model-based data from the authorities or because data for their specific location are not available. For EPAs this presents an opportunity as well as a challenge. Environmental protection agencies may profit from the high spatial and temporal resolution observations in the urbanised environment, if they find a way to assimilate these data in air quality and meteorological models to provide forecasts to the public. The involvement of citizens brings the prospect of having the dense coverage of observations needed for this purpose.

The next section describes several citizen science projects RIVM took part in and reflects on lessons learned and experiences gained. It also provides a brief description of innovations in environmental monitoring, and how these will help to ensure the continuity and effectiveness of citizen science measurements.

Collaboration with stakeholders and local initiatives

Measuring Ammonia in Nature

Since 2005 the Measuring Ammonia in Nature (MAN) network has monitored atmospheric ammonia concentrations in nature reserves in the Netherlands. The monitoring network is an example of citizen science even if it never was explicitly identified as such. Measurements are performed with commercial passive samplers, which are calibrated monthly against ammonia measurements of active sampling devices. The sampling is performed by an extensive group of local volunteers, mostly rangers, which minimises the cost and enables the use of local knowledge (Lolkema et al. 2015).

Fig. 23.1 A ranger exchanging a passive ammonia sampler. (Source: Erik Noordijk)

Without the unpaid help of the rangers, a monitoring network like this would not be affordable. The network provides countrywide coverage, crucial input for policy and a community of committed rangers.

Lesson learned: Including the voluntary help of societal partners may be a cost-efficient way to build a monitoring network on a scale that simply would not be feasible without trusting measurement devices to non-experts.

NO$_2$ measurements by Friends of the Earth Netherlands

One of the first citizen science projects RIVM had a small role in was led by Friends of the Earth Netherlands. Since 2012, Friends of the Earth and local community groups have been measuring nitrogen dioxide (NO$_2$) concentrations with Palmes tubes, a rather simple but well-established method to obtain monthly averages, at about 100 locations in the Netherlands. Friends of the Earth wanted to get an impression of local air quality and subsequently ask local authorities to take responsibility for good air quality. RIVM contributed in two ways. First, the Palmes tubes measurements were calibrated by mounting them next to official measurement stations of the RIVM, the Municipal Health Services (GGD) of Amsterdam and the Environmental Protection Agency of Local and Regional Authorities in the Rijnmond region (DCMR). Expertise from RIVM was used in the quality control of the measurements and subsequent calibration. Second, RIVM provided standardised procedures to calculate the air quality at the different measurement locations. These model results were compared to the measurements to independently assess the quality of the models used in the Netherlands.

The measurements show that NO$_2$ concentrations are still high in several locations in large cities, and sometimes exceed the legal limit for yearly averages. The NO$_2$ concentration measurements in the citizen science project were compared with the values calculated by RIVM using the official Dutch modelling system, and found to correspond well (Knol & Wesseling 2014).

Despite of the different roles of RIVM and Friends of the Earth, both parties benefit from this collaboration. Addressing the quality of the measurements together at the outset, sorting out differences in methodology and other potentially confusing issues, means that the final discussion is about the values measured rather than concerns about the *quality* of the measurements.

Lesson learned: Even if citizens, NGOs and EPAs have different goals, they all want reliable data, which is a good reason to work together.

Measuring air pollution with your iPhone – iSPEX

The iSPEX project (http://ispex-eu.org), initiated in 2012, played a decisive role in changing RIVM's views on the way that citizen science can contribute to environmental science. The project uses state-of-the-art technology and citizen science on an unprecedented scale for environmental monitoring, with more than 3,000 participants and over 10,000 contributed measurements. The iSPEX project propagated a relatively new type of citizen science, where a large group of participants turn their smartphones into measurement devices. Within this innovative type of citizen science, iSPEX distinguished itself by collecting and transmitting data to a central database. In the Netherlands, the data collection was organised in two large-scale, nationwide measurement campaigns (scaling up in 2015 to 11 major European cities). Alongside the scientific project partners Leiden University, Netherlands Research School for Astronomy (NOVA), Netherlands Institute for Space Research (SRON), the Royal Netherlands Meteorological Institute (KNMI) and RIVM, societal partners played an important role, especially the Lung Fund (a patient organisation for lung diseases), which supported publicity and the distribution of iSPEX add-ons.

The project measured the properties of particulate matter (aerosols) with iPhones supplemented with a small add-on for the camera. Together with a special iSPEX app that explains the measurement process and transmits the data, this add-on transforms the iPhone into an advanced measurement device. Only iPhones were used because of the uniform position of the camera and the calibration of the add-on. The participants' measurements were compared with, and complemented by, measurements from scientific equipment. One of the primary project goals was to find out how accurate the massive iSPEX measurements were, and what kind of additional information the measurements can provide. The experiment was successful and the scientific results have been published by Snik et al. (2014).

A study of the Dutch participants was conducted in close collaboration with the department of science communication of Leiden University (Land-Zandstra et al. 2016). The study aimed to examine (1) citizens' motives and conditions for (continued) participation in iSPEX; and (2) the impact of participation on citizens' understanding of both science and aerosols. An online survey showed that the project had attracted an older,

Fig. 23.2 The iSPEX add-on on the left; instructions for taking measurements on the right. (Source: iSPEX Team)

male, well-educated audience, typical for many citizen science projects. However, the project did attract people with limited previous experience with science and scientific research. There were two dominant reasons for participants joining the iSPEX project: (1) a desire to contribute to scientific research, the environment or health; and (2) an interest in science or more specifically in aerosols and their impact on health and the environment.

Respondents reported that their participation in the iSPEX project taught them how citizens can contribute to science and iSPEX was the first time many had participated in a citizen science project. Although there was agreement that they learned more about aerosols and their impact on health, understanding of the science behind the project was rather low. Respondents were primarily motivated by the prospect of contributing to a larger goal and liked that the measurements took a limited amount of time and could be done individually. Most importantly, the participants were motivated to contribute frequent measurements, including for longer projects. However, continuing the project would require significant investment in technology and operational costs. Therefore, additional funding would be necessary to establish a stable monitoring network for iSPEX observations.

Lessons learned: In principle it is feasible to have a large group of citizen participants performing measurements, and the results can be scientifically valuable and complement professional measurements (mostly in terms of spatio-temporal resolution). The limited amount of self-reported learning and understanding of the science imply that projects based on complex science need to find ways to ensure their participants understand what their measurements mean. Projects that use smartphones as measurement devices have the potential to attract a new audience to citizen science.

Waag Society Amsterdam Smart Citizens Lab

To experiment with more bottom-up citizen science approaches, the RIVM participated in the Amsterdam Smart Citizen Lab initiated by Waag Society in 2015 (Henriquez 2016). The idea behind the project was for citizens to develop tools and instruments that enabled them to register, measure and understand aspects of their direct living environment. The environmental focus was decided by participants themselves. Waag Society provided the facilities to build the tools in its Fab Lab, which included laser cutters and 3D printers. Other project partners included Wageningen University and the municipality of Amsterdam.

Over seven months citizens could participate in six workshops in which researchers from Waag Society and RIVM gave in-depth lectures concerning the large number of affordable DIY sensors and measuring kits available, and their differences compared to professional sensors. Participants were introduced to successful online citizen science platforms, technologies and additive manufacturing techniques that together make DIY sensing networks possible. Three teams were formed that focused on wind energy, air quality and noise pollution.

The outdoor-air-quality group included air-quality scientists from RIVM and Wageningen University, and the combination of both ordinary people and air-quality-sensor professionals helped the project progress (Jiang et al. 2016; Henriquez 2016). Initially, RIVM experts intended to be observers but soon became motivators and trusted sources of information. Their involvement increased confidence in the project and motivated participants; and their expertise helped participants assess ideas. The group succeeded in developing and testing new sensors and a sensing platform. The NO_2 sensor developed was significantly cheaper than those currently installed at official air quality–measuring stations. A flaw of the sensors was that they were over-sensitive to many factors and it was difficult to solve all hardware and software issues. An upgraded version of the same sensor was used in another citizen science project, Urban AirQ (see http://waag.org/en/project/urban-airq).

Lessons learned: The support of experts is often welcome and the chances of success increase if experts take citizen science seriously by providing support and information. Timing is crucial as participants need *enough* information at an *early* stage, when the plans can still be adapted and improved. Nevertheless, too much information limits their freedom: participants have different goals from EPAs and may want to measure with new technologies, or measure other pollutants for which there is not (yet) legislation.

Nijmegen Smart Emission project

RIVM is a partner on the Smart Emission project in the city of Nijmegen, initiated as a pilot project by Radboud University and the municipality. The project has its origin in Geographic Information Systems (GIS) and participatory mapping and planning, outside of RIVM's usual disciplines. The project aims to create and test the concept of a citizen sensor network, including a feedback loop of information from interested participants to sensors and back. It involved data analysis by students, information and communication technology companies providing the new sensor technology (hardware and software) and (geo-) professionals creating the necessary spatial data infrastructure.

As the pilot progressed, plans were continuously adapted to the needs of the project partners (with increasing enthusiasm from the municipality), technical possibilities (including the challenges of battery life and long-range data transmission) and wishes of participating citizens (through feedback processes). The first sensors were developed by SME companies 'Intemo' and 'CityGIS' and 34 were installed at the time of writing following advertisement for volunteers to accommodate a sensor (requiring a power-supply and Wi-Fi) in a local door-to-door magazine. The sensors not only register air quality, but also noise, light intensity, low-frequency vibrations, temperature, air pressure and relative humidity. Challenges remain, including need for the dataflow and data algorithms to work with the downtime of individual sensors, which may shut off temporarily when the interior gets too moist.

The project will compare local data from small and cheap sensors with the data from the RIVM air quality measurement network. A clear decision was also made for the project to be open and transparent. The research team shared the information portal and raw measurements with participants and other researchers as soon as they were available, and asked for direct feedback. In this way, the citizens were taken on board as co-working participants. For example, a digital forum for questions and answers was added to the portal, at the request of participants, so that discussions could take place between researchers and participating citizens in between meetings. In this way, the research team learned a lot about participants' needs and wishes, including the fact that there are large differences in their information needs: Graphs and pictures satisfy some, but others prefer the data behind the visuals. This shows the importance of transparency and flexibility in presenting information in various ways. Participants also learned about the process of gathering raw data, the construction of a data infrastructure, etc., and overall this feedback approach increased mutual trust (see also Hecker et al. 'Stories' in this volume on communicating with participants).

Lessons learned: Flexibility can be the key to success. It is possible to create a relatively low-cost network to monitor environmental parameters with citizen participants. Open data and transparency can create trust and a better understanding for both citizen participants and experts.

Ik heb last (I suffer now) app http://ikheblastapp.nl/

The Ik heb last (I suffer now) app is innovative because it directly uses the health complaints of individuals as indicators for environmental issues. The air quality in the Netherlands is continuously improving, but a significant number of people suffer from air pollution. Information about air pollution can help people to match their activities and medicine use with geographical locations. Estimates can be made for the Dutch population as a whole, but this may mean something different at the individual level.

The purpose of the 'I suffer now' app is to enable citizens to indicate that they have issues with their respiratory system. Reports can be matched with different air quality conditions and patterns derived using 'big data', which enables a forecast of sensitivity at the individual level. This forecast provides the user of the app the opportunity to plan activities and medicine intake. For RIVM and its partners, Friends of the Earth Netherlands, Lung Fund, Utrecht University and Hogeschool Rotterdam, the data gathered could improve the identification of causes and exacerbations of respiratory symptoms. Moreover, the project tests citizen participation and which incentives might drive this.

Lessons learned: Projects of this kind are innovative but complex, and participants require more feedback. However, they have the potential to directly measure effects on health, including on an individual level. The coming years will show how this approach may best be deployed.

Innovation for national EPAs

Recurring themes emerge from the various citizen science projects above that are relevant for an EPA and its societal partners:

1. Citizens are motivated to contribute to air quality measurements that complement existing measurement networks because of their high spatio-temporal resolution.
2. Data needs to be assimilated, for example, in models.
3. Low-cost sensors need developing.
4. Low-cost sensors require application calibration and validation.
5. Easily accessible expert information is needed.

The second point – about the assimilation of data in models – is perhaps the most challenging, and vital to ensure continuity in the measurements and motivation of citizen participants. Applying the data requires flexibility; this includes coping with data that do not meet the high-quality standards of official monitoring networks. Data science may help in dealing with cross-sensitivity or instability in measurements. Modellers may also be able to include data with a lower accuracy or of a different nature from official measurements (see Williams et al. in this volume). RIVM has concluded that citizen science measurements should be an integral part of the existing national monitoring network and employed in real-time modelling procedures. The current innovation of the monitoring network is intended to provide a stable basis for the testing, calibration and use of citizen science data.

A natural role for the EPA as a reference institute is to assess the quality of the data. In practically all the citizen science projects above, the quality of the data was a big issue. Although most relatively cheap sensors measure at least something, the relationship with official air quality measurements can be poor or even absent. The measurements by national EPAs can serve as a reference to aid the calibration of cheap sensors used in citizen science projects.

It is also important to recognise that citizens' need for information and data is diverse. In the Smart Emission project in Nijmegen, for example, some participants demanded more information and detailed insight into the underlying raw data, but others wanted less information. Although some wanted the data in a simplified form (e.g., good/bad rankings using colour codes or emoticons), making the underlying 'complicated' data available gave participants more confidence in the data and project. It is therefore crucial that data is open and available at a basic level, and that it is presented at different levels of complexity and in different forms (numbers, graphs, colour codes, etc.).

RIVM is currently developing an interactive knowledge and data portal for citizen science related to air quality monitoring (see also Brenton in this volume). This portal will be made in open collaboration with citizen participants and aims to connect with them by supplying knowledge and data in a way that is understandable to the interested public, and that has been adapted to their needs. The aim is to become an important source of information for air quality–related citizen science in the Netherlands. There is also a YouTube channel (Samen milieu meten) to experiment with short videos, for example to answer frequently asked questions, to introduce or explain different types of innovative measuring devices, or explain background information (e.g., What is particulate matter composed of? How can it be measured?). Tutorials will also be available showing how to build a sensor kit or download data). The focus is on air quality because of the interest of citizens, technological innovations and the authoritative role of RIVM in monitoring air quality. However, the generic knowledge and experience gained may be used for other environmental domains like noise, light, radiation, soil and water. Providing a knowledge and data portal will encourage and support long lasting data collection by citizen science projects. Consequently, citizen science will be an essential element of Dutch national monitoring networks.

Sensors for, among others, air quality and Internet of Things applications are developing at a rapid pace. Hence, RIVM expects that within five years, its network for measuring air quality could evolve from a network with a limited number of high-quality (reference) measurement stations to a hybrid system that uses a much larger number of sensors that are cheaper and of lower quality. Where possible, satellite data will also be integrated in this network and a limited number of high-quality measurement stations will still form the reference base of this system. Combining all these data of varying quality and levels of uncertainty with models is a cost-efficient way of monitoring. This results in a crowdsourcing system that provides local communities with trusted local environmental data and at the same time enriches the national system for air quality monitoring. In this evolution, RIVM has identified different phases to provide a roadmap for the innovation of environmental monitoring, as illustrated in figure 23.3. The different steps may overlap and timelines are not strict but help target innovation over five years.

Phase 1

The first phase, started in 2016 at RIVM, involves implementing efficiency actions (e.g., automating some steps in the validation process) and planning to decrease the number of measurements (and/or stations), which will create the capacity for innovation while still complying with international requirements (e.g., EU directives). This means difficult decisions about what measurements or stations may be discontinued. The RIVM innovation programme aims at balancing the effects of the foreseen reduction of the monitoring effort on monitoring quality by introducing new lower-cost sensor technology into the network. The goal is to keep or even increase the quality level of the monitoring programme in this way. Six official measurement stations belonging to RIVM and its monitoring partners will be equipped with a facility to test small sensors. These will be used by RIVM and its partner institutes to test sensors (such as the Air-SensEUR; see http://www.airsenseur.org), and the facilities are open to citizen science communities and sensor builders to test and calibrate their own sensors. The locations of the small sensor test stations represent a broad range of measurement situations. There is a rural site, an urban traffic station, two highway sites (one urban, one rural) and an industrial site. A test facility for small sensors will also be developed in a climate chamber.

Phase 2

In the second phase, around 2018 for RIVM, advanced yet relatively cheap (though still expensive for citizen science projects) sensors will be included in the national air quality monitoring system. The use of satellite data will also be included, where possible. Extensive tests will have to be performed to see how these sensors behave over longer periods such as a year. The practical effect of trading a limited number of reference measurements for (many) cheaper sensors on air quality monitoring will be determined.

Phase 3

In the third phase, envisioned to be fully implemented around 2020 at RIVM, a crowdsourcing platform will be developed to enable citizen science projects with low-cost sensors to participate in national monitoring.

The second and third phase will ask for creative solutions that enable the simultaneous use of data with different quality levels. Planning for the crowdsourcing platform will start with pilot projects with different degrees of citizen involvement. A limited number of reference measurement methods will act as the backbone of the national monitoring programme, supplemented with a flexible layer of alternative, low-cost sensor devices. Continuous validation cycles of these low-cost sensors with reference data will result in the gradual improvement of available sensor technologies.

Beyond phase 3

Health monitoring technologies are evolving at the same time as environmental monitoring, with technologies related to mobile health care rapidly maturing (e.g., GIS tracking to support individual exposure modelling, personal health measurements like heart rate or spatio-temporal tracking of medication use). They provide an excellent opportunity for environmental health research to become a key innovation partner in health transition technologies. Integrating environmental and health monitoring offers the potential for important follow-up innovation.

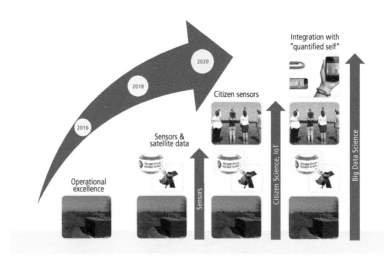

Roadmap innovation of environmental monitoring

Fig. 23.3 Innovating a traditional measurement network towards a hybrid crowdsourcing platform. (Source: Erik Tielemans)

Beyond phase 3, RIVM anticipates combining its static air quality monitoring network with personal exposure and health monitoring.

Conclusion

This chapter has explored the potential benefits and challenges of an EPA incorporating citizen science into its national monitoring. It has shown the important possibilities of this arena in improving the scale and scope at which data is available by supplementing limited and expensive monitoring equipment with widespread, low-cost sensors led by citizen monitoring. Participants in this field are highly motivated due to the societal importance of environmental issues, but EPAs and citizen science project leaders need to address issues of data quality and sensor calibration, and to provide appropriate feedback to reward and motivate participation. The trial projects RIVM has been involved in clearly point to the potential advantage of these methods for EPAs, such that it has now adopted a framework to further incorporate citizen science in its monitoring processes. This roadmap and lessons learned from the case studies may provide ways forward for other EPAs and official government agencies seeking to improve their traditional practices by engaging with the potential of citizen science.

Appendix A

Table 23.1   Claims about citizen science participant benefits

Citizen science participant benefit	Citation
Enhanced science knowledge and literacy (e.g., knowledge of science content, science applications, risks and benefits of science, and familiarity with scientific technology)	Braschler et al. (2010); *Brewer (2002); *Danielsen, Burgess & Balmford (2005); Devictor, Whittaker & Beltrame (2010); *Evans et al. (2005); *Fernandez-Gimenez, Ballard & Sturtevant (2008); *Jordan et al. (2011); Krasny & Bonney (2005); Sullivan et al. (2009)
Enhanced understanding of the scientific process and method	Bonney (2004); Bonney and Dhondt (1997); Braschler et al. (2010); Devictor, Whittaker & Beltrame (2010); Sullivan et al. (2009); *Trumbull, Bonney & Grudens-Schuck (2005)

(continued)

Table 23.1 (continued)

Citizen science participant benefit	Citation
Improved access to science information (e.g., one-on-one interaction with scientists, access to real-time information about local scientific variables)	*Fernandez-Gimenez, Ballard & Sturtevant (2008); Sullivan et al. (2009)
Increases in scientific thinking (e.g., ability to formulate a problem based on observation, develop hypotheses, design a study and interpret findings)	*Kountoupes & Oberhauser (2008); *Trumbull et al. (2000)
Improved ability to interpret scientific information (e.g., critical thinking skills, understanding basic analytic measurements)	Bonney (2007); Braschler et al. (2010)
Strengthened connections between people, nature, and place (e.g., place attachment and concern, establishment of community monitoring networks or advocacy groups)	*Devictor, Whittaker & Beltrame (2010); *Evans et al. (2005); *Fernandez-Gimenez, Ballard & Sturtevant (2008); *Overdevest, Orr & Stepenuck (2004)
Science demystified (e.g., reducing the 'intimidation factor' of science, correcting perceptions of science as too complex or complicated, enhancing comfort and appreciation for science)	Devictor, Whittaker & Beltrame (2010); *Kountoupes & Oberhauser (2008)
Empowering participants and increasing self-efficacy (e.g., belief in one's ability to tackle scientific problems and questions, reach valid conclusions and devise appropriate solutions)	*Danielsen, Burgess & Balmford (2005); Lawrence (2006); Wilderman, Barron & Imgrund (2004)
Increases in community-building, social capital, social learning and trust (e.g., science as a tool to enhance networks, strengthen mutual learning and increase social capital among diverse groups)	Bell et al. (2009); *Danielsen, Burgess & Balmford (2005); *Fernandez-Gimenez, Ballard & Sturtevant (2008); *Overdevest, Orr & Stepenuck (2004); *Roth & Lee (2002); Wilderman, Barron & Imgrund (2004)
Changes in attitudes, norms and values (e.g., about the environment, about science, about institutions)	*Danielsen, Burgess & Balmford (2005); *Ellis & Waterton (2004); *Fernandez-Gimenez, Ballard & Sturtevant (2008); *Jordan et al. (2011); *Melchior & Bailis (2003)

Source: Haywood 2014
* Studies that have empirically tested outcome hypotheses and reported results are noted with an asterisk.

24
Citizen science to monitor light pollution – a useful tool for studying human impacts on the environment

Sibylle Schroer[1], Christopher C.M. Kyba[1,2], Roy van Grunsven[1], Irene Celino[3], Oscar Corcho[4], Franz Hölker[1]

[1] Leibniz Institute of Freshwater Ecology and Inland Fisheries (IGB), Germany
[2] German Research Centre for Geoscience, Potsdam, Germany
[3] CEFRIEL, Milano, Italy
[4] Universidad Politécnica de Madrid, Spain

corresponding author email: schroer@igb-berlin.de

In: Hecker, S., Haklay, M., Bowser, A., Makuch, Z., Vogel, J. & Bonn, A. 2018. *Citizen Science: Innovation in Open Science, Society and Policy*. UCL Press, London. https://doi.org/10.14324/111.9781787352339

Highlights

- The alteration of light levels at night is a recent environmental change, which has become an increasing threat to nocturnal landscapes.
- Guidelines for illumination focus primarily on aesthetics, safety, security and energy efficiency. A policy shift towards considering the impact of light on ecosystems and health requires a sound transdisciplinary and supraregional approach.
- Citizen science projects could analyse changes in nighttime brightness worldwide, offer participation in various other scientific areas, and increase public awareness.
- Light pollution can be a unifying entry point for other environmental problems, connecting projects about the impact of human activities.

Introduction

In previous decades, the use of artificial light at night (ALAN) and the related brightening of the nightscape increased worldwide by more than 2 per cent per year (Kyba et al. 2017; Hölker et al. 'The Dark Side', 2010).

But compared to the global increase of temperature, the impact of alteration in ALAN has not yet been well studied. The research about ALAN is a showpiece example of how citizen science can assist collecting global data. To study the alteration of ALAN, measurements are needed from remote sensing, from the ground and about the impact of ALAN on the environment. Artificial light at night indicates hotspots of human activity and thus can be a unifying entry point for other environmental problems, potentially connecting citizen science for environmental monitoring (see also Owen & Parker; Peltola & Arpin; Danielsen et al.; Harlin et al., all in this volume).

As a result of increasing nighttime brightness, the measured range of night sky radiance is now often hundreds of times larger than it was before the existence of artificial light (Kyba et al., 'High-Resolution Imagery', 2015), depriving one-third of Earth's population of the possibility to enjoy a view of the Milky Way (Falchi et al. 2016). Light is the most important signal for circadian and seasonal rhythms, but ALAN can interfere with this signal, disturbing ecosystems (Hölker et al. 'The Dark Side', 2010; Schroer & Hölker 2016) and having adverse consequences on sleep performance and health (Reiter et al. 2011; Bonmati-Carrion et al. 2014). Furthermore, the transition to 'white' LED (light emitting diode) light sources increases the fraction of ALAN with short wavelength (blue) light. This is of concern for several reasons: Short wavelength light is more likely to scatter on clear dry nights (Aubé, Roby & Kocifaj 2013) and this part of the spectrum also has the greatest impact on the circadian system of higher vertebrates (Bailes & Lucas 2013; Brainard et al. 2015).

In the context of light pollution, citizen science is an indispensable tool for both data collection and knowledge dissemination, especially for collecting data on larger scales, such as at landscape or community levels (Kyba 2018; Kyba et al. 2013). Citizen science can mobilise people with various interests on a global level to measure the impact of ALAN (Schroer, Corcho & Hölker 2016). This chapter describes existing citizen science contributions to the highly interdisciplinary research field of light pollution. It discusses how raising awareness through citizen science may initiate changes in the use of illumination, and how citizens could become empowered to create a more social and sustainable environment. In contrast to other pollutants, reductions in light pollution could in principle be relatively easily achieved by increasing consumer awareness about the negative consequences and implementing guidelines for the use of ALAN: to use low light intensities, to shield lamps and to use lights with a low blue light content (Schroer & Hölker 2014).

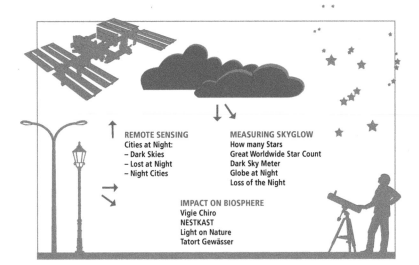

Fig. 24.1 Citizen science projects with focus on artificial light at night and changes of nightscapes. (Source: A. Rothmund)

Finally, this chapter shows how the ALAN research topic connects disciplines and interests, and how this issue can be used to network, for example, with social platforms focused on other pollutants such as air, water or noise pollution.

Citizen science projects on ALAN

Several citizen science projects focus on quantifying the effects of outdoor ALAN, which are summarised here to present both the broad ranges of used technologies (see also Mazumdar et al. this volume) and the diversity of disciplines engaged in ALAN research (figure 24.1).

Measuring ALAN with remote sensing tools

Members of the Group of Extragalactic Astrophysics and Astronomical Instrumentation from Universidad Complutense de Madrid use high-resolution images taken by astronauts from the International Space Station (ISS) to measure changes in ALAN. A lot of additional work is necessary to use the data scientifically. For this purpose, the project Cities at Night (http://citiesatnight.org/) involves citizens in creating a global map of satellite night views and artificial radiation (Sánchez de Miguel et al. 2014). Cities at Night uses three apps – Dark Skies, Lost at Night and Night Cities – to engage citizens to (a) identify images which qualify for mapping; (b) identify cities and communities; and (c) georeference the images.

The data obtained from remote sensing observes only radiation directed upward into the sky, but not the brightness that is experienced on the ground. Horizontal radiation and the degree of skyglow during overcast conditions are not recorded. Furthermore, satellite instruments show insufficient sensitivity to short wavelength light, such as for example to light emission by LEDs (Kyba et al. 'Worldwide Variations' 2015). Measurements from the ground are therefore needed.

Measuring ALAN and skyglow from the ground

Several citizen science projects assist in collecting data about the brightness of nightscapes from the ground:

- **How Many Stars** (http://hms.sternhell.at) encourages citizen scientists to classify how many stars they can see at their location in terms of the naked eye limiting magnitude (NELM), using a series of star charts on the Little Dipper and Orion constellations.
- **Great Worldwide Star Count** (http://www.windows2universe.org /citizen_science/starcount/) is part of Windows to the Universe and is designed to encourage learning in astronomy and identify changes in nighttime brightness, using star charts in a manner similar to How Many Stars.
- **Dark Sky Meter** (http://www.darkskymeter.com/) is an iPhone app that enables citizen scientists to measure night sky brightness by using the phone's camera as a photometer. The app also offers statistics about sunset, twilight and moon phases and provides weather forecasts based on satellite data.
- **Globe at Night** (http://www.globeatnight.org/) is a programme of the National Optical Astronomy Observatory, the national centre for ground-based nighttime astronomy in the United States. Citizen scientists are engaged in a similar way to How Many Stars but Globe at Night includes additional constellations for the Southern Hemisphere and invites participants to submit observations using a commercial sky quality metre (SQM).
- **Loss of the Night app** (http://lossofthenight.blogspot.de/) quantifies NELM, but instead of using star charts, participants are asked

to make decisions on whether individual stars are visible. Participants are interactively asked to make decisions on at least eight stars and the app checks participant data for self-consistency. The app also allows participants to submit SQM data.

Citizen science data measuring ALAN from the ground (e.g., from Globe at Night and the Loss of the Night app) have already proven to deliver fundamental contributions to academic publications and global maps about changes in ALAN (Falchi et al. 2016; Kyba et al., 'High-Resolution Imagery', 2015). This data can be valuable for projects about environmental monitoring. For example, light pollution data was shown to be a more useful predictor of bat activity than the proportion of impervious surface, a commonly used indicator for urbanisation (Azam et al. 2016). Comparing citizen science data of bat-monitoring with satellite data of light pollution showed that ALAN has a strong negative effect on bat activity.

Impacts on the biosphere

Wildlife responses to increasing ALAN and the loss of natural darkness in many nightscapes has far-reaching consequences for the environment, for ecosystems and their services, upon which human well-being relies (Hölker et al. 'Light Pollution', 2010; Gaston, Duffy & Bennie 2015). Various citizen science projects collect data on environmental conditions and put them into content of light pollution mapping. These projects do not always primarily focus on the effect of light pollution, but the data can still be used to analyse the impact of ALAN.

- **Vigie Chiro** (http://vigienature.mnhn.fr/page/vigie-chiro) is a French bat-monitoring programme and one example of how pre-existing citizen science data can be used to assess the impact of light pollution. In this programme, volunteer surveyors monitor bat activity along predefined transects. These results are compared with landscape characteristics such as the proportion impervious surface, intensive agriculture and radiance data from a polar-orbiting satellite.
- **NESTKAST** (https://www.vogelbescherming.nl/in-mijn-tuin/nest kasten) uses data from the project NEtwerk voor STudies aan nest-KASTbroeders, in which volunteers monitor songbirds breeding in nest boxes to analyse the impact of ALAN on the timing of breeding (de Jong 2016).

- **Light on Nature** (http://www.lichtopnatuur.org/) uses several experimentally illuminated semi-natural areas to monitor the responses of flora and fauna and analyse the impact of ALAN on the local species' community compositions (Spoelstra et al. 2015). Citizen scientists undertake the monitoring for some groups of animals, for example, moth populations are monitored by volunteers, co-ordinated by the Dutch Butterfly Conservation (De Vlinderstichting) and birds are caught and ringed by volunteers to measure bird populations, co-ordinated by the Dutch Centre for Avian Migration and Demography (Vogeltrekstation).
- **Crime scene freshwater** [*Tatort Gewässer*] (http://tatortgewässer .de/) is a German project developed to gain knowledge about the role of inland waters in the carbon-cycle and the effects of ALAN. The project is based on findings by Hölker et al. (2015), who observed changes in microbial community composition in freshwater sediments exposed to ALAN. Schroer et al. (2016) provided a questionnaire to the volunteers asking them to describe the local artificial light conditions. In just two weeks, more than 700 citizen scientists contributed to the project, providing an excellent data source for sediment and biodiversity pattern analysis in aquatic systems, far beyond the collecting capacity of scientist teams.

Many other citizen science projects about the occurrence of flora and fauna, such as monitoring wildlife, insects and plant species could in the future be examined in the context of light pollution (for wider examples see also Owen & Parker; Peltola & Arpin; Danielsen et al.; Harlin et al., all in this volume).

Collective Awareness Platforms for Sustainability and Social Innovation

The European Commission's initiative CAPSSI (Collective Awareness Platforms for Sustainability and Social Innovation) supports online platforms creating awareness of sustainability problems and offering digital networking between environmental and social platforms (funding framework of Horizon 2020). A platform like CAPSSI has the potential to link the multifaceted citizen science projects about light pollution and further to link existing projects monitoring environmental changes. Furthermore, CAPSSI has the prospective to recommend the principles as standard for future citizen science project development.

The CAPSSI project STARS4ALL provides a platform for citizen initiatives and activities to promote dark skies in Europe (http://www.stars4all.eu/). Light pollution initiatives are being developed to involve citizens, especially in cross-disciplinary areas such as energy saving, biodiversity and human health. The project expects to influence policy-making through the participation of citizens, by proposing specific measures for municipalities to protect dark skies in Europe.

Limits and opportunities for citizen science research on ALAN

This section presents efforts to increase the impact of citizen science involvement in scientific data collection about ALAN. It shows how data consistency and thus its impact can be improved and explains the benefit of enlarging networks and communities (see also Williams et al. in this volume).

Data reliability

Although citizen involvement in research about ALAN seems indispensable to allow data collection on a global scale, there are concerns about the reliability of the data for science. Citizens could in principle falsify data, or inappropriate handling of monitoring protocols could introduce errors that increase the uncertainty of the observations. Especially when using citizen science data for scientific studies, the measurements need to be approved and correlated with existing peer-reviewed scientific data. Kyba et al. (2013) and Schroer et al. (2016) have found a positive correlation of citizen science data with satellite measurements, demonstrating the usefulness of the data.

Another good example of how citizen science data can be evaluated and improved is the My Sky at Night project (http://www.myskyatnight.com). The web application was developed to allow participants of the Loss of the Night project to examine the observation data in detail, as well as measuring trends over time (including data collected by Globe at Night and the Dark Sky Meter app). The app visualises the self-consistency of the measurements and aims to motivate citizen scientists to improve their measurement technique (figure 24.2).

Georeferencing and the provision of automated apps may increase confidence in the data in the future. However, a lot of citizen science data has already been used without credit in academic publications, but its

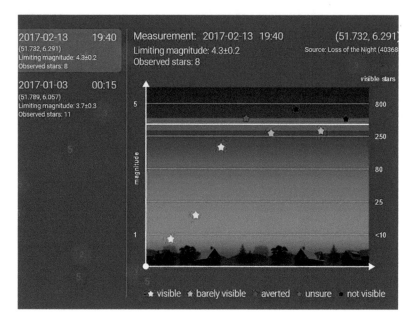

Fig. 24.2 Visualisation of citizen science observation data gives feedback to the user. (Source: www.myskyatnight.com)

acknowledgement may be the most important way to promote confidence in data from volunteers (Cooper, Shirk & Zuckerberg 2014).

Motivating citizen scientists

For the measurement of light pollution, it is critical to motivate people worldwide to measure sky brightness on a regular basis, especially in peripheral urban areas, where the most changes may occur. It is a challenge to engage new participants and raise awareness in young people (see also Harlin et al. in this volume). Motivation is needed for researchers and activists to step out of the circle of environmentalists and astronomers and to bring the relevance of light pollution to authorities, lighting planners, communities, the public and other stakeholders.

The STARS4ALL platform triggers the interest of citizens using a simple but powerful tool: broadcasting astronomy-related events such as solar and lunar eclipses, aurora borealis, meteor showers and so on, with the aim of involving citizens in experiencing a natural nighttime environment (http://www.sky-live.tv/). To encourage local engagement, the network uses participation portals in cities like Madrid (https://decide .madrid.es) and applications such as FarolApp4All (http://farolapp

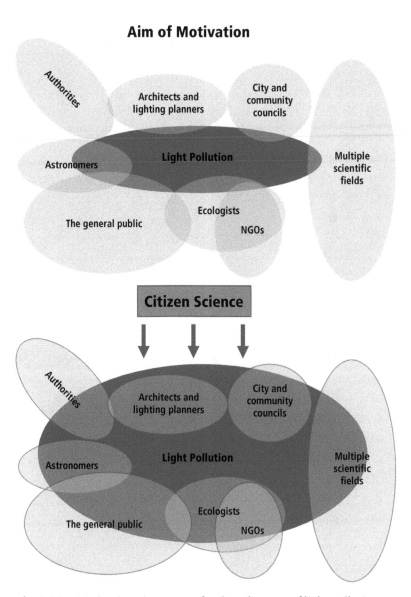

Fig. 24.3 Motivation aim to transfer the relevance of light pollution to a broader field of stakeholders

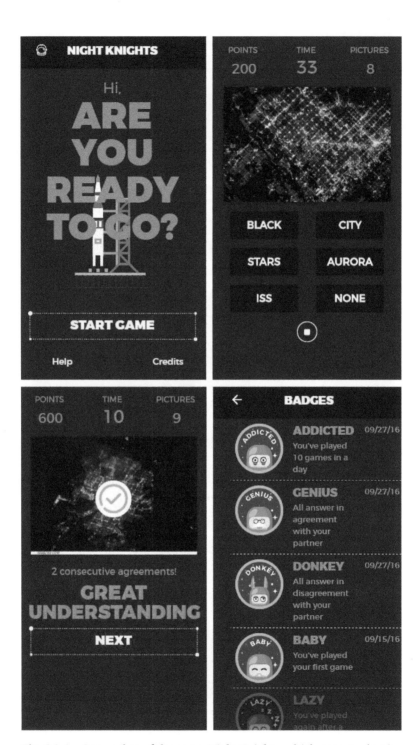

Fig. 24.4 Screenshot of the game *Night Knights*, which uses mechanics of output agreement, double player, contribution weighted by user reputation (measured against ground truth)

.linkeddata.es/) to allow citizens to detect and report cases where public lighting may not comply with guidelines for the sustainable use of ALAN.

Entertainment and gamification have increasingly been used to enlarge the number of participants in citizen science projects (Deterding 2012). For example, the Urbanopoly game involved players in validating and enriching OpenStreetMap data (Celino 2013). The STARS4ALL project has been experimenting with both the gamification of crowdsourcing initiatives, such as rewarding student crowdworkers with the prize of an international expedition to experience celestial phenomena, and pure entertainment apps to classify, for example, remote sensing images, implementing a 'game with a purpose' (von Ahn 2006). The game *Night Knights* (http://www.nightknights.eu) encourages volunteers to invest more time in categorising remote sensing photos. The game design follows recommendations for motivating sustained participation: The scoring mechanism provides personal milestone targets and feedback to maintain high-quality responses (Eveleigh et al. 2013).

Furthermore, interesting news and cartoons are disseminated by social media channels to revive attention within the light pollution community. The aim is to encourage the community to share personal experiences to foster a social aspect, and thus continued participation (Nov, Naaman & Ye 2009).

Establishing networks

The sheer range of environmental pollutants may discourage participants from becoming involved in monitoring a single environmental stressor. Networks could benefit each other, allowing citizen science data to be used for multiple purposes. Volunteers may be increasingly motivated by contributing to broader project about more than one environmental issue (Rotman et al. 2012).

Most pollutants are associated and mutually dependent, so sensors collecting data on pollutants can be combined. For example, air pollution sensors could be fused with photometers (for more on sensors see Volten et al. in this volume). Remote sensing data of ALAN may visualise human activity and locate high-priority areas for nature protection measures (Aubrecht et al. 2010). It is a matter of communication between project developers to make multiple sensor systems available and expand their networks. Platforms such as Pinterest can be useful tools to connect different environmental projects, share interesting content about various environmental actions and initiatives and distribute knowledge (Hansen, Nowlan & Winter 2012).

Integrating citizen science in policy

Guidelines for illumination currently focus primarily on aesthetics, safety, security and energy efficiency (Kyba, Hänel & Hölker 2014). This section discusses if citizen science has the transformative potential to reduce the negative impacts of light pollution through policy development (see also Nascimento et al.; Shirk & Bonney, both in this volume).

Light emission is mainly regulated on a regional or national level (Kyba, Hänel & Hölker 2014) and a common European policy, for example, for the reduction of light pollution is missing. The European Ecodesign Directive (Directive 2009/125/EC) aims to improve the energy efficiency of lighting products, but it does not address the adverse effects of ALAN on biodiversity or nightscapes. European Union regulations for infrastructure and outdoor activities recommend minimum lighting requirements but do not provide limits to brightness levels. For example, the European standard 'Road lighting' (EN 13201: 2015) is a non-binding recommendation for member states to implement minimum lighting requirements. Sound scientific justification for these minimum illumination levels, however, is missing. Many communities do not meet the recommended requirements on brightness and uniformity, and yet still appear to offer safe conditions. A recent study in the UK found no evidence of harmful effects on traffic casualties or crime when street lighting was reduced or switched off late at night in rural contexts (Steinbach et al. 2015). If more authorities felt compelled to implement EN 13201, this would result in higher energy consumption and a wider loss of natural nightscapes (www.cost-lonne.eu). The same applies to standards elsewhere, for example the ANSI/IES RP-8 in the United States.

A rising number of municipalities use planning instruments like lighting strategies or master plans to address the complexity of modern lighting technologies and the different dimensions for sustainability. These concepts require consultation with broadly trained lighting experts, experienced in the social, cultural and economic as well as the ecological impact of light. Currently, most municipalities avoid this investment and are often influenced by consultation with unbalanced expertise or industrial interests (Köhler 2015).

A policy shift will require a sound transdisciplinary approach to understanding the significance of the night and its loss for humans and the natural systems (Hölker et al., 'The Dark Side', 2010). Citizen science data from multiple disciplines may be able to support the process of scientifically determining requirements for sustainable lighting concepts.

With the help of citizen science comparison of lighting levels in various cities, exceedingly high light levels – and thus sites where lighting needs regulation – could be detected. Furthermore, engagement could lead to more participative democracy in empowering citizens to demand measures to improve the quality and level of public lighting. For example, the STARS4ALL project will empower citizens by aiming at the generation and presentation of a European Citizen Initiative to the European Commission in order to reconsider legislation for lighting in Europe (see Shirk & Bonney and Nascimento et al., both in this volume).

Conclusion

As nighttime brightness increases globally, well-distributed data collection is required to analyse its impact and determine the thresholds between beneficial to negative impacts of lighting. This data cannot be provided by single institutions or small groups of scientists working in isolation. Collective awareness of, and readiness for, volunteering is urgently required because changes in light at night is an experiment with unpredictable outcomes (Hölker et al., 'The Dark Side', 2010). Citizen science can help to understand the complexity of investment in public lighting technology and may increase willingness to invest in sustainable lighting. Collective awareness platforms can be useful for motivating and empowering citizens and offering collaborative solutions to change consumption trends. Social platforms can connect and build even larger communities. Light pollution can be a unifying entry point for other environmental problems and has the potential to create a global citizen science network increasing knowledge and awareness about the impact of human activities.

Acknowledgements

We acknowledge the support of the STARS4ALL CAPSSI project, which received funding from the European Union's Horizon 2020 research and innovation programme under grant agreement No. 688135. The European Cooperation in Science and Technology (COST) supported networking and allowed the initiation of several citizen science projects. ALAN research topic networking was supported by the COST Action ES1204 LoNNe (Loss of the Night Network).

Part V
Innovation in science communication and education

25
Science for everybody?
Bridging the socio-economic gap
in urban biodiversity monitoring

Taru Peltola[1] and Isabelle Arpin[2]

[1] *Finnish Environment Institute, Joensuu, Finland*
[2] *Université Grenoble Alpes – Irstea, Saint-Martin-d'Hères, France*

corresponding author email: taru.peltola@ymparisto.fi

In: Hecker, S., Haklay, M., Bowser, A., Makuch, Z., Vogel, J. & Bonn, A. 2018. *Citizen Science: Innovation in Open Science, Society and Policy.* UCL Press, London. https://doi.org/10.14324/111.9781787352339

Highlights

- Citizen science has the potential to bring societal benefits, but inclusivity is not an automatic outcome.
- The degree of inclusivity varies depending on the techniques used to involve citizens.
- Affective techniques can involve less experienced and less privileged participants.
- Addressing participants as individuals, with different learning abilities and skills, and the collective dynamics of learning are key to increased inclusivity.
- Successful techniques broaden the role of participants, address their concerns and support ownership of the learning process.

Introduction

In addition to scientific outcomes, citizen science often aims to achieve broader societal relevance and benefits, such as science education, empowerment or enhanced environmental citizenship (Edwards et al. in this volume). Citizen science is commonly presented as a way of opening science to everybody. The ECSA Ten Principles of Citizen Science also emphasise inclusiveness and societal benefits. However, the majority of participants in citizen science are well educated (Haklay in this volume). Many citizen science projects involve, for example, skilled amateur naturalists. Participation may also be biased towards the more affluent or powerful, leaving aside those whose lives could benefit most from the activities (Buytaert et al. 2014). Finding ways of engaging less educated and less privileged participants with less specialist backgrounds is thus an important goal if citizen science genuinely wants to move towards involving everybody.

This chapter explores the capacity of citizen science to foster responsive and inclusive science (see also Smallman in this volume) by drawing on a case study in the city of Grenoble, France. The Propage programme (Fontaine & Renard 2010) is a citizen science project about butterflies. It was implemented in Grenoble to make urban biodiversity more visible and meaningful for those who manage public urban green spaces. An in-depth investigation of motivations for, and modalities of, involvement enabled us to identify the conditions which facilitated the implementation of Propage (Arpin, Mounet & Geoffroy 2015). In 2014, we carried out 20 interviews with volunteer gardeners and their trainers, and observed their training sessions.

Insights from this study demonstrate that citizen science has the potential to provide a forum for social learning and the development of collective capacities among less privileged participants, provided specific conditions are met. Understanding these conditions requires the identification and evaluation of broader transformative outcomes of citizen science, which is often difficult. One reason for this is that the social and cultural benefits may be diverse, discrete and delayed. They are not limited to individual learning outcomes, but include social, cultural and institutional transformations (see Kieslinger et al. in this volume). These transformations are often much harder to document than instrumental outcomes, such as new knowledge and skills (Bela et al. 2016). This chapter starts by briefly discussing what kind of outcomes are relevant when evaluating the potential of citizen science to target groups with less scientific training or fewer skills. We then move to lessons from the implementation of the Propage programme and demonstrate the importance of specific affective techniques in involving participants.

Social learning and transformative outcomes in citizen science

In its simplest form, participants in citizen science are considered as the 'crowd' or mere 'data-drones' (Ellis & Waterton 2004) who have the time, equipment and skills to provide good-quality data (but see especially Ballard, Phillips & Robinson; Haklay; Mahr et al., all in this volume). Other outcomes, such as learning or empowerment, are not always goals for data-driven projects (see Keislinger et al. in this volume). Instead, learning-oriented projects may aim to improve participants' knowledge or enhance their scientific literacy and skills (e.g., Makuch & Aczel; Harlin et al., both in this volume). A French community biodiversity project, for example, is tellingly named 'ABC' (*Atlas de la Biodiversité Communale*). Socially and politically oriented citizen science projects, in turn, may focus on transforming participants' personal and collective identities and capacities. They aim to change participants' lives and careers, and even promote the knowledge of citizens, seldom taken into account in decision-making.

This multiplicity of participatory imaginations implies that learning may have a different role in different types of citizen science projects. Learning outcomes can also be diverse. In addition to the acquisition of new knowledge and skills, learning can be understood in terms of shared perspectives, clarification of arguments, enhanced dialogue, development of social capital (e.g., trust and partnerships) or adaptive capacities (Buytaert et al. 2014). According to Bull, Petts & Evans (2008) the notion of social learning also broadens understanding of learning processes as a whole. Social learning not only means that individuals adopt new knowledge or skills through social interaction, but also that learners become members of a community of practice who learn to collaborate, reflect on what they are doing and make collective judgements. Outcomes of social learning include understanding motivations for knowledge acquisition and moral development.

From the social learning perspective, the societal relevance of citizen science is linked to its ability to support problem-solving and the generation of actionable knowledge (Franzoni & Sauermann 2014). Collaborative research should generate new knowledge that matters not only scientifically but to all participants. Ideally, it opens up new roles and identities for the participants and even triggers new concerns or questions (Hinchliffe, Levidow & Oreszczyn 2014). Learning to approach problematic situations and the development of collective problem-solving capacities are key elements in evaluating the societal relevance of citizen science. From the social learning perspective, citizen science not only involves a generic crowd, but also deals with the public – citizens who have concerns of their own or adopt an active role concerning their living environments (see, e.g., Marres 2007).

The following section highlights how capacity for social learning can be developed among a less-educated group of citizen scientists (see also

Richter et al. in this volume). In particular, we demonstrate how collective reflection, new roles and identities have been built, and participants' concerns and questions met, in a project that seeks to make biodiversity visible and meaningful for green space workers. We focus on techniques of affecting the participants in ways that enable them to develop individual and collective capacities.

Propage: Biodiversity for green space workers

The city of Grenoble employs some 140 green space workers, predominantly male, of various age groups and with various educational levels. Some qualification is required to work at the city's green space department. Most workers hold a professional certificate (mainly a CAP, *certificat d'aptitude professionnelle*, or BEP, *brevet d'études professionnelles*, less often a *baccalauréat professionnel*), and only few have a higher degree (BTS, *brevet de technicien supérieur*). Eleven informants in our research had a BEP, whereas one had a *baccalauréat professionnel*. They were significantly younger (from 30 to 45 years old) than the green space workers' average.

All the city's green space workers are familiar with cultivated flowers but, until recently, knew little about insects except those well known to be useful (e.g., ladybirds) or harmful (e.g., aphids) to cultivated flowers and plants:

> I remember having bought a booklet during my initial training, which presented the main useful insects, such as ladybirds, hoverflies, green lacewings. So there was already some vague interest in integrated pest management at that time. But we spent 95% of the time learning how to use chemicals and which product to use against harmful insects or fungus diseases rather than when to release ladybirds or green lacewings   (green space worker, 45 years old).

Neither biodiversity in general nor butterflies in particular were addressed in the workers' initial training. Their approach to butterflies was restricted to some leaf-eating caterpillars, which they had learned to fight. In 2014, 12 workers agreed to participate in the Propage programme, which aims to collect data about common butterfly species. These are relatively easy to detect and identify, so participation in the project does not require high naturalist skills.

Launched in 2010, Propage is one of the three butterfly monitoring programmes developed by the French National Museum of Natural History (MNHN). Propage targets the staff of the park and garden departments of cities and transport institutions while the two other projects are aimed at the general public or skilled naturalists. Propage was designed to respond to concerns about the impact of city management practices on biodiversity. The idea was to create a cost-effective and easy way to implement a protocol initially developed by an MNHN PhD student studying butterflies in urban parks and gardens. Propage resulted from close collaboration between the MNHN and an environmental non-governmental organisation (NGO), Noé Conservation. The MNHN is responsible for the data and website management, while Noé trains participants, communicates with them and disseminates written instructions. Before Propage was implemented in Grenoble, the MNHN and Noé made no particular effort to recruit participants as they were encountering serious technical problems with the programme website. At the time, Propage had two main contributors, the city of Nantes and the *département* of Seine-Saint-Denis (Paris), and a set of minor contributors. In Grenoble, the vice-head of the service responsible for the city's park management, David, knew about Propage and informed the MNHN that the city was willing to participate.

Addressing participants' concerns

Like many other cities (see Ernwein 2015 about Geneva), Grenoble had decided some years previously to move from lawn-oriented and pesticide-based management of its parks and gardens to biodiversity-oriented and insect-based management (see Tollis 2012). This shift changed the outlook of urban parks and triggered some criticism among the inhabitants who felt 'abandoned', especially in disadvantaged areas (see also Menozzi 2007). The green space workers, in turn, found that they had to work harder, for instance, to remove weeds by hand, while the result was less satisfactory and socially contested. The workers also suspected that this shift had hidden economic grounds beyond the official ecological reasons. At the time of our survey, the department was indeed being reorganised with a substantial decrease in the number of teams (from 19 to 13). David wanted to demonstrate to his staff that the new management practices did have positive effects on biodiversity. Therefore, contributing to citizen science was not the main objective of encouraging the green space workers to become involved in Propage or the main motivation for workers to participate. Instead, participation in Propage was a means to address David's concerns – showing his staff that the new management practices had positive effects on biodiversity – and the concerns of the workers – how to answer residents' questions and responses.

Affective techniques of educating attention

In the post-pesticide era, the green space workers saw weeds thriving in the city but not necessarily all the life, and in particular, all the insects that were thriving, too. Seeing is neither obvious nor spontaneous (see also Peltola & Tuomisaari 2016) but requires 'education of attention' (Ingold 2001), resting on specific techniques aiming to conduct the conducts of individuals. In Grenoble, several techniques were simultaneously adopted to teach the green space workers to become sensitive to insects in general and butterflies in particular. They were given the leaflets prepared by Noé Conservation explaining the Propage protocol and showing the butterfly species they were expected to recognise (figure 25.1). In addition, they were given nets to catch butterflies. 'Referent' workers for biodiversity were designated among the participants and training sessions were organised.

Training was carried out by staff of local environmental NGOs specialised in entomology and by the local museum of natural history. Before Propage, David had invited an entomologist, Édith, to come to Grenoble once or twice a year to teach the green space workers to detect and identify useful and harmful insects. This proved to be a major factor in the success of Propage. Édith had studied entomology first at university and then in a public research institute. In 1997, she founded a small company to develop integrated pest management in French cities and was later hired by a gardening company. She also writes articles for gardening journals and disseminates information about integrated pest management on the website of her company.

> Basically, my job consists of monitoring urban cultures. I work a bit like a doctor: I go to places, I observe plants and their health and we decide to manage the parks so as to facilitate the arrival of auxiliary insects, or to release insects, or possibly to use chemical treatments, when there is no other option. But I also have a pedagogical approach: I train the staff, which is not so good for my job, because then I don't come so often as they can cope by themselves.

David and Édith had met in the mid-2000s in the glasshouses of a city where she had been called to implement integrated pest management. David invited Édith to audit the gardens and gardening practices in Grenoble, and then to train the staff in integrated pest management. Since 2006, she has facilitated collective outdoor sessions, gradually showing the green space workers how to identify the insects likely to be

Fig. 25.1 Propage butterfly protocol. Instructions and leaflets were a technique for educating attention, but to become effective, they required other techniques helping the participants attune to insects. (Source: MNHN and Noé Conservation)

found in the parks. This work was ongoing when Propage was implemented in 2014.

David participated in the training sessions whenever he could and the workers could email Édith questions and send her insect photographs. Édith and David's enduring commitment played an important role in establishing trust with the green space workers. During the training sessions, Édith introduced short, varied games and used humour:

> They collect specimens, we usually look at a specimen together, normally I find out quite quickly what it is and then I ask them to identify. I help them to implement an identification method by giving them some clues and advice in a humorous tone. This helps them to memorise. Generally it's quite interactive, we play and have fun. I start with a few jokes and then we have a sort of contest, or a vote, regarding a specimen: who sees legs? Who doesn't?

This quotation demonstrates that educating attention can operate through affects and emotions. The bodily capacities of individuals, including their feelings, become 'object-targets' for action (Anderson 2014). Édith, for example, used humour and playfulness to influence the green space workers' sensitivity to particular elements of their environment. Gamification is a popular technique in citizen science (Bowser, Hansen & Preece 2013; Eveleigh et al. 2013) and is often discussed as a means to motivate citizens who might not be interested in science otherwise. In the current case, games played a crucial role in making the green space workers 'see' and relate to the insects. The playful atmosphere they helped create (see Anderson 2014 for a discussion on affective atmospheres) increased the participants' ability to become attuned to the presence of insects, enabling them to 'think as insects' (see Lorimer 2008).

Another technique facilitating learning was the feedback given to the gardeners. After each session, Édith sent a detailed, positive and reassuring progress report. Her attitude helped the green space workers to overcome their initial anxiety and doubts about their capacity to navigate the huge and creepy world of insects. The reassuring atmosphere was fundamental to the ability of the workers to use the Propage guides and leaflets. They gradually overcame their fear of being unable to identify the butterflies in flight and became more confident in their learning capacities.

Édith also carefully observed the conditions under which the green space workers were willing to learn. For instance, during the first training sessions, she observed how long the participants would stay indoors listening to theoretical explanations before showing signs of boredom or annoyance. The fact that most training sessions took place outdoors and were collective was very important for their learning. Sharing experiences with colleagues gave the workers a feeling of togetherness and of being on a fairly equal footing.

The green space workers' involvement in Propage was therefore supported by a long phase of collective education of attention using various techniques: humour and playfulness, reassurance, symbolic rewards and training based on long-term relationships with instructors. These techniques targeted creating a learning collective. However, this was complemented by an individualised approach. Distancing themselves from the conception of participants as a crowd, David and Édith paid close attention to the participants as individuals, called them by their first names and expressed genuine interest in them. They also identified a range of attitudes, knowledge and skills among the green space workers. For example, some were good at finding insects and catching butterflies during the training sessions but would not suggest names or respond to questions, while others were eager to offer suggestions and participate actively in the games. One participant, Christophe, had already acquired knowledge of butterflies before Propage started, so he was encouraged to take an active role in the learning process. Some participants were motivated from the beginning of the process, whereas it took some time before others started showing signs of deeper interest. David and Édith were patient and accepted the diversity in the rhythm and extent of the green space workers' involvement. Engaging them in Propage was thus based on seeing them not as a homogenous group of undifferentiated participants but as a complex, dynamic set of individuals endowed with diverse skills and characteristics that could all contribute to the collective learning process, albeit in different ways and at different paces.

Involving the green space workers also required learning by the instructors. David and Édith paid attention to the participants' social characteristics and the influence of these characteristics on the inclusivity of the learning collective. Most of the participants had become leaders in their teams without necessarily holding a high school diploma. The fact that only team leaders volunteered was seen as a potential obstacle to the involvement of workers without any hierarchical position or responsibility. When a female gardener who did not lead a team finally decided to participate, it was hoped that this would convince others that Propage was genuinely accessible to all and encourage them to participate.

Empowering outcomes

The use of computers and simple mobile applications has been presented as an effective means of including groups with limited literacy or numeracy skills in citizen science (Bonney et al. 2014). Our case study offers other possible avenues for encouraging participation from groups with limited skills. Importantly, the techniques of educating attention were linked to the previous knowledge and experiences of the participants, sometimes dating back to their childhood or teenage years. Propage in Grenoble was not only aimed at learning new knowledge about insects, but also involved sharing, dialogue, new partnerships, trust and community building. Participation in the programme also influenced the participants' careers, as they acquired naturalist knowledge to reinforce their professional status and legitimacy. It even had far-reaching consequences on some participants' personal lives. For instance, 45-year-old Christophe, who had become interested in insects and butterflies before Propage started, related this change to his decision to quit smoking and reorient his life:

> After I quit smoking, I had to find something good to do to occupy myself. I felt I had never stopped, never looked up since I started working until I was 40, 41, 42. So at one point, I thought: this must change. I don't know why, perhaps because I quit smoking. So now, as a hobby and passion, I go hiking, I've got my backpack with my cameras. I go out every day after work, two to three hours, and take pictures to identify insects, principally in spring and summer.

Stéphane, 44 years old, began organising open evening sessions in his neighbourhood to show how to tend gardens in a more biodiversity friendly manner:

> I'll try to organize meetings in my neighborhood to show people what I know. Because I have a strong professional background now.
> **Q:** Will this be part of your work? Or of an environmental NGO?
> **A:** No, just myself. I'll start with neighbors, people I know, and expand gradually. For instance start with a small meeting at somebody's place and explain alternative methods to tend gardens. And gradually people ask questions. I wanted to start last year but I didn't have enough time with small children at home but, yes, I will do it. It's something I want to do. I want to share what I know. I'm really keen on this: changing practices by drawing on my experience and knowledge.

Laurent, 30 years, spent time during his holidays looking for butter-flies and teaching his young daughters how to identify them. The Propage project therefore played an unanticipated role in some volunteers' per-sonal lives, families or neighbourhoods, beyond their professional sphere.

Conclusions

Based on a French case study, this chapter outlined some key lessons about the conditions that facilitate the implementation of citizen science. In particular, we focused on the conditions under which citizen science sup-ports more responsive and inclusive forms of learning. The degree of inclu-siveness varies according to how citizens are involved in scientific research. Providing participants with well-designed documents, information and appropriate tools were definitely crucial to the success of citizen science. However, other techniques also affected participants' bodily responses and created a positive, reassuring atmosphere for learning, which were crucial for involving less educated and non-specialist participants.

If citizen science projects are to involve not only skilled experts but also wider target groups, it is important that participation is meaningful and adjusted to participants' own interests, histories and ways of think-ing and learning. If the sense of meaning is lacking, potential participants may refuse to be involved or withdraw rapidly, leaving only the 'usual-suspects' (see also Jupp 2008). This may also lead to a situation in which transformative outcomes are limited or absent. For example, citizen sci-ence may fail to support collective reflection or the development of more versatile roles for participants. Techniques accommodating individual spe-cificities and fostering ownership and responsibility of the process and its outcomes may be more effective in producing transformative outcomes than techniques based on participants' instrumental roles. However, such conclusions have been challenged by previous studies pointing out that also instrumental, data-driven approaches to citizen science can have far-reaching effects in participants' lives (Lawrence 2006). Detailed studies about transformative effects can illuminate how they emerge from vari-ous kinds of citizen science initiatives and how they influence participants in different contexts.

Our case study of the Propage project in Grenoble demonstrates the value of such studies. It also appeared that our in-depth interviews with the volunteers helped David, the project's initiator in Grenoble, to discover the far-reaching effects of this citizen science project in the green space workers' professional and personal lives. Our study also enabled him to reflect on the social learning process and strengthen commitment to the project as a result. The changes triggered by citizen involvement in citizen science projects can remain invisible to the promoters and practitioners of citizen science, even the most considerate and mindful ones. This underlines the potential role of the social sciences in highlighting and reinforcing these changes (see also Mahr et al. in this volume).

Based on our study, the close and continued attention to participants as individuals and to their diverging learning abilities and skills, on the one hand, and attention to the collective dynamics of learning within the group, on the other hand, proved to be key factors for increased inclusiveness. This required instructors to learn about participants' professional and personal trajectories, concerns and motivations, and ways of learning. It also required reflection on how these could support a collective learning process. Similarly, it was important to recognise factors otherwise external to the citizen science project, such as the conditions pre-dating the project and wider social dynamics of participation. In the case of Propage in Grenoble, the long and passionate commitment of a few people was crucial in getting green space workers interested in insects. Where and when this is not the case, developing close personal interactions in the field is likely to be all the more important.

Other factors might also play important roles in different contexts. While it is clear that inclusiveness is not an automatic outcome of citizen science projects, but depends on the techniques and practices of involving participants, studies exploring both successful and unsuccessful examples can further help to understand how openness, inclusiveness and broader diffusion of the benefits and learning outcomes can be fostered within citizen science.

Acknowledgements

We are grateful to David Geoffroy, who initiated the Propage project in Grenoble, for offering his insights and comments for this paper. Taru Peltola's contribution has been supported by the Strategic Research Council at the Academy of Finland (decisions 313013, 313014).

26
Learning and developing science capital through citizen science

Richard Edwards[1], Sarah Kirn[2], Thomas Hillman[3], Laure Kloetzer[4], Katherine Mathieson[5], Diarmuid McDonnell[6] and Tina Phillips[7]

[1] *University of Stirling, UK*
[2] *Gulf of Maine Research Institute, Portland, US*
[3] *University of Gothenburg, Sweden*
[4] *University of Neuchatel, Switzerland*
[5] *British Science Association, London, UK*
[6] *University of Stirling, UK*
[7] *Cornell Lab of Ornithology, Ithaca, NY, US*

corresponding author email: r.g.edwards@stir.ac.uk

In: Hecker, S., Haklay, M., Bowser, A., Makuch, Z., Vogel, J. & Bonn, A. 2018. *Citizen Science: Innovation in Open Science, Society and Policy*. UCL Press, London. https://doi.org/10.14324/111.9781787352339

Highlights

- Increased attention is focused on how to support and evaluate participation and learning through citizen science.
- The dimensions of science capital provide a new framework through which to consider participation and learning.
- The links between volunteers' prior level of educational qualifications and disciplines studied, and the learning they report from contributing to citizen science are not uniform across projects.
- The levels and dimensions of volunteers' engagement and learning do not always reflect the intentions of citizen science project designers.

Introduction

Inclusiveness and learning are two concepts underpinning the principles of citizen science put forward by the European Citizen Science Association (ECSA). The learning of volunteers in citizen science and its educative potential have been much discussed in recent years, as have the educational backgrounds and qualifications of those contributing to such projects (e.g., Bonney et al., 'Citizen Science', 2009; Garibay Group 2015; Haklay in this volume). There has also been growing exploration of how project design can affect the educational profiles of volunteers, the learning potential of projects (e.g., Phillips et al. 2014) and how projects may be designed to widen participation in citizen science (Novak et al. and Mazumdar et al., both in this volume). Overall, however, the educational impact of participating in citizen science has remained under-researched (see also Peltola & Arpin in this volume). Evidence is often anecdotal or based on evaluations rather than up-to-date learning theory and systematic research (Falk et al. 2012).

This is beginning to change and this chapter offers research evidence on learning through citizen science based on work that has been developed in the United States and Europe. This is ongoing, so broad conclusions would be premature. Research on learning through citizen science is in its infancy and, while it can draw on wider research traditions in informal science learning (e.g., Falk et al. 2012) and informal and experiential learning more generally, this is not yet fully the case. The chapter also seeks to make the case for considering learning through citizen science within a broader conceptual framework, that of science capital (Archer et al. 2015). The developing concept of science capital points to the iterative relationship between people's dispositions towards science, participation in science-related activities and science-related outcomes, including learning (DeWitt, Archer & Mau 2016). Basically, the more one is part of a culture of participation in science-related activities, the more one is likely to develop science learning outcomes and the disposition to participate further in science-related activities. In other words, developing science capital means developing a culture of participation in, and learning from, science-related activities, including citizen science.

The concept of science capital is relatively new within research on science learning in general and at the periphery of research and practice in citizen science (Edwards et al. 2015). It provides a broader framework to consider issues of participation and learning in citizen science. It contrasts with many preliminary explorations of learning through citizen science, which focus primarily on what people learn and how best to evaluate learning outcomes while trying to draw connections to peoples' motivation to participate. These outcomes can be identified narrowly or broadly. Narrow science learning outcomes may embrace areas such as domain knowledge, for instance in relation to specific fauna or flora, or specific scientific methods. Broader learning outcomes may embrace areas such as environmental stewardship and the development of science identities.

While there has been a widening of ideas about what volunteers may learn from participation in citizen science, less attention has been given to how they learn; that is, the practices in which they participate that enable learning when contributing to citizen science projects. Exploring how people learn focuses on the people and resources with which volunteers interact and how they engage with them to learn, if indeed they do learn. Better understanding how people learn can enable practitioners to better design projects or develop curriculum, training materials or professional development materials for teachers to enhance the educative potential of citizen science projects. Learning is not simply cognitive, but also social and cultural (Fenwick, Edwards & Sawchuk 2011), hence the interactions among volunteers or between volunteers and project coordinators, and facilitation thereof, should be carefully considered.

If it is important to develop broad science-related outcomes, including learning, then exploring the social and cultural aspects of volunteer participation – the nature and extent of their science capital – and how these can be developed becomes important. Some citizen science practitioners are becoming interested, therefore, in how citizen science might enhance the building of science capital among volunteers – developing a wider culture of engagement in science-related activities – as well as their specific science learning through individual projects (e.g., Bailey 2016; Kirn 2016).

This chapter suggests that an approach to developing citizen science projects that seeks to develop science capital could have positive benefits on the educational profiles of those who participate and enhance the educative potential of citizen science.

Science capital

The concept of science capital has been developed from the work of the French sociologist Pierre Bourdieu. Science capital refers to the educational qualifications, social networks, dispositions and behaviours among those working in, or engaged with, sciences (Archer et al. 2015). It is a subset of the social and cultural capital that accrues to individuals unequally in society and results in the reproduction of those inequalities. In other words, inequality is not only economic, but is also social and cultural.

The existence of individuals and families with higher or lower levels of science capital, therefore, can be utilised to explain inequalities in participation in science-related activities and the unequal learning of science. It can also help to shape practitioner responses to this situation. Science capital can be seen as a resource to support the development of science learning and identities as part of a culture of engagement with science-related activities. Individuals and families may develop more or less science capital and the children of those families with most science capital are more likely to consider science education and a scientific career as options for their futures (Archer et al. 2015). Science capital helps explain the ways some people engage with and learn sciences, while others do not, and can also be considered an outcome of participation in science-related activities. In other words, the more one develops science capital, the more one is likely to participate in science-related activities, thus further enhancing one's capital.

Archer et al. (2015) identify eight dimensions of science capital:

- scientific literacy;
- scientific-related values;
- knowledge about transferability of science in the labour market;
- consumption of science-related media;
- participation in out-of-school science learning contexts;
- knowing someone who works in a science-related job;
- parental science qualification; and
- talking to others about science outside the classroom.

Some of these dimensions may be used to design citizen science projects and develop pedagogical and other interventions that can build science capital and change current patterns of participation. Science capital also suggests that broadening participation in citizen science and enhancing its educative potential is not simply an educational issue, but also social and cultural. Citizen science projects may become a means to enhance volunteers' science capital, but, at present, they seem to draw largely from populations with higher pre-existing levels of science capital. Refocusing attention on potential volunteers with lower science capital means addressing the wider cultural factors that influence what and how people participate in science-related activities in society.

Little research has yet been done to investigate how citizen science participation may increase science capital and the concept itself is still in development. More rigorous research drawing on the notion of science capital is required before stronger claims can be made. This might involve new studies or the re-analysis of existing datasets. Some existing studies are discussed in the next section.

Science capital and volunteer demographics

Understanding who currently contributes to citizen science and their educational and wider backgrounds is an area of concern. While there are significant attempts to widen participation and encourage diversity among volunteers contributing to citizen science projects, to date most surveys show that it is those that are older, more highly qualified and from higher socio-economic backgrounds who are most likely to participate (e.g., Garibay Group 2015). In addition to these factors, gender and race are also significant in who volunteers in what types of citizen science project. In general, it is the already advantaged – those with the greatest social and cultural capital – who are most likely to volunteer. This is a pattern to be found in volunteering more broadly (European Foundation for the Improvement of Living and Working Conditions 2011). Determining the extent to which this is also related to higher levels of science capital, however, means examining the specific scientific disciplines previously studied by volunteers and the nature and level of their engagement in wider science-related activities.

[handwritten margin note: and w/ more science capital too!]

In two ornithology citizen science projects in the UK studied by Edwards, McDonnell and Simpson (2016), 83 per cent of respondents were male, 98 per cent were white and the largest proportion was in the 61–70 age range. As a proxy of their higher socio-economic status, 67 per cent of respondents had a university-level qualification. In other words, the majority of volunteers might be argued to have high social and cultural capital. Exploring further, the study found that large numbers of volunteers had gained either school and/or university-level qualifications in the sciences. Therefore, a majority of volunteers could further be argued to have higher levels of science capital as the basis of their participation in these citizen science projects than the wider population. To explore the impact of citizen science participation on the development of science capital, the study also explored volunteers' enjoyment of participation in a wider range of science-related activities, such as scientific hobbies or watching science television programmes as a result of participation in the projects. Little overall affect was found. The building of science capital was not an explicit goal of the two projects studied, nor does it appear to be a significant implicit outcome.

[handwritten note: what would a citi sci project look like if building science capital was a goal? would it still have scientific merit?]

Working with schoolchildren offers an opportunity to engage a population with more diverse levels of science capital than would be the case through volunteer-based projects (see Makuch & Aczel; Harlin et al., both in this volume). In these cases, the student citizen science participants are not volunteering out of interest, but rather participating in a compulsory curriculum. Increasingly, citizen science programmes are designing experiences and curriculum that engage students in both practising the skills of science and interacting with the broader community of volunteers and scientists also participating in the project. For instance, the Gulf of Maine Research Institute's Vital Signs programme (Kirn 2016) has developed novice-friendly protocols, standards-aligned curriculum, and professional development support and coaching for schools and teachers to facilitate the successful engagement of children in scientific investigations and provide an opportunity for increasing science capital. Through Vital Signs, students practice scientific skills to explore their environments, collect rigorous observational data, conduct peer review of one another's work, share data online, and engage in public discussion through the programme website with the scientists and natural resource managers using their data. Kirn notes how resources and protocols designed explicitly for novice volunteers, as well as interaction with experts, helps to encourage and sustain engagement and learning, contributing to an increase in science literacy. Additionally, these interactions between experts and novices give science novices the opportunity to get to know scientists and/or science enthusiasts with high science capital.

[handwritten margin note: proximity to high levels of science capital builds it for others?]

Other providers of science-related activities, such as urban ecology centres and museums, link with citizen science projects to promote wider engagement and learning. Some citizen science projects, such as eBird at the Cornell Lab of Ornithology, have developed curriculum and professional development materials for teachers to support engagement and learning. While valuable initiatives, the extent to which they continue to engage those with pre-existing higher levels of science capital rather than provide bridges for those with lower science capital remains unknown.

How and what volunteers learn

In addition to increasing knowledge in science content areas, some citizen science projects aim to increase science learning in the broader sense and include cognitive, affective, practical and behavioural outcomes (Bonney et al. 2016). Here, learning is not simply focused on the knowledge and skills relevant to the scientific goals of the specific project, but extends to a wider engagement with science as a whole. For instance, learning outcomes intertwined with environmental science knowledge include interest in science and the environment; efficacy to do and learn about science and engage in environmental activities; motivation to participate in science and environmental learning; understanding of the nature of science; acquisition of science enquiry skills such as data collection, analysis and interpretation; and involvement in environmental stewardship practices outside of project activities (Phillips et al. 2014). It is in developing these broader learning outcomes and how they are enhanced that citizen science might be said to contribute to the building of science capital and a culture of engagement with science-related activities in society more generally.

However, although many citizen science projects have successfully demonstrated an increase in participants' understanding of science content and processes, fewer studies have examined wider outcomes. For instance, in their study of ornithology citizen science projects, Edwards, McDonnell & Simpson (2016) found that large percentages of volunteers identified themselves as learning something across a range of science-related outcomes. However, it was only in relation to 'learning about the topic' and 'learning about data collection' that volunteers identified themselves as learning a lot. Prior level of educational qualification, one marker of science capital, was significant here, as there were statistically significant differences between volunteers with or without a degree. Overall, the less *only so* qualified the volunteers, the more they evaluated themselves as learning *much* across most of the outcomes. This suggests that those with a higher level of *science* qualification are not being extended or are not extending themselves in *capital to* contributing to projects – they are simply drawing on their existing levels *be gained* of science capital. There are indications from this study that citizen science *from* participation can enhance the learning of those with less science capital. *citi sci*

However, existing research is not entirely consistent on this point. For instance, Kloetzer, Schneider & Jennett (2016) researched learning and creativity in nine online citizen science projects. Unlike Edwards, McDonnell & Simpson (2016), they found a very low correlation between level of education and self-reported learning. However, as with other studies, they found also different degrees of participation among volunteers with a minority being more active than the majority. The degree of active participation was linked to the level of learning outcomes reported. The extent to which that participation was linked to prior levels of science capital remains unknown. However, there are some indications that high engagement enhanced science capital as higher-order learning was related to active and social learning, and 37.5 per cent of the participants claimed that participation in a citizen science project helped them discover a new field of interest. This shows the importance of examining not only who is participating in citizen science but also how they are participating and examining the impact of citizen science on volunteers within the framework of a wider culture of participation in science-related activities.

Kloetzer, Schneider & Jennett (2016) identified a number of ways in which people learn through participation in citizen science: contributing to the project; using external resources; using project documentation; interacting with others and personal creations. These point to the relational and material ways in which people learn, and the heterogeneity of learning outcomes and processes: people learn different things in different ways within the same project. In other words, how learning is designed into citizen science projects does not guarantee that volunteers will learn what is intended or in the ways planned. The mismatches between planned and actual outcomes is found elsewhere. Drawing on a study of six citizen science projects across a spectrum of contributory to co-created (see also Ballard, Phillips & Robinson; Novak et al., both in this volume, on these different types of participation), Phillips (2016) identified four different dimensions of engagement – behavioural, affective, effort and social – and various indicators of each. Significantly, levels of engagement were not directly related to the type of project as more co-created projects did not necessarily have a larger proportion of participants identifying themselves as having higher levels of engagement. This suggests that what is planned and designed as learning does not necessarily result in the anticipated outcomes; volunteers engage with and make use of projects in unplanned ways and, as a result, learn different things.

Evidencing participants' existing expertise and how peer support occurs is another important element of citizen science practice. For instance, Hillman and Mäkitalo (2016) studied the learning of contributors to Galaxy Zoo, an online international project to classify images of galaxies (see also Haklay in this volume). They argue that online citizen science projects that focus on classification tasks tend to deliberately require relatively low skill levels since their goal is often to enrol as many volunteers as possible and render all their contributions equal in relation to the scientific protocol. Communities of volunteers were identified as developing around classification tasks and it was activities in these communities that provided a rich source for learning. It was also in online discussion among these communities that the resources volunteers drew on became visible through, for instance, moving from using URLs to newspaper articles or popular science websites to referring to more research-focused resources such as astronomical databases. Drawing on a sociocultural conception of learning as the appropriation of cultural tools or resources, Hillman and Mäkitalo (2016) used changes in resource use in online forums as an indicator of learning and scientific literacy. In the discussion forums, those with less scientific literacy moved from the use of popular to more scientific resources, from curating content to formulating arguments, and from soliciting advice to providing guidance as contributors developed more familiarity and expertise. In particular, the authors argue that the appropriation of scientific resources is a strong indicator of scientific literacy and that progression along learning trajectories is visible for new members of citizen science communities as they successively appropriate these resources. While such shifts are difficult to track in the more ephemeral interactions of face-to-face citizen science projects, the technologies of the internet often render them readily chartable in relation to online citizen science projects. Tracing the activities of citizen science volunteers as they discuss online means that data produced can be argued to reflect trajectories in the building of science capital and reveal some of the means through which it can be built.

Issues for the future

It is clear from both the research and practice worlds that learning is occurring among volunteers in citizen science. Yet exploring how that occurs as well as what is learnt remains in its the early stages, and the picture emerging is complex, full of tensions and highly influenced by context. Prior qualifications, volunteer demographics, project design, participation and engagement are all significant. Examining these issues through a social and cultural framework and drawing on the dimensions of science capital could enable a better understanding of the dynamic interrelationship of these and other issues.

At a broad strategic level, building a stronger international research base on learning in citizen science; relating it more clearly to wider educational research; and engendering stronger relationships between research and practice are clear priorities for the future. More specifically, questions remain about the correlation between project design and learning outcomes; variations in the prior science capital of participants; and what and how resources are used in citizen science projects. Much research and evaluation in citizen science to date relies on self-reported learning processes through surveys and less often interviews (see Kieslinger et al. in this volume). There is a need for more refined, ethnographically informed studies to examine more closely how volunteers learn (see, for example, Peltola &

Arpin in this volume). Exploring the extent to which patterns emerge in relation to prior science capital, project design, participant recruitment, engagement and learning would be helpful for practitioners. The forms of project support for learning, and the possibilities for peer learning within the context of projects, are also of interest. The possible contribution of citizen science to enhancing science capital has also yet to be addressed fully, as has how best to support those contributing to projects with different levels of science capital. These are only a few of the issues emerging for research and practice as the field of citizen science expands.

Conclusions

At a policy level, the potential of citizen science to engage citizens in more informed debates on science and scientific issues as they relate to broader social, economic, environmental and cultural questions is becoming clear. In relation to education policy specifically, the continued growth of the links between citizen science projects and formal educational institutions is to be encouraged (see also Wyler & Haklay in this volume). Here citizen science can be rethought of as itself a form of pedagogy, and one with the capacity to increase learners' science capital. The extent to which citizen science can build science capital and enable wider engagement with science-related issues, such as the impact of climate change, deserves further experimentation and investigation.

In relation to the management of citizen science projects, a more explicit engagement with the issues of learning and science capital would be welcome when designing and resourcing projects. This entails more and greater systematising of relevant research, and developing more and better models of research-practice interactions. As the field of citizen science grows, there will no doubt be a related growth in the diversity of projects and scientists seeking to engage participants or understand the dynamics of participation. Supporting that growth while enhancing the diversity of participants in citizen science and their learning remains a challenge, but one for which there is a growing evidence base.

Acknowledgements

Funding for the research projects reported in this chapter is gratefully acknowledged from the British Academy, European Union, National Science Foundation and Wellcome Trust.

27
Children and citizen science

Karen E. Makuch[1] and Miriam R. Aczel[1]

[1] *Imperial College, London, UK*

corresponding author email: k.e.makuch@imperial.ac.uk

In: Hecker, S., Haklay, M., Bowser, A., Makuch, Z., Vogel, J. & Bonn, A. 2018. *Citizen Science: Innovation in Open Science, Society and Policy*. UCL Press, London. https://doi.org/10.14324/111.9781787352339

Highlights

- Children can both learn from and contribute to citizen science. Scientific learning can develop children's environmental citizenship, voices and democratic participation as adults.
- The quality of data produced by children varies across projects and can be assumed to be of poorer quality because of their age, experience and less-developed skill set.
- If citizen science activities are appropriately designed they can be accessible to all children, which can also improve their accessibility to a wider range of citizens in general.

Introduction

To date, a cursory examination of the literature tells us that a large number of citizen science projects have been, or are, in the environmental domain. It is thus on environmental citizen science that we focus this work[1]. This chapter suggests why children ought to be involved in citizen science – largely through environmental projects, highlights some case study examples to show positive and negative outcomes of child participation in said projects, comments on the potential contributions to science education and environmental awareness, and highlights some practical considerations of child involvement in citizen science. This work is thus premised on the two-way benefits of engaging children in environmental citizen science:

1. Children can both learn from and contribute to environmental knowledge, education and scientific enquiry; and
2. Where activities take place outdoors, child involvement in citizen science provides access to the environment, enabling children to develop environmental awareness, responsibility, emotional and physical benefits.

As the European Citizen Science Association (ECSA) assert in their formative 'Ten Principles of Citizen Science'[2], 'Citizen science is a flexible concept which can be adapted and applied within diverse situations and disciplines'[3]. It is exactly this adaptability and promotion of diversity which we embrace in this chapter, as we argue that such approaches can open up opportunities, outlined below, for *child* participation, in the *environmental* field. Furthermore, the involvement of individuals, (thus including children), in citizen science is advocated in ECSA Principle 3, which states that 'learning opportunities, personal enjoyment, social benefits, satisfaction through contributing to scientific evidence e.g., to address local, national and international issues [. . .] and influence policy', *inter alia*, may be some of the gains of participation in citizen science projects, and this is very much aligned to the work of environmentalism.

Why citizen science?

Cheng and Monroe (2012, 32) assert that '[h]uman behaviour is implicated in a number of environmental problems. In addition to solutions that can be offered by experts and policy makers, citizens' conservation actions are needed'. Thus, *citizens* and benefactors of the earth need to be responsible for the planet and all that is sustaining and enriching. A 'citizen' can be defined broadly as someone who has a stake in the future of the *global* environment. This chapter also adopts a more localised definition of a 'citizen' as someone who has a stake or interest in their *local* community.[4] Principle 1 of the ECSA Principles[5] avers that 'Citizen Science projects actively involve citizens in scientific endeavor that generates new knowledge or understanding [. . .] Citizen may act as contributors and have a meaningful role in the project'.

Further, citizen science, on a practical level, has the potential to:

1. Educate individuals about the environment in a broad sense, and ecology, species and scientific concerns, among others, in a narrower sense;

2. Be efficient, as local citizens undertaking data collection with qualified and trained experts can save time and money for regulators;
3. Engage at the local level, physically and temporally, rather than being remote or detached. Citizens live and work in their local environment and are more likely to notice, or be affected by, environmental change;
4. Be participatory and contribute to environmental justice. This is particularly pertinent in situations where local authorities or politicians are not willing or able to act on environmental matters.

Based on the above definitions, all children are current and future citizens and have a stake in the natural world.

Developing environmental citizenship

Children have an innate curiosity and desire to experience and learn that can enhance their citizen science experience (Jenkins 2011). Working with children brings insightful questions, new ideas and fresh perspectives on how scientific information is presented and interpreted. Research shows that children are naturally 'exploratory, inquiry-oriented, evidence-seeking' in their learning (National Research Council 2009, 67). The communication and exploration of science in a way that is primarily directed to children will arguably benefit children (*and* laypersons) involved in a citizen science project (Bonney et al. 2016). This child-focused form of communication can thus have multiple positive benefits: Difficult concepts are clearly explained, understanding of environmental problems is arguably made easier, knowledge can then be shared within communities and preconceptions of science and environment can be challenged and 'corrected' (Kambouri 2015a; Kambouri 2015b), particularly as 'the core components of initial science learning are (1) accurate observation, (2) the ability to extract and reason explicitly about causal connections and (3) knowledge of mechanisms that explain these connections' (Tolmie et al. 2016, 2). Citizen science offers the opportunity for children and young people to undertake research and ask questions from their unique perspectives, which may lead to a different understanding of issues, alternative solutions (Wells & Lekies 2006) and learning of distinct skills.

In addition to contributing to scientific enquiry, exposure to positive experiences as a child can have a profound effect in adulthood, particularly in developing responsibility or positivity towards an issue (Jones,

Greenberg & Crowley 2015; Edwards et al. in this volume). Positive exposure to the environment as a child is shown to create positive attitudes towards the environment as an adult (Wells & Lekies 2006; Cheng & Monroe 2012). Children are more receptive to specific aspects of the natural world at certain ages, with particular developmental stages crucial in engaging the child citizen scientist as an emerging supporter of science and of the environment (Kellert 2002; White & Stoeck 2008).

Noting the above, an additional argument for the inclusion of children in citizen science is that children are gradually becoming disconnected from nature and the environment (Louv 2005; Miller 2005; Kahn, Severson & Ruckert 2009). There is an abundance of literature examining childhood behaviour (Wells & Lekies 2006; Cheng & Monroe 2012), education (Littledyke 2004; White & Stoeck 2008), psychology (Kellert 2002) and participation (Hart 1997; Wells & Lekies 2006), and the benefits to be gained from exposure to nature through citizen science (Purcell, Garibay & Dickinson 2012). If children find their participation on a nature-focused citizen science project exciting, and the experience of the outdoors stimulating, it could help develop self-confidence, connection to the environment and responsibility and empathy for nature and others. Furthermore, the practical tasks involved, such as preparation of the experiment and data collection and monitoring, can help to develop a sense of responsibility particularly for the work and for the environment and/or species with which they are engaged.

Filling a regulatory and democratic gap

Fluctuating political, social and, *inter alia*, economic circumstances, can arguably have an impact on government budgets and investment in environmental monitoring (Conrad & Daoust 2008). Here, citizen science can potentially fill a regulatory gap through the contributions of volunteers (Shirk & Bonney; Volten et al.; Owen & Parker, all in this volume). Furthermore, it may be easier to organise child participation and engagement as there are ready-formed pools of schoolchildren, scouting groups, activity clubs and so on (Wells & Lekies 2006). Advocates comment that creating a fun and engaging project is central to recruiting volunteers, particularly children (Dickinson et al. 2012). Moreover, an approach to participating in science and environment that is not solely adult-centric could promote inclusivity and a democratic approach to public participation in environmental decision-making (Hart 1997), very much in line with ECSA Principles 1 and 3[6], with adults making a deliberate decision to extend participation and inclusivity in citizen science projects to children. As Silvertown (2009, 467) asserts, 'the characteristic that clearly differentiates modern citizen science from its historical form is that it is now an activity that is potentially available to all, not just a privileged few' (but see also Mahr et al., in this volume, on the history of participatory science). However, commentators observe that we still have a long way to go in making citizen science truly democratic and diverse (Tweddle et al. 2012; Smallman in this volume), though including children is positive step towards these goals.

Citizen science ought not to be limited by age, geographical, racial, economic, (dis)ability, gender or other boundaries, and can be a group or solo activity (Liebenberg 2015; Stevens et al. 2014). Projects can be designed to be appropriate for children and can take place in urban environments or places not typically associated with the exploration of nature, such as schools, yards or windowsills. Projects can also be designed to be inclusive with respect to learning or physical disabilities.

This chapter argues that:

1. Citizen science by definition should be inclusive across gender, ethnicity, class, disability, level of education and so on. A diverse mix of participants contributing to scientific enquiry means that a broad range of perspectives can inform the research. Including children in citizen science further broadens the scope of the research due to their ways of viewing and enquiring about the world.
2. Citizen science may improve access to STEMM fields (science, technology, engineering, mathematics and medicine) for marginalised groups that have been historically excluded. Citizen science projects can be designed in ways that aim to overcome gendered, racial and other biases often associated with STEMM (Ceci & Williams 2011); and,
3. Generating interest or opportunity for engagement in citizen science among groups that might traditionally have been excluded from STEMM fields, or groups that might suffer from environmental discrimination or inequity, has social, educational, health and developmental benefits (European Citizen Science Association 2014, see Robinson et al in this volume; Dickinson & Bonney 2015).

Benefits for child citizen scientists

Physical and emotional

Children in citizen science can profit in their emotional and physical development and well-being. Box 27.1 highlights some of the key benefits of participation.

Interpersonal and social

Understanding and communicating science is arguably vital to the development of a sustainable world. Moreover, interpersonal and social skills are needed to prepare children for a healthy and productive future. Outdoor learning experiences, and particularly citizen science projects, can give children the courage to try new activities with new people, which ultimately have a positive effect on their self-esteem and confidence to

---

Box 27.1. Physical and emotional benefits of child participation in citizen science and engagement with nature

Sense of inclusion with nature is associated with understanding how an individual identifies his or her place in nature, the value that he or she places on nature, and how he or she can affect nature [. . .] Connectedness to nature, caring for nature and commitment to protect nature are core components of inclusion within nature (Cheng & Monroe 2012, 34, citing Schultz 2002).

Citizen science can:

- engage children with a purposeful and positive activity, which can help improve mental and physical growth;
- get children outdoors and help children to connect with nature;
- help children understand the environment and the important role of ecosystems;
- assist children in claiming some ownership of their environment and provide them with the ability to participate in its guardianship;
- teach children scientific concepts and provide information and data that can be used both to further develop understanding of science and safeguard the environment.

---

participate in further collaborative opportunities (Dillon et al. 2006). Projects such as community gardens can give children a 'context for learning that addresses multiple societal goals, including a populace that is scientifically literate, practices environmental stewardship, and participates in civic life' (Krasny & Tidball 2009, p.1). For example, the intergenerational Garden Mosaics programme[7] has a variety of activities in urban settings, and seeks to 'integrate learning from the "traditional" or practical knowledge of community gardeners with learning from science resources produced at Cornell University', allowing children both to learn about environmental science and engage in *civic ecology* (Krasny & Tidball 2009, 5) community garden restoration and management initiatives (see further, Tidball & Krasny 2007). Further, the Little Seedlings phenology project (The Conservation Volunteers, online, undated) at a garden centre in Scotland[8] worked monthly (April–August 2014) with children aged 4–12 (some with parents) using The Woodland Trust's Nature's Calendar website survey and recording sheets[9] to record and view seasonal events and the impact of climate change on wildlife. Organisers taught child participants about seasons and the changes they bring to nature, though acknowledged the limitations of younger children (under 7) in collecting data (The Conservation Volunteers, online, undated[10]).

Moving on, we comment below in box 27.2 on some key benefits to children, the environment and to citizen science from child participation in a plastic debris sampling project, while box 27.3 draws out some pros and cons of data analysis by children engaged in a citizen science project.

Box 27.2. National sampling of small plastic debris, supported by children in Chile

1. A wide pool of capable child citizen scientists
   This inclusive project, which combined environmental stewardship with child citizen science, highlights the 'two-way' benefits of citizen science projects. In this case, children from all around Chile and Easter Island both filled gaps in data on the accumulation and abundance of debris on Chilean beaches, and their personal development benefited from engagement in an environmental activity (Hidalgo-Ruz & Thiel 2013, 14). A pilot study was used to first test the protocols and data reporting forms so that adjustments could be made to guarantee data quality (Bonney et al., 'Citizen Science', 2009, 979; Hidalgo-Ruz

(continued)

Table 27.1 Results from the final evaluation survey applied to students (Hidalgo-Ruz & Thiel 2013, 14)

Question	Majority response	Percentage (%)
On a score from 1 to 7, how much fun was this small plastic debris project?	7	61.1
Had you heard about small plastic debris before this project?	No	73.1
Had you participated in an activity related to the environment before this project?	No	61.7
Did you read the story "The journey of Jurella and the microplastics"?	Yes	83.0
On a score from 1 to 7, how interesting did you find the story of Jurella?	7	51.8
Would you like to participate in other environmental activities in the future?	Yes	96.1
What was your favorite part of the project? Please mark one option	Field work on the beach	76.2

& Thiel 2013, 13). The programme ran for two months and involved nearly 1,000 students (from 8 to 16 years old).

2. Positive experiences

Students completed a survey (see summary in table 27.1) to evaluate their overall satisfaction with the programme and children 'rated the activity with an average grade of 6.3, in which 61% of all students qualified it with a 7 (the best possible grade). The favorite part of the activity was the field sampling (76% of the students)' (Hidalgo-Ruz & Thiel 2013, 14). Roughly three-quarters of the students had never heard of 'small plastic debris' before the programme, and for 62 per cent, this was their first 'environmental activity' (14). Yet, 96 per cent of all students said they wanted to participate in future similar activities (14).

3. Outcomes

'To validate the data obtained by the students, all samples were recounted in the laboratory. The results [. . .] showed that the students were able to follow the instructions and generate reliable data' and that children who take part in citizen science projects can engage in 'a scientific thinking process' (Hidalgo-Ruz & Thiel 2013, 15) through learning about impacts of pollutants such as plastic debris, in addition to experiencing positive changes regarding their attitudes towards science (Lawless & Rock 1998, 7–8) and potentially the environment (Phillips et al. 2012, 92–93). Further, this case study demonstrates that child citizen science can help in the collection of 'large-scale spatial [. . .] data on the occurrence and abundance of small plastic debris' (Hidalgo-Ruz & Thiel 2013, 17).

Box 27.3. Conducting ecological research: Analysing data collected by German schoolchildren

1. Limitations in collection of data by children
   In conjunction with a project investigating dispersal and predation of seeds in rural and urban ecosystems, Miczajka, Klein and Pufal (2015) conducted a study to determine whether children could contribute to an ecological experiment by collecting data qualitatively. In Hamburg and Luneburg, Germany, 14 classrooms with a total of 302 children aged eight to ten years old, with 'no comparable experience or training in conducting scientific experiments', were taught 12 lessons by scientists, of which four were dedicated to the citizen science project (5). Six experiments were devised, with different conditions. The children 'used pre-designed field protocols' to measure conditions including weather; vegetation cover (using words, such as 'lots of cover') and height (using a ruler); treatment and colour of seeds; number of seeds exposed; and number of seeds recovered at the end (4). The scientists conducted the same measurements as the children to compare their data. The study found that 'only in five classes out of 14, children and scientists provided similar cover estimates' (5), and the measured range of vegetation heights for scientists was 0–40 cm and for children was 5–800 cm (5).

(continued)

2. Outcome dependent on task

On the other hand, from a total of 1,680 seeds, the children recorded 83.9 per cent of seeds compared to 88.7 per cent recorded by the scientists. The authors demonstrate that 'seed count data from children and scientists was mostly similar' (and differed significantly only for one particularly small type of seed), while on the other hand there was 'only little concordance in the estimation and measurement of vegetation and height data' (Miczajaka, Klein & Pufal 2015, 5, 7). Therefore, the results show that measuring height and conducting estimates is 'difficult' for children with little experience (5). Conversely, as 'counting is an innate skill for children aged eight to ten because they learn it early', they achieved mostly similar results to the scientists. The authors conclude that it is 'possible to integrate elementary school children as citizen scientists in [ecosystem science] projects . . . if these projects require skills that the children are already familiar with' but that citizen science experiments requiring skills beyond their level 'would require intensive preparation and training' (6). Thus, this case study illustrates that children can contribute as scientists, but it is arguably important to first assess the skills and knowledge required to ensure valuable and more accurate data.

Educational

Citizen science has been credited with 'hold[ing] much promise' (Jenkins 2011, 501) for making classroom-based/laboratory-based learning less boring and science more accessible to children (Jenkins 2011, 505–6; Rodríguez 2015, 14; and see Wyler & Haklay in this volume re: motivating university students). Corrigan (2006, 51) asserts that science educators 'need to have a clear purpose of what they hope their students will learn' for science education to better engage students. Cherry and Braasch (2008, 1) argue that there is a demonstrated need to increase both formal and informal science and climate literacy, and show that citizen science 'works because data collection stimulates experiential and cognitive ways of learning'. Citizen science is also credited with promoting scientific and ecological literacy and offers the potential to develop a lifelong inter-est in science (Jenkins 2011, 502; Rodríguez 2015, 13–16; Miczajaka, Klein & Pufal 2015, 2).

Pike and Dunne (2011, 494–5; 498–9; 487) state that often UK stu-dents are not motivated beyond the compulsory school curriculum to study science after age 16, finding the curriculum fails to enthuse students or they see science as irrelevant to their lives. Jenkins (2011, 504) con-tends that students might find science inaccessible because they cannot relate to it. Thus, citizen science can be pitched at varying levels of aca-demic ability and experience, and can 'translate' abstract topics into ones that can be visualised (Johnson, Hart & Colwell 2014a, 12–13).

Citizen science can also help to address inequalities in *access* to edu-cation (Gommerman & Monroe 2017), particularly in developing coun-tries. Working in groups and using the natural world as the laboratory resulted in a low-cost educational model. The National Research Council (2009, 3) found that there are benefits to learning science through 'informal' environments and that non-school programmes 'may posi-tively influence academic achievement'.

Oberhauser and Prysby (2008, 104), commenting on their Monarch Larva Monitoring Project, observe that from an educational perspec-tive, volunteers, including many children, have learned data collection protocols and had the opportunity to be engaged in authentic research. Many teachers, parents and other youth leaders use this programme to engage children in the scientific process'. A 2015 study by Wells et al. (2015, 2873) documented a 'modest, positive effect on science knowledge among elementary school children in low-income communities' through randomised controlled trials. In addition, school garden science projects benefit the development of the 'whole child' by 'contributing to social, academic, cognitive, and health outcomes' (Wells et al. 2015, 2874). Fur-ther, Trautmann et al. (2012, 179) observe that citizen science provides 'meaningful connections to the natural world' for children 'through observation, data collection and [. . .] investigation'. Yet, though there are broader benefits to be obtained from data collection and child participa-tion in citizen science, not all child-centred citizen science projects will yield wholly accurate results or data (see box 27.2 and box 27.3, above).

On a positive note, citizen science allows for children to learn about nature and the environment in an immersive and structured way, bene-fiting from the solid disciplines and underpinnings of scientific inquiry, while being engaged in an experience that will impact their role as future custodians of the world and also as potential future scientists (Krathwohl,

Bloom & Massia 1964; Harlin et al. in this volume). It is clear though that the scientific tasks being undertaken should be tailored to the ability of the participants and potentially employ 'skills that the children are already familiar with' (Miczajaka, Klein & Pufal 2015, 6). Therefore, in spite of some limitations, children who had the chance to undertake actual scientific research and work with 'proper' scientists had a learning experience 'shown to be more effective than education by teacher-centered teaching in other studies'. (Miczajaka, Klein & Pufal 2015, 6).

Curriculum enhancement

Child involvement in citizen science projects can also play a role in formal education and curriculum enhancement. For example, the Greenwave project (1997), an initiative of the Discover Science and Engineering Programme at Science Foundation, Ireland, was 'the longest running phenology network in Ireland in which school children were the main participants' (Donnelly et al. 2014, 1239). According to Donnelly et al. (2014, 1241), over 150 schools participated in this project, which fed into the Primary Science/Social, Environmental and Scientific Education curriculum. Participation in Greenwave was one of the criteria applied to awarding the Science and Maths Excellence mark to schools.

Formally connecting citizen science projects to national curricula helps realise its potential benefits. Where there is not the political expediency to formally adopt or integrate citizen science into the curriculum there is a strong potential for citizen science projects to develop their own school-centred learning materials and lesson plans, which may well feed informally into national learning schemes (see, for example, the Imperial College London/Open Air Laboratories [OPAL] project[11].)

Jenkins (2011, 501) states that 'participation in citizen science projects moves scientific content from the abstract to the tangible involving students in hands-on, active learning. In addition, if civic projects are centred within their own communities, then the science becomes relevant to their lives because it is focused on topics in their own backyards'. However, some guidance is needed, from a teacher, parent or other leader (with some form of scientific expertise) because '[c]itizen science, by definition, relies on co-operation between a range of experts and non-experts, which in many cases, involves some sort of public engagement, education, and data collection' (Jordan et al. 2015, 208).

Participation, engagement and children's voices

Bultitude (2011, 2) comments that '[a] recent major review [. . .] within the UK identified four key cultural factors that have influenced the separation of science from society, resulting in an increased need for scientists to engage with public audiences: 1. The loss of expertise and authority of scientists; 2. A change in the nature of knowledge production; 3. Improved communications and a proliferation of sources of information; 4. The democratic deficit'.

Bultitude acknowledges that these issues might also be relevant outside the UK and the following points in turn suggest how they might be addressed by child citizen science:

1. Loss of expertise. Citizen science engagement with children can expose children to the joy of science, equip them with some key knowledge and enthusiasm for science, and introduce them to the crucial roles of scientists in society. When children are included in citizen science projects, they learn about the rigour of scientific experiments and the importance of scientific integrity, and how to question valid research and evidence. Citizen science can thus increase awareness of the particular areas of scientific study being undertaken (Gommerman & Monroe 2017).

2. Nature of knowledge production within the context of an increasing variety of actors and collaborators producing 'science'. There is clearly a role for children in science and the more they learn how to do science, the better. Citizen science participation can thus be seen as a supplementary form of learning, beyond the curriculum.

3. Proliferation of communication channels and sources of information. In Bultitude's (2011) view, this is positive and can be further nurtured within the context of children and citizen science. Children can be actively engaged in citizen science through games, apps, computers, Geographical Information Systems and other technologies.

4. Democratic deficit. Bultitude (2011) comments on the disenfranchisement of citizens and their disconnectedness from decision-making and participatory processes (see also Smallman in this volume). She points out that 'recent changes in the nature of decision-making processes have created a "democratic deficit", whereby political-scientific decisions are increasingly made outside of the public arena' (Bultitude 2011, 3). As active participants in environmental

citizen science, children are more likely to be informed, understand the local – and global – issues at hand, and ultimately have more to contribute to a discussion or decision-making.

Engaging children in citizen science now – as they will be the future guardians of the environment – is a useful way to teach them about wildlife and habitats, engage them in conservation efforts and attain useful monitoring data and evidence on biodiversity and population health, and other environmental impacts which contribute towards effective environmental management. Hart (1997, 3), cautions, however, 'that all children can play a valuable and lasting role [in environmental protection] but only if their participation in taken seriously and planned with the recognition of their developing strengths and unique competencies'. As Kellett (2005, 10, section 6) asserts, children contribute to research through addition of their 'genuine child perspective', and their ability to communicate with their peers and disseminate information from a 'child voice' Kellett (2005, 10; 16; 19). Furthermore, on a positive note, as Cornell Professor John Losey explains, 'kids are high energy' and their lack of training may arguably lead them to search in places experts may overlook.[12]

Education also develops awareness and the ability to generate an informed opinion. Referring to Corburn's (2005) analysis of citizens in New York who educated themselves about neighbourhood environmental risks and successfully rallied against polluters because their children continually suffered from health issues, Jenkins (2011, 507) observes that:

> if these citizens had citizen science experiences during their science education, just imagine how much more empowered they would feel when facing such challenges. They may have to learn the specifics of the pollution to which they are being exposed, but they would already have authentic science experiences that they could build upon. Science becomes a tool of many that can be used to address concerns in people's everyday lives.

Data quality, ethics and practical considerations

Data quality is a key issue in citizen science (Fowler et al. 2013; Wiggins et al. 2011; Kosmala et al. 2016) and it is also perhaps assumed that children are more *likely* to obtain inaccurate data, due to their age, over-excitement or lack of attention (Miczajka, Klein & Pufal 2015; see also box 27.2 and box 27.3, above). There are not yet many studies comparing the quality of child citizen science data like-for-like with data collected by adults, though Burgess et al. (2017, 116) suggest that there is a 'higher probability' of professional scientists using data collected by retirees for primary research purposes. They are significantly more likely to use data collected by college students or adults with college degrees, and are ultimately more likely to use data collected by college students or adults with college degrees than younger individuals (Burgess et al. 2017, 116–7). As noted above regarding the two case studies (see boxes 27.2 and 27.3), data collection by children will vary from project to project due to variability in project design and goals, scope of the research being undertaken, prior knowledge and age of the children. However, we assert that citizen science projects can be designed to work around the abilities of the children involved (see further, Miczajaka et al. 2015; box 27.3).

Related to the data collection issue, this chapter has also alluded to the need for science to be legitimate, rigorous and accurate (Bultitude 2011), which is also a concern for academics involved in citizen science projects (Riesch & Potter 2014; Wyler & Haklay in this volume). A sensible approach for projects with children is to keep research methods simple, which will produce simple results more likely to be fairly accurate (Riesch & Potter 2014). Other scientists researching citizen science projects have commented that 'there is no such thing as quality of data, it's what you use the data for. Different uses will require different quality' (Riesch & Potter 2014, 112; Williams et al. in this volume). Although quality depends on the age and level of development of the child participants, children can still make valuable contributions to a project, particularly ones that require extensive monitoring over time and space (Miczajka, Klein & Pufal 2015). Furthermore, involving large numbers of children and changing the pool of researchers will increase accuracy. Anecdotal evidence from an OPAL[13] event in 2014 conducting a group-level and species-level identification exercise (trees and bumblebees) indicated that parents tended to 'jump in' and make a species identification based on existing knowledge, whereas children were more methodical, followed the guidance and came to the correct identification more often than their parents (OPAL 2016). To this end, advocates have produced guidelines and methodologies for including children in research (Johnson, Hart & Colwell 2014b). Johnson, Hart & Colwell (2014a) suggest six steps for engaging young children in research (see figure 27.1), including the development of ethical protocols for working with young children[14]. Further below, box 27.4 also suggests some child-centred approaches, following Piaget (Wadsworth, B.J. 2004) and

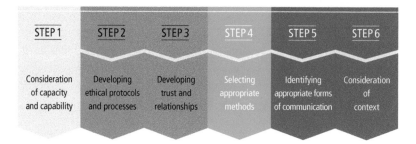

Fig. 27.1 The six steps for engaging young children in research. (Source: Adapted from Johnson, Hart & Colwell 2014a)

Box 27.4. Child-centred learning approaches

Child-centred learning approaches, as promoted by Piaget, fit well with citizen science, which:
- offers an environment within which to facilitate learning, rather than providing direct tuition;
- focuses on the process of learning, rather than the end product of it;
- promotes 'active methods' of learning that require 'discovery' and investigation;
- allows children to learn from each other and from their leaders/facilitators, working collaboratively and in a group;
- means children can be included based on the level of their individual development and ability. Tasks can be set accordingly so all children can be included.

---

box 27.5 includes some common sense approaches, for the inclusion of children in citizen science.

Going forward

This chapter has outlined reasons for the explicit inclusion of children as a distinct group in environmental citizen science projects. Reasons for inclusion have focused on the contribution that children can make to a

Box 27.5. Common sense policies for engaging
child citizen scientists

- Always obtain prior informed consent from parents/guardians **and** children (if there are specific vulnerable children, their school is likely to be aware of this so working with schools is sensible);
- Do not post photographs of child participants or name child volunteers (even if you are thanking them for their involvement) unless consent has been obtained;
- Do not give specific details as to ages, names, addresses, etc. Precise data can include using codes and generic information, so that what needs to be made public will be anonymised.

citizen science project in terms of data collection and monitoring, and also the insights to the project that emanate from their unique childhood perspectives and enthusiasm for learning as a result of being included. In turn, there is acknowledgement that participating children can be from a diverse mix of backgrounds, experiences and ages, and can gain the benefit of formal and informal scientific and environmental education and awareness raising, that can subsequently enhance their interest in and access to STEMM at school and in the wider community, foster the development of positive attitudes towards the environment, promote physical activity and assist in further developing their potential for inclusion in environmental decision-making.

The conclusions, so far, above, have drawn together some of the potential positives of including children in environmental citizen science but the question also needs to be asked as to whether children can really actually contribute to citizen science. Are they too young? Do they lack capacity and experience? Can they do the science 'properly'? Will they behave well? In response to these questions, it is noted in the work that there are shortcomings in relation to including children in citizen science projects, largely centred on the accuracy of data collection. Perhaps, however, the broader benefits of inclusion, outweigh the dis-benefits associated with data collection, and this work has suggested ways to address some shortcomings, for example, through setting scientific tasks for children that are aligned to their unique abilities and skill sets, pitching at an appropriate level of academic ability and experience, and being realistic about what the data will be used for.

Further, just as adults may have colleagues that are difficult to work with, some children might be 'challenging'. This does not mean that they should be excluded – if a citizen science project is engaging and interesting, children are likely to contribute well. Children like to be given tasks and to be productive, and they like to explore and learn when in a stimulating environment (Kellert 2002). To this end, there are many positive contributions that children can make to citizen science and that citizen science can contribute to children. To facilitate this, tasks need to be age-appropriate and with adequate supervision, explanation and guidance. A citizen science project with children is about developing the citizen and also developing the science. For project leaders, respectfully communicating in a way that is aimed at children, building their self-confidence, developing a sense of responsibility and ownership of the work can greatly assist in developing the child and the project.

Children arguably view the environment and their place in it differently from adults. Including children in citizen science means they will learn substantive skills, develop as individuals and hopefully, go on to be custodians of the natural world. Enabling – and encouraging – children to participate in science research projects 'is an empowering process', leading to a 'virtuous circle of increased confidence and raised self-esteem resulting in more active participation by children in other aspects affecting their lives' (Kellet 2005, 10). Furthermore, as demonstrated throughout this chapter, if projects are designed with regard to children's specific skills and abilities, they are able to contribute valuable data and research as citizen scientists. There is, therefore, a double reason for including children in citizen science.

Acknowledgements

The authors thank Susanne Hecker, Jonathan Silvertown and an anonymous reviewer for their comments and feedback. All errors and omissions remain the authors' own.

Notes

1   Though our experience is in the environmental field, we note that citizen science is not limited to the environment. The SciStarter database (https://scistarter.com/citizenscience .html – accessed 28 November 2017) contains environment-related citizen science projects, and some related to psychology, social sciences and computers and technology. There are also projects related to *inter alia*, language, literature, health, data processing, disasters, cybersecurity, war, etc., listed on the Scientific American Citizen Science database (https://

www.scientificamerican.com/citizen-science/ – accessed 28 November 2017) – another example of an excellent repository.

2 European Citizen Science Association, Ten Principles of Citizen Science, https://ecsa.citizen -science.net/sites/default/files/ecsa_ten_principles_of_citizen_science.pdf – accessed 28 November 2017.

3 See note 2, above.

4 Environmental citizen science can also positively contribute to objectives advanced under Local Agenda 21 schemes to achieve sustainable development at the local level. See further, *Constructing Local Environmental Agendas: People, Places and Participation*, edited by Susan Buckingham-Hatfield & Susan Percy, 2005, Routledge.

5 See note 2, above.

6 See note 2, above.

7 Cornell University Civic Ecology Lab https://civicecology.org/outreach/garden-mosaics / – accessed 30 May 2017.

8 Dobbies, in Kinross, Perthshire, Scotland, https://dobbies.com/events/little-seedlings / – accessed 25 May 2017.

9 Woodland Trust http://www.woodlandtrust.org.uk/visiting-woods/natures-calendar/ in collaboration with the NERC Centre for Ecology & Hydrology https://www.ceh.ac.uk / – accessed 25 May 2017

10 See: *Phenology Recording with Young Children,* https://www.tcv.org.uk/sites/default/files /172/files/CSR_Dobbies.pdf – accessed 30 May 2017.

11 The UK Open Air Laboratory Project: https://www.opalexplorenature.org/schools – accessed 30 May 2017.

12 Kids Count: Young Citizen Scientists Learn Environmental Activism: Student researchers become the eyes and ears of environmental scientists, By Evantheia Schibsted, October 2, 2007 George Lucas Educational Foundation, *Edutopia*: 'A comprehensive website and online community that increases knowledge, sharing, and adoption of what works in K–12 education'. https://www.edutopia.org/service-learning-citizen-science – accessed 30 May 2017.

13 'Open Air Laboratories is a UK-wide citizen science initiative founded in 2007 that allows people to get hands-on with nature while contributing to important scientific research.' http://www.imperial.ac.uk/opal/about-us/ – accessed 04 December 2017

14 In terms of getting 'free labour' and 'free data' from children participating in citizen science projects, ethical questions have been raised. Riesch and Potter (2014) comment that many citizens are willing to participate, *pro bono*, in citizen science projects in exchange for the learning and engagement opportunities. Here we will add then, that guardians, school-teachers and parents have to act legitimately in the best interests of the child when consenting to the participation of minors in citizen science projects. We contend that the larger benefits arguably outweigh the pitfalls. Further, any concerns over 'ownership of data,' the role of participants, and safeguarding of precise or personal data can be ironed out at the onset of the project, as seems to be the case in practice, and should not be a barrier to the participation of children in citizen science (Bowser 2014).

28

Turning students into citizen scientists

John Harlin[1], Laure Kloetzer[2], Dan Patton[1], Chris Leonhard[1], and Leysin American School high school students[1]

[1] *Leysin American School, Switzerland*
[2] *Université de Neuchâtel, Switzerland*

corresponding author email: jharlin@las.ch

In: Hecker, S., Haklay, M., Bowser, A., Makuch, Z., Vogel, J. & Bonn, A. 2018. *Citizen Science: Innovation in Open Science, Society and Policy*. UCL Press, London. https://doi.org/10.14324 /111.9781787352339

Highlights

- Schools can introduce vast numbers of citizens to participatory science.
- Students feel more engaged in their learning by participating in genuine scientific investigations where they are contributing to world knowledge.
- Citizen science projects offer opportunities for teacher professional development.
- Teachers have many opportunities to merge their curriculum with citizen science projects.
- Teachers need support in efficiently finding projects that fit their immediate classroom needs.

Introduction

Citizen science is growing in popularity, but most attention focuses on adult volunteers and their potential contribution to science and society. This disregards the millions of children studying science in school as they learn the skills of citizenship. Would hands-on involvement in real science projects simultaneously teach them about the scientific process and engage them with the world around them? Could these budding scientists contribute actual data and knowledge that adds value to science and society? Experience in education suggests that the answers to these questions are a resounding yes (see also Edwards et al.; Makuch & Aczel; Wyler & Haklay, all in this volume).

This chapter explores citizen science in schools. It highlights key learnings from the scientific literature, then explores a large teacher-developed citizen science project at the Leysin American School in Switzerland called LETS (Local Elevation Transect Survey). It includes the voices of students who participated in the LETS project. Finally, it shares ideas on how to integrate citizen science into teaching at schools. The conclusion discusses connections to the Ten Principles of Citizen Science.

How and why to embed citizen science into schools

Young people spend a large part of their lives in school. Some engage easily, whereas others struggle to see the 'relevance' of what they are studying. This can be especially true in the sciences, where concepts often feel remote from a young person's life. This is not helped by the fact that experiments they do in class and the data they collect are later thrown away. What if the data they collect could be preserved because it contributes to scientific knowledge, and maybe even helps to solve real problems?

Connecting citizen science and schools seems like a natural step. The promise of citizen science as an educational tool would appear to be a win-win game: teachers and students get authentic access to science in action, including scientists, scientific research questions, processes, data and data analysis, all of which promotes engagement with science and learning opportunities. Meanwhile, scientists get many enthusiastic volunteers (the students) along with team leaders and data quality filters (the teachers), while also expanding public awareness of their research topics and findings. A careful reading of the emerging scientific literature that explores citizen science projects in schools partly supports this hypothesis. It also highlights a few critical challenges, suggesting a 'trade-offs' model (Zoellick, Nelson & Schauffler 2012, p. 310).

benefits for all parties

Educational, motivational and transformative outcomes

What do students learn from participating in citizen science projects? There is considerably more literature on learning outcomes from general public participation in citizen science than there is from student participation. Research on public participation shows that learning outcomes are widespread but difficult to evaluate and highly differentiated. In environmental projects, Jordan, Ballard & Phillips (2012) identify the following learning goals from public participation: understanding ecology; understanding the science process; engagement with, and interest in, science and nature; motivation to participate; skill development in the scientific process and inquiry; environmental stewardship behaviours; and science and ecological identity. This typology has been extended to other scientific fields and to online citizen science projects with an alternative typology covering six levels of potential learning outcomes for individual participants: project-specific learning outcomes directly related to the tasks, concepts and mechanics of the project; disciplinary knowledge related to the topic of the project (for example, synthetic biology, philosophy or meteorology); scientific literacy; other knowledge and skills unrelated to the main topic of the project; personal development, including expanding interests and social networks; and identity change (Kloetzer et al. 2013; Jennett et al. 2017).

Citizen science in formal education, including primary schools, secondary schools and higher education, might be expected to bring similar individual learning outcomes. However, the material and social context of the classroom, as well as its social dynamics, are different from what can be observed in the general public. Specific research is needed to evaluate these learning outcomes but remains limited at the present time. The existing papers highlight three main things. First, citizen science projects indeed seem to teach disciplinary knowledge and increase scientific literacy (Zoellick, Nelson & Schauffler 2012), as well as positively alter attitudes towards science (Vitone et al. 2016). However, secondly, and most importantly, their value may go beyond these science-specific learning outcomes. The main outcomes of these projects may be motivational and transformative.

Participation in citizen science projects in college classrooms is reported to increase the sense of meaning of school learning and science courses. Considering Cell Spotting, a cell biology project, the authors write: 'besides helping students to consolidate and apply theoretical concepts included in the school curriculum, some other types of learning have been observed such as the feeling of playing a key role, which contributed to an increase in students' motivation' (Silva et al. 2016). In the classroom, teachers often struggle to find a balance between strict curriculum requirements and the desire to find new and interesting ways to engage and motivate students. Participation in the collection and analysis of real-world data is engaging for both students and teachers (Trautmann et al. 2012). By having actual value, citizen science imparts a sense of meaning in learning. Engagement with nature at a younger age may provide a means

give the learning contexts

for engagement with science: 'Connecting young learners to the natural world through a citizen science approach provides a meaningful context for learning about science in the primary/middle age of schooling' (Paige, Hattam & Daniels 2015).

Thirdly, going beyond students' individual learning outcomes, citizen science projects may also benefit teachers and change the nature of schools themselves. Benefits to the educational process itself may be observed through increased engagement of the students and social relevance of the topics. Vallabh et al. (2016) report the following example: studying a river-monitoring project, they suggest that shifting the emphasis of the project from scientific testing to matters of concern for the local community serves as a driver of learning and change by emphasising situational motives and lifeworld contradictions.

The need for careful design and support for teachers

The existing literature also identifies critical challenges for the success of citizen science projects at school. The primary challenge is the balance between scientific and educational goals: 'Citizen science program leaders and scientists must clearly define the desired balance between learning goals and scientific goals. If broader learning goals are a priority, then that should be reflected in the activities of participants, and these goals should be stated explicitly'. (Jordan, Ballard & Phillips 2012, 307). The tasks offered should be consistent with learning goals, which are largely defined for teachers by the school curriculum (see also Makuch & Aczel in this volume). This requires the careful design of tasks offering both scientific interest and educational potential, which might be difficult, as 'the questions of interest to the scientists [are] not aligned with student learning outcomes specified in state educational standards' (Zoellick, Nelson & Schauffler 2012, 312). Keeping this balance between scientific goals and educational goals may therefore require third-party mediation (Houseal 2010, cited in Zoellick, Nelson & Schauffler 2012).

science vs. learning: does one always have to suffer?

The second challenge is that of supporting teachers. Citizen science programmes need to offer relevant teaching material to ease the work of teachers in connecting them to school curriculum. However, 'simply offering project support materials, such as leaders' guides, to individual groups or teachers rarely leads to project adoption' (Bonney et al., 'Citizen Science', 2009, 980). Even more importantly, these programmes request and offer opportunities for teacher professional development. Reporting on the Acadia Learning Project, a collaboration with 11 schools, 20 teachers and thousands of students to investigate spatial variations in mercury in macroinvertebrates, Zoellick et al. (2012, 310) analyse the original impetus for working with teachers and students, which was 'a need to undertake long-term sampling and a desire to engage students in authentic scientific research', and noted that the project identified 'a need for teacher professional development'. As Zoellick et al. wrote, 'we had teachers and students who needed additional support to undertake basic scientific work but who valued the engagement with a real and complex project' (Zoellick et al. 2012, 312). This was solved with further professional training for teachers through regular online and occasional in-person access to scientists and summer institutes. Similarly, Paige et al. (2015) present two citizen science programmes developed as part of larger teacher professional development projects. In these cases, 'teachers realized the benefits for their students and their own professional learning' (Paige, Hattam & Daniels 2015, 11). This happened mostly through 'long term participation in small professional learning communities supported by university academics' (Paige et al. 2015, 12). Therefore, citizen science projects for classrooms should consider the needs of both teachers and students (Zoellick et al. 2012).

Support for teachers and careful design for educational purposes are important because otherwise it may be difficult for overworked teachers, constrained by busy curriculums, both to engage themselves in new, complex activities and to engage their students in activities with no clear connections to the required curriculum. Teachers often feel pressured for time and unsupported by administration when it comes to the extra effort needed to try a new form of teaching. Consequently, there are recurrent difficulties in recruiting classrooms into citizen science programmes.

Three models for embedding citizen science in schools

Three models for embedding citizen science in schools, which offer different resources to overcome the challenges identified above, are listed below.

Type 1: Adoption and adaptation of an existing programme
Type 2: Autonomous local development
Type 3: Local partnerships between scientists and teachers

These three models are summarised below before the chapter turns to present a case study of a type 2 project.

Type 1 projects take advantage of hundreds of school-friendly citizen science programmes worldwide, which may bring the difficulty of knowing which projects might fit a school's region or curriculum. For example, CITI-SENSE (www.citi-sense.eu) used high school students as citizen scientists in indoor air quality research. This international effort (across nine European cities) had the dual mission of gathering and analysing data, while also exploring how citizen science projects can best work with students and schools. In his report, SINTEF senior scientist Sverre Holøs stated that 'Results from the collaboration so far indicate that students and teachers are motivated to engage in these environmental studies, and able to perform studies of good quality' (Holøs 2016).

CITI-SENSE also found that while each city was successful at recruiting a school, considerable attention had to be paid to fitting the research into narrow windows of time during which the needs of the curriculum matched the needs of the science investigation. Schools also had concerns about privacy, misuse of data and how to navigate school policies on technology and internet access. CITI-SENSE therefore found that while recruiting schools was often successful, it requires significant time and effort.

challenges

Rather than actively recruit schools, it is more common for citizen science projects to simply make themselves available online for teachers to discover. Some projects have developed supportive resources, from teacher's guides to specific protocols and individual lesson plans. Perhaps the oldest and most widely used citizen science programme for schools is the GLOBE Program (www.globe.gov). Launched by NASA in 1995, GLOBE (for Global Learning and Observation to Benefit the Environment) is now used in over 100 countries and has over 100 million entries in its international database. Developed explicitly for schools, its teacher support materials are extensive and tied to American standards. A number of regional GLOBE offices have sprung up worldwide to serve local needs. Ironically, teachers in the GLOBE programme stated in personal communication that the sheer quantity of material that they offer is overwhelming for many time-strapped teachers.

Various hubs are also developing where teachers can learn about projects they might want to participate in. Some are highly regional, such as Tous Scientifiques (www.schweiz-forscht.ch), which promotes citizen science projects in Switzerland. For others, the earth is not large enough – Zooniverse (www.zooniverse.org) grew out of the popular Galaxy Zoo project, where anyone with a computer can help scientists to classify galaxies. The Zooniverse now offers photo-based identification and classification projects as wide-ranging as counting penguins in Antarctica and identifying endangered condors in California. Many of their projects offer supportive materials for teachers to use with their students.

The broadest citizen science project finder is SciStarter (www.sci starter.com), which offers over 1,600 projects. Users can narrow their searches by activity, location, whether projects are school-based and whether they offer teaching materials. SciStarter is currently biased towards the United States, which stems largely from its origins at Arizona State University's Center for Engagement and Training in Science and Society. However, according to the project's management, SciStarter also features an ever-increasing number of projects from outside the United States and is working to further develop its support for international education.

Type 2 projects are suited to especially motivated teachers who want to design their own projects relevant in their local environments. Key considerations in this context include, first, the choice of a research question, along with developing a connection to the scientific community. Next is the professional training of the teachers, if possible in a group where they can discuss how to guide students, as well as the ethical, scientific and practical issues of the research. Finally come the practical issues, including the choice of data collection and data analysis tools. Entering data on a website or app custom-built for someone else's project might not be useful. Several services now offer completely customisable data entry forms that are simple to use. MyObservatory (www.my-observatory.com), for example, allows users to create forms on their website that can be filled out in the field on a smartphone. The data goes to the MyObservatory site, which among other things offers the ability to create graphs and export the data in multiple formats, including universal comma-separated values (csv).

Type 3 projects involve deliberate partnering between scientists and educators. This type of project has been tested by Cornell Lab of Ornithology (CLO) over the last 20 years, where it has been extremely productive but requires careful planning and significant efforts from both teachers and researchers (Bonney et al., 'Citizen Science', 2009). It also requires interdisciplinary collaboration, in CLO's case between experts 'in education, population biology, conservation biology, information science, computational statistics, and program evaluation' (Bonney et al., 'Citizen Science', 2009). The co-construction of the research project facilitates connections to the curriculum, as in the BirdSleuth project, which was 'developed over three years with extensive input from more than 100 middle-school teachers across North America. Teachers helped to develop, pilot, and field test the curriculum so it covers subject matter (e.g., diversity, adaptation, and graphing skills) that teachers can easily integrate into their lessons' (Bonney et al., 'Citizen Science', 2009, 981). These partnerships require significant funding for both the teachers' and scientists' time to be able to reach a productive balance of scientific and educational goals.

In all cases, further research is needed to evaluate the outcomes and challenges of each of these types of citizen science projects in schools (see Kieslinger et al.; Richter et al., both in this volume). The next section presents a case study of a type 2 project.

LETS Study Leysin: An annual school-wide citizen science project

Presentation of the LETS Study

In 2014 the Leysin American School (LAS) in Switzerland decided to get involved in citizen science. Its teachers closely examined the local environment, considered how best to study it and developed a long-term research project appropriate to their locality and school. The school's individual experience has been broadened into a roadmap to starting a citizen science programme at school (see later section), which is intended to inspire other schools to develop their own long-term research projects.

The LETS Study Leysin project emerged from teachers' belief that getting kids outside and into the local forest would excite them about learning science; two years into the programme, they believe this more strongly than ever. Reaching across the curriculum to other departments, the study also engages the whole school, including nearly all the teachers. Following a strict set of data-collection protocols, students feel that they are contributing valuable information to experts who can use it to model the impact of global climate change on the forests of the Alps (the project, including its growing set of protocols, is described in depth at www.lets-study.ch.)

The town of Leysin is perched on a steep mountainside in Switzerland (figure 28.1). The town itself spans an altitude range from 1200 m to 1700 m. The hillside drops below the town to the valley floor at 450 m and rises above the town to a limestone peak at 2300 m, well above the local timberline. Thus, the obvious environmental characteristic of Leysin's geography is elevation. The main questions students are exploring through LETS are (1) How does altitude affect life? (2) How will climate change affect altitude distributions of species? The first question can be partially addressed during twice-annual days of research. The second question can only be addressed over a longer time period, but the

Fig. 28.1 The LETS Study Leysin plots studied by the Leysin American School span an elevation range from approximately 600 m to 2,300 m. Students visit the 30 m × 30 m plots that are not covered in snow twice per year, once in May and once in October. The plots are displayed in Google Earth.

research is expected to continue for decades, eventually turning into a serious longitudinal climate study, because it is being institutionalised in the school.

The transect itself was set up by LAS teachers in consultation with Dr. Christophe Randin, an ecologist from the nearby Université de Lausanne who specialises in the Swiss Pre-Alps, including Leysin (Randin et al. 2009). Teachers have so far established 14 fixed plots of 30 m by 30 m at altitudes from 600 m to 2,300 m (plus a dozen smaller meadow plots). These were chosen for their consistency of aspect, slope and forest cover, though there is diversity in forest type. Inside these plots, trees are identified, measured and mapped; species are inventoried with the iNaturalist app (which photographically records and geolocates species that can be corroborated via a social network of professional and amateur taxonomists); and students are given the opportunity to carry out their

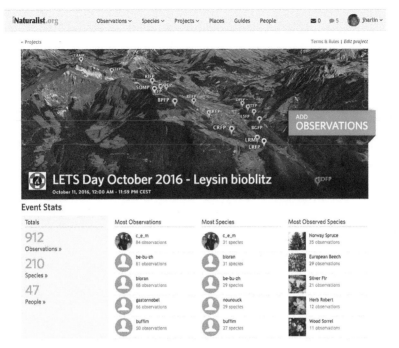

iNaturalist.org Observations ∨ Species ∨ Projects ∨ Places Guides People ✉ 0 💬 5 jhartin ∨

« Projects   Terms & Rules | *Edit project*

ADD
OBSERVATIONS

LETS Day October 2016 - Leysin bioblitz
October 11, 2016, 12:00 AM - 11:59 PM CEST

Event Stats

Totals	Most Observations	Most Species	Most Observed Species
912 Observations »	c_e_m — 84 observations	c_e_m — 31 species	Norway Spruce — 35 observations
210 Species »	be-bu-zh — 81 observations	bloran — 31 species	European Beech — 29 observations
47 People »	bloran — 68 observations	be-bu-zh — 29 species	Silver Fir — 21 observations
	gastonnobel — 66 observations	nounouck — 29 species	Herb Robert — 12 observations
	buffim — 50 observations	buffim — 27 species	Wood Sorrel — 11 observations

Fig. 28.2 LETS Days' happen twice a year: in October, about 130 7th- to 10th-graders fan out in groups of 10, one group to each site (with teachers for supervision); and in May, over 100 11th graders do the same as part of their International Baccalaureate (IB) Group 4 Day (Group 4 is a mandated co-operative science research in the IB programme).

own investigations. All accessible altitudes are investigated on the same day (snow-cover permitting).

Before heading out into the field, students write a journal entry recording their thoughts in response to the prompt, 'Describe the forests of Leysin'. In education terminology this is known as activating prior knowledge: asking students what they *think* they are going to see creates the mental space for them to absorb what they *actually* see. Laden with picks, tape measures, thermometers and cameras (among other tools), the groups then walk to their assigned plots. Once on location, students set up the boundaries of their study site then divide it into nine 10 m × 10 m subquadrats with string. Smaller teams measure, photograph and dig to collect data (figures 28.3 and 28.4).

During LETS Day in October 2016, LAS students were joined by about 50 students from the Université de Neuchâtel, along with a few

Fig. 28.3 The highest LETS plot with trees lies at 2,000 m on the Tour d'Aï. Here students have laid out strings to divide the 30 m × 30 m plot into nine subquadrats that are used for mapping the tree cover. (Source: John Harlin)

Fig. 28.4 The highest tree discovered by LAS students was found at 2,090 m on the Tour d'Aï during LETS Day 2015. (Source: John Harlin)

Fig. 28.5 Students practice tree measurements near campus at 1,390 m. During LETS Days the circumferences of all mature trees in most LETS plots are measured. (Source: John Harlin)

PhD candidates and their professor. Their mission was to conduct a more thorough BioBlitz (species inventory) of each plot by utilising iNaturalist and their skills with a taxonomy book. The school students finish each LETS Day by creating a quick poster based on their research.

Today we were scientists: Students recount their experiences during LETS Day

Overview of the experience: LETS Day report by the students

The following report has been compiled from students' own words as written in their afternoon reflection following LETS Day. The writing is lightly edited for continuity between the multiple authors.

> 'LETS Day was amazing. I was like a scientist. It was harder than I imagined, but an exciting experience. Learning about your community is very interesting.'

'Over time this project will help us understand the changes that are happening to the forests around Leysin. From that information people will learn what to expect. Climate change is a big issue, and Leysin, being on top of a mountain, could be very affected by it. Our studies could help our town to prepare for and adapt to the coming changes.'

'Citizen science is collecting data, analyzing it, and putting it out there for scientists to use. The data can be collected by anybody: students, teachers, workers, and many others. On LETS Day we collected data about trees, temperature, and other factors. It was quite interesting to feel like a specialist in tree identification. We entered the data into a document that can be looked at by scientists so they can observe climate change. This data will show differences when compared to data five or ten years later. Scientists can learn how the plants and animals start to move up the mountain because of climate change.'

'We saved lots of time for these scientists. We did part of the work on the forest and now they have to do the other part.'

'Here is what we did on LETS Day. First, we hiked for a long time. Many students were so tired. I like hiking, but I'm usually too lazy to walk anywhere, so I was happy to have this experience. I fell a lot of times, but it was okay because I learned how to hike in the forest.'

'Our first job was to find the orange buttons [these mark the corners of each plot]. It was very difficult because there were so many trees. But we found all of them, then we put strings between the buttons so it became a square. Then we put strings again and it became like a grid with nine subquadrats. Then we divided into groups and measured and mapped the trees inside the subquadrats. The highlight of our exploration was putting the white strings on the hill.'

'My group was in charge of "baby trees" and had to measure, take photos, and identify little trees. Others recorded the temperature every 30 minutes. Others wrote down the circumferences.'

'It was kind of confusing to do all the things at first but then we got it. We made mistakes but we fixed them easily and carefully. The data we collected today was pretty accurate. Our group members were working together and we got everything done fast and with high quality.

Then we ate our sandwiches under a huge tree with mud.'

'At the end of the day we put our measurements into the computer and it gave us statistics like average circumference. We input the data that we collect every year where everybody can see them.'

Highlights of the day

Interestingly, the highlights of the day vary greatly from one student to the other, even in the same group, highlighting that the real benefits of the project depend mostly on the subjective experience of the students. We identified three main types of reported highlights, which we call acting in nature, acting together (teamwork) and discovery.

Most LAS students come from large urban environments and their strongest impressions were simply of being outside in the forest, which was new to them. Some memorable quotes are included below.

'It is so cool to be in the forest. You get to run around, take pictures, and help other scientists. I will invite friends to come here and see what I just saw: a magical forest. Overall, it was a very wonderful and memorable experience, one that you have to have once in your life.'

'The highlight was eating and laughing with my group. We were all really cold and it was funny. I was really badly dressed but overall it was fun and interesting.'

'We went to the highest point of the mountain and I liked it very much. The view was amazing and in my opinion the exploration we did will help to find the difference in forest climate within the next years.'

'The highlight was being able to hold an earth worm in my hand for the first time. It was great!'

Some students remarked on how they felt empowered by teamwork and leadership:

'My favorite part of the day was having the opportunity to be a leader and help my community. Even if it was just a small part, small parts can have big jobs.'

'I enjoyed the leadership opportunity. I think it inspired me to try more leadership activities at LAS.'

'The team work made us close to each other. I really appreciated how teammates helped each other, all united in order to contribute to the ecosystem study.'

It suggests that various roles in teams may trigger different experiences and outcomes.

For others, the highlight was discovery, including how much they enjoyed field science:

'The highlight was meeting the university students and exploring the forest with them. Finding mushrooms and plants, observing them, and looking at differences.'

'I liked talking to the college students because that gave me knowledge on why climate change is a real issue that affects all of us directly.'

'How great of a school I am in to be able to physically study climate change and understand nature!'

'When I go to a university, I want to research forests.'

Roadmap to starting a citizen science programme at school

While LETS Study Leysin is an altitude transect and is thus not universally applicable, the teachers who invented it hope that the concept of transects will be picked up by other schools and adapted to their local environments. Transects are well established in ecological research and are thus a good concept to teach students. Even more important is the concept of long-term research, which is especially vital in climate studies. If schools can establish long-term observations of their local environment and collect the data, they can simultaneously teach basic biology and contribute to the advancement of scientific knowledge.

20 design claims for educational citizen science at school

Based on the literature and our experience in the LETS project, we would like to suggest some 20 design claims, which are organised into five categories: curriculum, resources, official support, teacher training and networking, and community. These design claims may orient overall project designs (in the broad sense of designing the tasks and the pedagogy of the project, but also its human, social and technical context).

1. **Curriculum**
   1.1. Tie project tasks to the curriculum, even at the textbook level.
   1.2. Create adaptable lesson plans and support teachers creating their own lesson plans.
   1.3. Design assessment tools that match local standards.
   1.4. Scaffolding: offer levels of advancement, both within projects and between projects.

2. **Resources**
   2.1. Plan resources for the teachers to support the extra effort required to engage in citizen science.
   2.2. Create and moderate a system for peer-to-peer sharing, discussing and learning.
   2.3. Provide flexible tools to create citizen science projects.
   2.4. Provide technical support from experts.

3. **Official support (support from school administration and education departments)**
   3.1. Integrate citizen science into the school philosophy and recognise citizen science as an educational tool in the school policy.
   3.2. Support citizen science training as professional development.
   3.3. Encourage flexibility at higher levels, including administration, education boards and curriculum developers.
   3.4. Provide official recognition for innovative pedagogy in science education.

4. **Teacher training and networking**
   4.1. Provide hands-on interactive training.
   4.2. Develop sources of fresh ideas for teachers who want to try something new each year.
   4.4. Develop library of how-to videos for using citizen science in schools.

4.5. Encourage peer-to-peer sharing, provide a user-friendly platform for the teachers to connect with each other, share experiences, get feedback and cooperatively develop lesson plans.

4.6. Develop a platform for teachers to connect with scientists who support school projects.

5. Community (connection with local community)

5.1. Team up with museums and cultivate them as key allies.

5.2. Develop public spaces (such as elegant websites) for presenting school projects.

5.3. Build in opportunities for parental involvement.

A process to launch a first citizen science project at school

One of the great wonders of citizen science in schools is that there are so many possible directions to take. Ironically, this cornucopia of choice can be daunting for a teacher. How to choose the right project for one class or school? How to launch a first citizen science project in a specific school context? Based on the literature and experience, we suggest the following 10-step roadmap to help schools launch their own citizen science programmes.

Conclusion

Citizen science engages school students in many of the same ways that it is known to engage adults. Although literature on citizen science and schools is still emerging, there are some rich experiences to draw on. This chapter shows that such engagement adds significant value to formal education.

One challenge lies in merging the scientific value of projects with their educational value and, when necessary, prioritising goals. The Ten Principles of Citizen Science puts a strong emphasis on the scientific value of projects, including making the data publicly available. Science teachers, by contrast, have a clear professional priority to fulfil their educational mandate and have neither the tradition nor external motivation to achieve meaningful (e.g., accurate) data, nor to share their data outside the classroom. Without shared meaningful data, the projects have no direct scientific value. Thus, for citizen science to achieve its potential within formal education, it will need significant teamwork between practising scientists and practising educators. Such teamwork that benefits both professional

Box 28.1. 10 steps roadmap for launching a citizen science programme at school

1. Listen to your stakeholders. What questions excite the teachers and students? What are the talents of the people around you? Do you know any local scientists to discuss this with?
2. Consider your environment. What is available locally that you could research?
3. Hatch an idea. Think of engaging research topics. Are there any environmental or social hooks you can bring to your project? (e.g., water quality, garbage, air pollution, biodiversity changes, habitat conservation and so on.)
4. Build institutional support. Present the idea to administration departments and other stakeholders. Be sure to understand the details well enough to respond to concerns.
5. Cultivate connections to the scientific community. Universities, museums, science centres and other community groups often include community education in their missions. Use these human resources whenever possible – they add meaning to the project and help with student engagement. Ideally get them involved in steps 1–3.
6. Use good pedagogy. Be sure to tie your project to your curriculum. The project must support student learning at their level. Consider safety and privacy issues.
7. Follow the Ten Principles of Citizen Science as best you can, but recognise that your bottom line as a teacher is to educate, which loosely falls under number 9, 'participant experience and wider societal or policy impact'.
8. Launch your project. Expect something to go wrong.
9. Ask for feedback and adapt accordingly.
10. Think long term. The first time you try a new project might not yield great science, but student learning is at least as valuable as the data you gather. If you are doing worthwhile research, repeat it year after year, improving the results over time and gradually building a long-term study that offers real value to science as well as to education.

and citizen scientists ('citizens' in this context being students and teachers) is in fact one of the Ten Principles in its own right.

Another challenge is working within the difficult constraints faced by teachers, including time, training and curriculum. These challenges can be overcome by motivated educators with the help of developing technologies (apps, knowledge-sharing platforms) and a support structure built around integration of citizen science into the curriculum that includes a recognition of its value for young citizens and for science.

29
Citizen science and the role of natural history museums

Andrea Sforzi[1], John Tweddle[2], Johannes Vogel[3], Grégoire Lois[4], Wolfgang Wägele[5], Poppy Lakeman-Fraser[6], Zen Makuch[6] and Katrin Vohland[3]

[1] Museo di Storia Naturale della Maremma, Italy
[2] Natural History Museum London, UK
[3] Museum für Naturkunde Berlin, Germany
[4] Muséum National d'Histoire Naturelle Paris, France
[5] Zoological Research Museum Alexander Koenig, Bonn, Germany
[6] Imperial College, London, UK

corresponding author email: direzione@museonaturalemaremma.it

In: Hecker, S., Haklay, M., Bowser, A., Makuch, Z., Vogel, J. & Bonn, A. 2018. *Citizen Science: Innovation in Open Science, Society and Policy*. UCL Press, London. https://doi.org/10.14324/111.9781787352339

Highlights

- Historically, natural history museums (NHMs) have a long history of collaboration with the amateur-expert naturalist community. A tradition of two-way knowledge sharing that continues today.
- Over time, NHMs have renewed their functions within society and assumed a relevance not only for the conservation of collections, but also for engaging society in the generation of new scientific awareness and understanding of the natural world.
- Natural history museums now deliver a wide range of field-based and online citizen science projects and play a central role in supporting the development of citizen science and citizen scientists.
- Natural history museums have also taken a central role in establishing the European Citizen Science Association (ECSA) and are well-placed to both promote the field of citizen science and support capacity building within critical subject areas such as taxonomy.

Introduction

The study of nature and the environment by amateur natural historians pre-dates both the professionalisation of science and modern definitions of the term 'citizen science' (e.g., Silvertown 2009; Bonney et al., 'Public Participation', 2009). In some parts of Europe, volunteer-led gathering of observations of wildlife ('biological records') and specimens has a long and illustrious history and this continues today (Silvertown 2009; Roy et al. 2012; Pocock et al. 2015; and see also Mahr et al. in this volume). A significant level of taxonomic expertise is contained within this community: a recent study demonstrated that over 60 per cent of the 770 new species discovered on average each year in Europe since the 1950s are described by non-professional taxonomists (Fontaine et al. 2012).

Since their establishment, NHMs have worked closely with these amateur-expert communities, with many NHMs being founded – and their scientific collections subsequently developed and maintained – with the invaluable support of enthusiastic, highly skilled amateur naturalists. Similarly, long before citizen science became the widely publicised concept of today, NHMs, other academic organisations and amateur naturalists worked together to build understanding of the breadth and dynamics of a country or region's biodiversity (see Miller-Rushing, Primack & Bonney 2012). One of the earliest examples of such collaboration is the Christmas Bird Count: initiated in 1900 by the American Museum of Natural History ornithologist Frank Chapman, this project has subsequently developed into the longest-running active citizen science project (National Audubon Society 2017). Since the initiation of this formative project, NHMs and amateur naturalists have collaborated on diverse programmes, from co-ordinating national bird ringing initiatives to the production of national species distribution atlases. This long history of amateur-professional collaboration through the spontaneous evolution of sound partnerships continues today.

Over the past two decades, however, NHM-led citizen science programmes have undergone a rapid and marked increase in both number and diversity, with a wide range of project types and collaborations being developed alongside these more traditional activities. In part, this change mirrors the wider expansion of the global field of technology-mediated citizen science (e.g., Roy et al. 2012; Pocock et al. 2017; Novak et al. in this volume). Critically, though, it also reflects the two synergistic and shared goals of museums and citizen science: the generation of new scientific understanding and education (Ballard et al. 2017).

Many NHMs have recently undertaken the ongoing, and in some cases quite profound, transformation of the organisation of their core functions and the ways in which they interact with visitors and local communities. Alongside their traditional roles of conserving and providing access to specimen collections, NHMs are increasingly looking to actively engage members of the public in projects that seek to build understanding of the natural world. They have expanded and diversified their public-facing work to encompass the development of new educational approaches and tools that seek to engage broad sectors of society with the science of natural history.

This increased emphasis on societal engagement reflects two key factors:

1. Museums – and national museums in particular – are increasingly driven by the need to maximise their public value and impact across society and over large geographic areas. This includes reaching and engaging non-traditional audiences and people that cannot physically visit the museum itself (e.g., Robinson et al. 2016).
2. Growing recognition of the importance of actively engaging wider society with the key biodiversity challenges of our time, from understanding the impact of environmental change, to building awareness and knowledge of the critical role that nature plays within society, and developing effective conservation practices.

Citizen science is a central approach through which many NHMs are seeking to tackle these challenges. As a methodology, it is an effective way to combine the scientific and educational remits and expertise of museums. It makes full use of their wide audience reach and trust (Ballard et al. 2017) by directly involving people with museum collections and cutting-edge research on key socio-scientific challenges.

This chapter illustrates how European NHMs are seeking to support the development of citizen science and citizen scientists as well as some of the key forms of NHM-delivered citizen science. It concludes by summarising factors influencing the success of citizen science in a museum environment, the challenges and next steps.

The following sections outline four of the most common ways NHMs are actively contributing to the field of citizen science.

Supporting the development of amateur-expert naturalists

Citizen science includes a broad range of potential participants, from untrained members of the public with limited subject knowledge, to practitioners whose expertise is equivalent to that of professional researchers. Whilst NHMs often look to support this spectrum, there is often emphasis on supporting amateur-expert naturalists and taxonomists to develop and share their faunistic and floristic knowledge, such as identification and field skills. These areas are otherwise facing a profound skills crisis (Cutler & Temple 2010; e.g., Boxshall & Self 2011) at a time when the demand for the biological monitoring and conservation assessment of habitats and species is increasing (Collen et al. 2013; Owen & Parker in this volume).

Museums directly support amateur-expert citizen scientists in diverse ways, from providing access to resources such as reference specimen collections, libraries, meeting rooms and technical equipment; to archiving personal specimens, herbaria and book collections; to direct scientific support, training and mentoring in species identification, field survey and research methods. Creating opportunities for amateur and professional naturalists to interact and share their skills is key to these activities, to mutual benefit.

Non-governmental organisations (NGOs), such as regional natural history associations, are common in many European countries but many face recruitment problems, arguably because school biology classes – and increasingly universities – rarely include teaching units on species identification or local biodiversity. Young scholars have few opportunities to experience the joy of discovering biological phenomena or interesting organisms in nature, and schoolteachers rarely receive training in faunistic and floristic studies. The demography of many natural history societies is ageing as a result (e.g., Hindson & Carter 2009). In more specialist societies (e.g., entomological groups) in Germany, the majority of members are often older than 50 years of age and male. Programmes are therefore needed to train the next generation of naturalists and diversify this demographic. In the UK context, NHM London's Identification Trainers for the Future programme (www.nhm.ac.uk/take-part) is one such project.

Natural history museums can help to develop an out-of-school/ university curriculum for interested scholars. They can provide space, books and microscopes, while NGOs contribute experts for taxonomy and environmental assessments where this expertise is not available in-house.

In Bonn, the Zoological Research Museum Alexander Koenig (ZFMK, a member of the Leibniz Association) runs a series of junior research clubs. There are four age classes for young people aged between 8 and 18, who work in the museum at weekends. Initially they (playfully) learn the difference between animal groups (e.g., reptiles, amphibians, insects) and take environmental samples (e.g., in creeks and lakes). The older age groups work on comparative morphology and taxonomic tasks. As well as imparting biological knowledge, this direct experience of taxonomy and fieldwork can help to demonstrate the relevance of natural history as a pastime and career (www.zfmk.de/en/research/education).

Hosting biological recording schemes and developing species monitoring projects

A key issue for launching biological recording schemes – long-term species observation and recording initiatives – is ensuring that they are financially and practically supported over a substantial period of time by key stakeholders. In the UK and Germany, many of these schemes are entirely run by volunteers or NGOs, who lead on identification training and data collection, collation and verification. Data storage and analysis support is often provided by the national biological records centre (e.g., Pocock et al. 2015). In France (where national NGOs were not available to manage national monitoring initiatives), the Muséum National d'Histoire Naturelle Paris (MNHN) started to support national recording schemes as a significant component of its citizen science programme, playing a lead role under its obligations to the joint authorities of the Ministry of Research and the Ministry of Environment. Monitoring started with birds, as the museum holds the French bird ringing scheme. In 1989, the MNHN launched a national Breeding Bird Survey (BBS) based on point counts, and a constant effort site-based, capture-mark-recapture study led by amateur ringers (Julliard, Jiguet & Couvet 2004).

In Paris this model was subsequently adapted to encourage a less experienced audience and incorporate public surveys of garden butterflies, snails, bumblebees and birds (see also Peltola & Arpin in this volume on an identification project with city gardeners in Grenoble). SpiPoll, a photographic survey of flower-dwelling insects, and a survey of wild plants in city streets were also introduced in 2010 and 2011, respectively. All of these schemes are co-ordinated by the same scientific team based at the museum, but each also relies on a specific NGO partner to play a key role in supporting participants.

As well as gathering scientific data, these schemes provide learning opportunities for participants. Within the SpiPoll project, participants' identification accuracy was found to significantly increase over time, with web tools supporting learning about the key identification features of invertebrates (Julliard 2015, pers. Comm.). There were 1,300 participants, of which only 43 per cent declared 'to know about insects prior to commit in this program', and a recent study has shown that identification accuracy rates increase with experience (Elise Elwood, pers. comm.). Photographs of 630 insect taxa, identified by participants and checked by experts, started with a poor success rate of 50 per cent but subsequently increased by 7 per cent for every 10 identifications a participant made.

The NHM London has also developed a wide range of mass participation biodiversity monitoring surveys, which have collectively engaged over 64,500 people since 1996 (Ballard et al. 2017). Between 2009 and 2014 NHM London was a key partner in the highly successful OPAL programme, which launched seven national citizen science surveys and held a number of BioBlitz events across the UK (see also Makuch & Aczel in this volume). A BioBlitz is a public-facing event that aims to discover as many species of living organisms as possible, within a set location over a defined time period, usually 24 hours. The national surveys studied a range of taxa including earthworms, lichens and invertebrates as bioindicators of environmental condition. Surveys were developed in collaboration with 13 OPAL partner organisations and a range of supporting organisations, including voluntary natural history societies, research institutes and government agencies. OPAL is funded by the UK National Lottery and in line with funder priorities has a strong focus on engagement and participation, especially with traditionally less-engaged audiences (Davies et al. 2013).

Data from many of the national surveys have been published in peer-reviewed journals indicating the wider research contribution of this kind of citizen science data (Fowler et al. 2013; Seed et al. 2013; Bone et al. 2014; Bates et al. 2015). OPAL also introduced many new people to the process and value of biological recording, in particular through online resources and data entry, and a country-wide network of community scientists who delivered training and events, and supported new participants (Davies et al. 2013; Barber et al. 2016; Davies et al. 2016). As a trusted voice of authority, nationally recognised brand and much-loved institution, NHM London played a key role in OPAL as the public face of the project, leading on media and publicity, public events and exhibitions, and taxonomy.

A key outcome of the OPAL project was open source software designed to enable naturalists to create their own biological recording website (Indicia 2016). Inspired by the OPAL model, this software was subsequently used by the Maremma NHM (Grosseto, Italy) to set up a recording website for public surveys (www.naturaesocialmapping.it). The platform is used to collect fauna and flora sightings, as well as to develop national species inventories. A leading example is the national survey of the crested porcupine (*Hystrix cristata*), a large rodent currently expanding its range in Italy. The national survey collected public sightings to reconfirm the presence of the species within its historical range and to map newly colonised territories. Contributors included wildlife enthusiasts, first-time citizen scientists and wildlife professionals. Most sightings were accompanied by pictures of road-killed animals, quills, faeces or tracks. Given its peculiar body structure and distinct tracks and signs, the crested porcupine is easily recognisable and almost impossible to misidentify. Nevertheless, data are verified by project experts before being added to the main database. The data has allowed the creation of the first live map of the species, educating people about native and non-native species (Mori, Sforzi & Di Febbraro 2013). Feedback to participants also includes access to a private area of the website containing the lists and pictures of all the sightings and a personal, automatically updated map. There are plans to engage repeat participants by communicating the importance of regular monitoring.

A key example from Germany is the German Barcode of Life Project (GBOL) (www.bolgermany.de), which co-ordinates an inventory of all species in the country to develop genetic markers (DNA barcodes) for rapid, automated species identification from environmental samples. The programme is financed by the Federal Ministry for Research and Education and supports 300 voluntary experts. Contributors help to collect and identify a wide range of taxa for barcoding, from spiders and insects, to fungi, diatoms and flowering plants. These samples are sent to participating NHMs where DNA samples are extracted and sequenced, and associated voucher specimens (any specimen that serves as a basis of study and is retained as a reference) stored. Sequence data are made available via free and easily accessible databases (BOLD, GBOL: Geiger et al. 2016) and the contribution of all partners is shown on the project website. Citizen scientists can access training on how to evaluate sequence data and receive a small expenses allowance for providing specimens (figure 29.1).

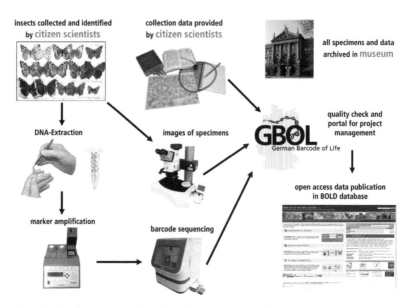

Fig. 29.1 Flow chart showing the structure of the German Barcode of Life

Museum-led BioBlitzes

BioBlitz events provide the opportunity for professional scientists, amateur naturalists and local communities to explore and learn together (see also Gold & Ochu in this volume on citizen science events). They raise awareness of biodiversity and the importance of biological recording, while generating a biodiversity 'snapshot' inventory for a given site (see Robinson et al. 2013).

MNHM has organised several 24-hour BioBlitzes (Sforzi 2017), each located within a Natura 2000 site – a European network of core breeding and resting sites for rare and threatened species and natural habitat types. The choice of sites facilitates promotion of the Natura 2000 network among local people while contributing to the gathering of knowledge for the implementation of EU Bird and Habitats Directives' reports. On average, 30 different wildlife surveys were carried out in each BioBlitz, with more than 1,200 participants contributing over the last five years. This level of participation is encouraging, especially considering the geographic position and low population density of the survey areas (Southern Tuscany, Italy). Final species lists ranged from 450–700 terrestrial, freshwater

Fig. 29.2 Snapshots of some activities carried out during the BioBlitzes organised by the Maremma Natural History Museum. (Source: Andrea Sforzi)

and in some cases marine species. Rare native and invasive alien species were also recorded to address specific conservation issues and reports synthesising the main outputs were distributed to all participants (Sforzi et al. 2013) (figure 29.2).

Many other NHMs have focused on BioBlitzes as a way of doing citizen science. For example, NHM London has led or convened events that have attracted between 500 and 8,000 members of the public and helped to develop understanding of UK biodiversity, as well as supporting site management practices (see Ballard et al. 2017). In Germany, BioBlitzes are typically organised in co-operation with the journal *GEO*. In 2014, the MfN Berlin, with citizen science partners ORION and naturgucker.de, organised a BioBlitz in the president's garden and presented the results during a large public event. As well as generating biodiversity data, the event aimed to raise awareness of urban biodiversity. This is a common theme across NHM citizen science activities, reflecting the urban location of museums and their visitors.

Digital technology–mediated citizen science

Digital technologies are increasingly enhancing both citizen scientists' experiences and research leaders' ability to access and process collected data. In the context of NHMs, digital participation is a highly scalable approach that can extend project reach and engagement nationally and internationally (Robinson et al. 2016). The extraction of data from specimen archives is a key area where this is being applied (see www.idigbio .org). Natural history museums house vast collections on the world's biodiversity and geodiversity, with biological material collected across more than three centuries. The majority of specimens are accompanied by a written or typed label detailing the species name, its collection location and date; a treasure trove of information with the potential to enhance scientific understanding of species range dynamics, population genomics and responses to environmental change (e.g., Johnson et al. 2011; Bi et al. 2013; Willis et al. 2017).

The digitisation of this specimen data – which is often handwritten – cannot yet be fully automated and specimens held across institutions cannot be easily accessed or searched. National history museums across Europe and the rest of the world are initiating mass digitisation programmes to image and catalogue their collections, and make the information freely available and searchable for researchers and the public. MNHN

Paris' project, Les Herbonauts, and NHM London's involvement in projects, including Notes from Nature and Herbaria@Home, demonstrate that involving the public in transcribing information from specimen labels and registers into digital databases can be highly successful, but not without its challenges. For example, nineteenth-century handwriting can be extremely difficult to interpret, particularly when it comes to unfamiliar place and species names.

Previous activity of this sort was undertaken on site by museum volunteers and required computers, desk space and supervisor time. Online crowdsourcing is a relatively new activity, enabled by widespread access to technologies and fast download speeds for accessing high-quality images. Undertaking crowdsourcing on a mass scale moves museums into a new sphere of interacting with online, geographically dispersed digital citizen scientists. NHM London is currently exploring how lessons learned from traditional citizen science activities can be applied to this online environment, how participant motivations and online engagements differ (or are similar to) those of outdoor citizen science projects, and how the benefits gained by volunteers working alongside curators on site may be transferred to the geographically distributed model of crowdsourcing. The Orchid Observers project is one route through which these questions are being investigated (figure 29.3).

The digital revolution is also influencing how field-based NHM citizen science projects are being delivered. Improved access to the internet and availability of mobile phones with built-in cameras and GPS receivers is rapidly expanding the digital element of many pre-existing or 'traditional' recording activities (see also Mazumdar et al. in this volume). For example, MNHN Paris' SpiPoll project asks participants to photograph every different kind of invertebrate visiting the flowers of a chosen plant for 20 minutes. The participant then selects one photo per 'morphospecies' and uploads it to the website. An online-identification tool based on a learning neural network then compares the image with a database that currently contains data on 630 species, and estimates the probability of a correct identification. Together with a photo of the plant and its immediate surroundings, this list of photographs makes a 'collection' which is then shared with other participants for validation. Over six years, 1,500 participants have contributed 27,000 collections and an incredible 270,000 pictures; a significant volume of scientific data that could not have been gathered by other means.

The OPAL Bugs Count app, developed by NHM London as part of the OPAL programme, enables participants to photograph the target species

Fig. 29.3 The Orchid Observers project at NHM London combines the online transcription of historical specimen data with contemporary biological recording. It investigates how the UK's flora is responding to climate change by building a 200-year record of flowering times for 29 species of UK orchid. Over 1,900 people have participated, extracting scientific data from 3,700 specimens and generating 1,700 new field observations. The project is a collaboration with Zooniverse and the University of Oxford. (Source: Orchid specimen – Natural Museum [NHM] London. Screenshots of Orchid Observers website – NHM/Zooniverse)

of invertebrate and immediately submit a georeferenced photograph to the survey team for verification. The latter step streamlines data upload and enables data quality to be managed. Significantly more photographs were submitted via the app than through the project website (Robinson 2015, pers. comm.), reflecting the ease of use and growing familiarity with app-based systems. MfN Berlin is currently developing an app platform called Anymals, which can be adapted to any taxonomic group or research question. Data from the app can be automatically made available via the Global Biodiversity Information Facility (GBIF), to the benefit of global research and conservation communities (see also Ballard, Phillips & Robinson in this volume for more on the GBIF).

Discussion

This chapter has described some of the principle ways through which NHMs are actively employing citizen science to meet their joint educational and scientific research goals. Alongside more traditional mechanisms of support for amateur-expert naturalists, the benefits of involving large numbers of members of the public (who may have had no prior subject experience) in gathering and curating species observations and extracting data from historical specimens is increasingly recognised (e.g., Ballard et al. 2017). In the context of NHMs, citizen science can be thought of as a disruptive approach that is actively shaping how museum-based biodiversity research and engagement are being delivered. This final section summarises some key conclusions relating to the application and future potential of citizen science, through the lens of European NHMs.

Factors that influence the role and success of citizen science within NHMs

National history museum missions generally span collections development and access, advancing scientific knowledge about the natural world, and inspiring and educating the wider public about the wonder, diversity and importance of nature. A key reason for ongoing investment in citizen science by major NHMs is because it can simultaneously deliver each of these core, mission-related priorities. This has been critical in securing the necessary resourcing from scientific and public-facing teams. NHMs are also particularly well-placed to deliver and facilitate citizen science, due to their long experience in science communication and collaboration with amateur-expert naturalists, and their combination of scientific, education and engagement expertise. Other relevant factors are institutional longevity, public profile, reach and trust, all of which aid project publicity and uptake (see also Richter et al. in this volume).

A number of published resources review and outline best practice steps delivering citizen science projects based on the success of individual projects (e.g., Roy et al. 2012; Tweddle et al. 2012; Pocock et al. 2014b, Robinson et al. in this volume). Collectively, these highlight that citizen science is most effective when a) there is clear benefit for both researchers and citizen scientists; b) the project aims are clearly stated from the outset; c) researchers have (or can benefit from colleagues who have) experience in public engagement and outreach; d) evaluation is built in and used to improve the programme over time; e) citizen scientists are valued and adequately supported; f) the quality of the scientific data generated is measurable; and g) the project is appropriately resourced. In addition, Devictor, Whittaker and Beltrame (2010) advised that projects should manage their resources sustainably to ensure that data are adequately stored, analysed and published. The most successful NHM-led projects take all of the above elements into account. However, this has clear resource implications and is not always fully achievable. Working effectively across the institution can be a particular challenge, as museum departments often manage time and resources in different ways.

Finally, as respected and politically neutral institutions, museums are particularly well-placed to act as platforms and conveners for citizen science co-ordination, exhibitions, discussion and debate. Indeed, this type of activity directly helps museums to demonstrate their societal relevance and value (see Wyler & Haklay on the university context too). Citizen science requires institutional support, such as trained project personnel and an investment of researchers' time and resources (Novacek 2008; Richter et al. in this volume). NHMs are favourably placed to provide such support at the science-society interface. Germany is an excellent example of how this is being implemented as the Federal Ministry for Education and Science supports a consortium of research institutions, including the MfN Berlin, to hold workshops, develop guidelines and develop a strategy (Bonn et al. 2016) for citizen science within Germany. This has been achieved through a broad consultative process and investment in supporting networking between initiatives, for example, via the web page (www.buergerschaffenwissen.de) and events. In the UK, the Angela Marmont Centre for UK Biodiversity at NHM London is a free resource centre where amateur naturalists can develop their skills and acts as a hub for citizen science. A core strand of their citizen science programme is dedicated to supporting other citizen science practitioners.

Challenges

In keeping with other organisations, securing sufficient resources to establish and maintain projects is not always straightforward. Resources (e.g., staff time) for community management to deliver project evaluation is often most affected. In keeping with the field of citizen science as a whole (e.g., Pandya 2012), significant work remains to develop truly inclusive NHM-led projects (see for example Haklay; Smallman, both in this volume). The broad visitor demographic of many NHMs indicates that they can play an integral part in tackling this critical challenge (Ballard et al. 2017) but if citizen science is to become a truly accessible field of practice, sustained work is required in this area and that of open data (e.g., Groom, Weatherdon & Geijzendorffer 2017).

Arguably the most pressing concern in the specific context of NHMs, however, is the decline in numbers of both amateur and professional taxonomists. Volunteer biodiversity recording has also declined and a new generation of enthusiasts needs to be recruited (Hopkins & Freckleton 2002). In Germany, Frobel and Schlumprecht (2016) document a 21 per cent reduction in the number of taxonomists over the past 20 years, with only 7.6 per cent of experts being younger than 30. This is especially problematic for nature conservation. It has been argued that specialist amateurs are declining, while more generalist volunteers and environmental enthusiasts are on the rise (Lawrence 2010). More recently, other authors have warned about the decline, death or 'impending extinction' of natural history as both an academic subject and amateur activity (Tewksbury et al. 2014). Should this be confirmed, it would be a source of great concern, not least as the need for biodiversity knowledge is increasing (Dayton 2003). An adequately trained group of amateur and professional taxonomists is central to knowledge of the world's biodiversity and how it is responding to pressing environmental changes (e.g., Davies et al. 2016). Increasingly, citizen science shows the supporting role that can be played by people of all ages, backgrounds and subject knowledge. However, continued support is needed for the development of individuals and communities with high levels of taxonomic knowledge and the motivation to observe and document changes within nature over long timescales. National history museums, universities, academic researchers and NGOs can cooperate to fill this gap, developing solutions especially designed for particular demographics such as youth audiences (see also Harlin et al.; Wyler & Haklay, both in this volume).

The future of NHM citizen science

The examples described above demonstrate how museums are continuing to support traditional citizen science activities such as biological recording, while embracing new approaches and technology developments that were unimaginable a few decades ago. Collections-based projects will remain a central area for innovation, as will the development of ever-more engaging citizen science gallery-based displays and interventions.

More broadly, NHMs are ideally placed to continue to lead by example, delivering projects with strong outcomes for both science research and environmental education and showcasing the work of other practitioners.

This role is likely to continue, including through the sharing of knowledge and experiences between citizen science practitioners via practitioner-based associations. In recent years, NHMs have taken a central role in establishing the ECSA, a network of people and institutions (research institutes, universities, museums and civil society organisations) aimed at sharing best practice, building capacity for citizen science across Europe and advocating for it as a participatory research methodology with relevance to both researchers and decision-makers. European NHMs have been involved in the development of ECSA from the inception of the idea to its incorporation as an NGO and charity, now co-ordinated by its headquarters at MfN Berlin. All NHMs that contributed to this chapter are key organisations in the association. The European Citizen Science Association aims to establish closer links between museums, other research institutes, civil society organisations and citizen groups. Strategic associations like ECSA help to mainstream environmental citizen science as an approach for gathering data to improve environmental policy, as well as to monitor compliance with existing regulation, thus increasing opportunities for participatory environmental governance.

30
Stories can change the world – citizen science communication in practice

Susanne Hecker[1,2], Monique Luckas[3], Miriam Brandt[4],
Heidy Kikillus[5], Ilona Marenbach[6], Bernard Schiele[7],
Andrea Sieber[8], Arnold J.H. van Vliet[9], Ulrich Walz[10] and
Wolfgang Wende[11]

[1] Helmholtz Centre for Environmental Research – UFZ, Leipzig, Germany
[2] German Centre for Integrative Biodiversity Research (iDiv) Halle-Jena-Leipzig, Germany
[3] Futurium Berlin, Germany
[4] Leibniz Institute for Zoo and Wildlife Research, Berlin, Germany
[5] Victoria University of Wellington, New Zealand
[6] Rundfunk Berlin-Brandenburg, Germany
[7] University of Quebec at Montreal, Canada
[8] Alpen-Adria-Universität Klagenfurt, Austria
[9] Wageningen University, Netherlands
[10] Dresden University of Applied Sciences, Germany
[11] Leibniz Institute of Ecological Urban and Regional Development, Germany

corresponding author email: susanne.hecker@idiv.de

In: Hecker, S., Haklay, M., Bowser, A., Makuch, Z., Vogel, J. & Bonn, A. 2018. *Citizen Science: Innovation in Open Science, Society and Policy*. UCL Press, London. https://doi.org/10.14324/111.9781787352339

Highlights

- There has been a paradigm change from a one-way transfer of science information to a paradigm of exchange that demands adequate science communication.
- Communication in citizen science projects is key to motivating and retaining participants and exchanging information.
- Stories can play an important role in translating the abstract and logic scientific discourse into a concrete, emotion-related narrative of societal relevance.
- Communication and media coverage improves the chance of scientific expertise and knowledge influencing policy-making.
- Innovative collaboration between science and the media can benefit both partners – attracting participants to citizen science projects and generating media stories.

- Storytelling and visualisation are powerful communication tools, affecting the brain and emotions more effectively than words.

Introduction

Societies are facing global environmental and social challenges, such as the loss of biodiversity, environmental damages and climate change (Owen & Parker; Ballard, Phillips & Robinson, both in this volume). Simultaneously, social transformations mean that global society is connected in new ways (Novak et al. in this volume). Facing a world of uncertainties and threats, people need to redefine their way of life and ways of living together in society and with nature.

At the same time, people also dispose of knowledge and scientific evidence in new ways (see Mahr et al. in this volume). Global human knowledge currently increases at high speed, but this is not solving the various environmental and societal challenges. Millions of scientific papers are published every year without leading to a significant change in behaviour. Scientific knowledge is needed to make sound political decisions and enable meaningful conversations, but there is a lack of communication from the scientific community to wider society (Shirk & Bonney; Nascimento et al., both in this volume). Innovative methods are needed to bridge the gap from science to society and policy, generating changes in everyday lives and behaviour (Smallman in this volume).

The communication of science to diverse audiences and the engagement of scientists with all parts of society, including policymakers, are key factors in this process. Studies have shown that scientists communicating outside of the scientific community, for example, with the media, have a higher chance of being noticed and taken seriously by society and policymakers (Peters et al. 2008). Research is legitimised when the perception of its social relevance is reinforced, making it more likely to be used in the policy-making process. The media – as part of the process of science communication – influences policy agendas by communicating and discussing deliberately chosen topics. Media and news programmes are influential as they not only report problems, but also seek to analyse them and present possible solutions, thus turning into advocates for particular policy solutions (Howlett & Ramesh 2009). The process of making news means that journalists and editors choose stories or issues that they consider important. This selection follows its own rules of interest and means that not all stories are told.

From a scientist's point of view, communicating science to audiences is not enough. New forms of interaction between scientists and society are needed, going beyond the so-called 'deficit model' of science communication (Snow 1974). The deficit model assumes that the wider public lacks knowledge, interest and the ability to think scientifically and process data (Bauer 2016). Still, these deficits could be overcome and the public educated, gaining some scientific literacy through science communication (Snow 1974). The deficit model also assumes one-way communication, transmitting information and knowledge from science to the public. It is important to state that science and society are not two worlds apart with science being 'elsewhere', but rather that science is part of society's culture (Irwin 1995; Trench 2006; Schiele 2008).

Citizen science can be part of the larger process of engaging people in new forms of interaction, challenging scientists and citizens whilst enlarging scientific knowledge and providing learning opportunities for all parties involved (Bonney et al. 2014; Bela et al. 2016; Peltola et al., all in this volume). These diverse interactions require innovative forms of multiway science communication.

Stories can play an important role in bridging the discourses between science and society, and translate the abstract and logic scientific rationale into a concrete, emotion-related narrative for non-specialists that can easily be linked to existing knowledge and experience (Constant & Roberts 2017). This chapter discusses the innovative potential, limits and opportunities of science communication with a focus on storytelling through the framework of citizen science, illustrated with case studies and best practice examples.

Changing paradigm in science communication

A paradigm change (Kuhn 2012) can be observed in the field of science communication as basic scientific concepts and practices are challenged. In recent history, scientists were legitimised by creating and possessing knowledge and the public was deemed unable to understand scientific concepts, methods and findings (see Mahr et al. in this volume). However, there is now a shift from this simplistic understanding of one-way knowledge transfer, which is reflected in the change from a deficit paradigm to an engagement paradigm (Schiele 2008). The engagement paradigm understands science communication as also involving identity, democracy and scientific citizenship (Davies & Horst 2016). At the same time, the reference frame is rapidly changing from passive to active knowledge, influencing science communication and the interaction of science and society.

Engaging volunteers in research both requires and offers multiway communication between different actors in the scientific process. Actors in citizen science projects can be scientists, citizens, mediators or communicators, and they take on different roles, such as research managers, information providers or data providers (see Haklay in this volume). To make this collaboration successful, there must be adequate communication. The aim of communication in citizen science therefore goes beyond outreach and the one-way diffusion of information. Communication can have multiple aims and must be considered and designed according to the project goals. Long-term citizen science projects need ongoing collaboration and, therefore, communication to inform, motivate and engage participants. Other projects might want to raise awareness of their project issue at a local level, introducing it to the policy agenda. They therefore need to address a different target group and use other methods of communication. Educational projects need to consider training tools and might need to communicate them to different age groups (see Harlin et al.; Makuch and Aczel; Wyler & Haklay, all in this volume).

The border between *science communication*, understood as external communication from scientific actors to the public, and *scholarly communication*, understood as communication between those involved in the scientific process (usually only the scientists), is becoming increasingly blurry as citizens are becoming part of the process in multiple ways. Issues of democracy, local empowerment and community identity should be included in the concept of citizen science communication, thus enlarging the framework of science communication (Smallman in this volume). Communication can develop into multiway exchanges driven by technical tools but also by feedback possibilities.

The success of communication in citizen science is more relevant than in conventional science because it might motivate people to get or stay involved (or not) and thus contribute to the project's scientific success. More engaging formats of communication are needed to make these complex interactions possible. In addition to traditional channels such as broadcast media and newspapers, social media are gaining in importance, allowing science to reach its audience directly (Mazumdar et al. in this volume).

Science-media exchanges and co-operation

Citizen science projects often attract media attention as they deal with topics of societal relevance and provide good stories; and citizen science projects profit from media presence for the recruitment of participants. Innovative collaborations between science and media can therefore benefit both partners (see box 30.1) with the potential of even changing the journalistic perspective on how to explore a good media story (box 30.2).

Box 30.1. Foxes in Berlin: Science meets media for mutual benefit

In 2015, the public broadcasting corporation rbb (Rundfunk Berlin-Brandenburg) started a media campaign on red foxes in the city of Berlin, Germany, designed to operate as the starting point for a citizen science project on the ecology of foxes run by the Leibniz Institute for Zoo and Wildlife Research (IZW). The topic 'Foxes in Berlin' was covered in numerous TV and radio shows, and audiences were asked to send in photographs, videos and narratives of fox encounters, which were published on a dedicated website. The response exceeded all expectations: More than 1,000 'fox watchers' submitted contributions. The resulting publicity was used to kickstart a citizen science project asking volunteers to take over various tasks in the IZW's research project on foxes.

Fig. 30.1 A citizen photograph for the 'Foxes in Berlin' campaign. (Source: Margit Schröter)

(continued)

The joint campaign benefited both partners. For the IZW, it provided access to a large audience and a valuable starting point for the recruitment of citizen scientists. For the rbb, the combination of scientific information and the 'human dimension' in the interesting and amusing stories people contributed generated excellent audience ratings.

Box 30.2. Media needs a novel approach

I got interested in citizen science some time ago. We started by producing a broadcast of 30 minutes about citizen science and placed several features in the live programme of the Berlin-Brandenburg broadcast rbb. Then we started the foxes in the city programme with the Leibniz Institute for Zoo and Wildlife Research. We wanted to find out whether people were actually willing to participate continuously. Nowadays when planning a science feature, I always try to find a citizen science approach for the very research question, for example, for the Mars mission. In conventional media reporting we would tell the story from the scientist's perspective. But now I ask editors to go find citizen science approaches and participating citizens in the region to let them talk about it. That way we not only have a regional approach but also a level reaching our audience. With topics like the death of bees or soil quality, our colleagues already ask for the expertise of non-scientific experts. But when it comes to more complex and not everyday life topics, we have not done this so far. So my main message from a media point of view is not only that science needs to better communicate. But we as media also need to redefine our premise.

Ilona Marenbach, Rundfunk Berlin-Brandenburg (rbb)

Working together with scientists and citizen scientists also led to self-reflection on the part of the media partner, who reconsidered their involvement in, and practices of getting, science stories (see box 30.2). Thus, communication may improve not only from the science side, but also from the media side. Instead of only including scientist perspectives, media may also try to actively involve citizen scientists.

Connecting policy and citizen engagement

Members of society are likely to participate in citizen science projects when they address issues of relevance to their lives and experiences (see box 30.3). Policy, on the other hand, deals with concerns and problems of society that call for solutions. Collaboration between science, policy and society in citizen science can fulfil multiple needs for all actors (Shirk & Bonney in this volume) and produce meaningful results.

Box 30.3. Cat Tracker: International citizen science project exploring the movement and management of cats

Cats are one of the most popular pets worldwide, providing their owners with enjoyment and companionship, but they can also be a nuisance to neighbours and may have a negative impact on native wildlife. A citizen science project, Cat Tracker, was initiated to collect both environmental and social data exploring the relationship between cat management and cat behaviour. The project combines a social survey and GPS tracking to turn cat owners into researchers. It aims to help better understand cats' home range, how much time they spend in different kinds of habitat, and how owners can manage pet cats to reduce their impact on wildlife. This project was established in the United States and has expanded through collaborators to other countries. In New Zealand, information collected via the Cat Tracker project – from public attitudes towards the management of cats to distances travelled by individual felines – has been used to inform local government. The social survey of the New Zealand public indicated a high level of support for mandatory microchipping of pet cats. In August 2016, the Wellington City Council voted to make microchipping of pet cats compulsory – a first in New Zealand. It is hoped that the public involvement in this project will encourage responsible pet ownership and improve the welfare of both domestic cats and native wildlife.

Source: www.cattracker.nz

(continued)

Fig. 30.2 *Image A* – A domestic cat wearing a GPS tracker on a harness as part of the Cat Tracker New Zealand citizen science project. *Image B* – an example of a cat's movements over a one-week period. (Source: Heidy Kikillus)

Connecting and saving worlds

Science communication can help to connect complex issues like climate change with the changes in nature that people can see in their own backyards. Connecting complex issues with concepts that are familiar to people can result in lots of media attention – if communication is planned and managed in the right way (see box 30.4). Researchers in citizen science have a good chance of reaching larger audiences and communicate during the whole process.

Communication plays an important role in citizen science projects that apply an oral history approach with the aim of safeguarding cultural heritage. In this approach, participants communicate and interview other participants about historical events, traditions or daily life. Oral history is a way of giving history back to people in their own words (Thompson 2017). Involving members of society in this process and the subsequent research, leads to a citizen science project. Communication can thereby lead to social cohesion, connect generations, encourage the valuing of cultural heritage and create cultural resources, as case studies show (see box 30.5).

Box 30.4. *De Natuurkalender*: New communication tools and technologies

New communication tools and technologies allow individual scientists to more easily take the responsibility and initiative to communicate to society at large or to specific stakeholders. *Nature Today* (*De Natuurkalender*) experienced this by letting experts write stories on topical developments in nature in an understandable way and by actively informing journalists and interested people about these stories, generating mass outreach. This would not have been possible if the initiative of writing and publishing these stories had been the responsibility of communication officers alone. Citizen science projects are well suited to this type of frequent communication as they often continuously produce interesting, newsworthy data. The fact that already many volunteers are involved makes the topics even more attractive. Communication fulfils multiple objectives from which scientists and society/stakeholders benefit.

Box 30.5. Intergenerational dialogue as a research tool to save cultural heritage

Cultural heritage in all its components is a valuable, if not vital, factor in the reorganisation of societies on the basis of dialogue between cultures, respect for identities and a feeling of belonging to a community of values. BreadTime (2015–2017) focuses on the cultural sustainability and the agricultural and manual practices of the cultivation and processing of grains and the production of bread in the rural region of Lesachtal, Austria. The goal of

FIELDWORK ON LOCAL KNOWLEDGE AND PRACTICE

Fig. 30.3 Enabling process conducting research through fieldwork on local knowledge and practice. (Source: Andrea Sieber)

the project is the analysis, protection and documentation of local knowledge and practice related to the immaterial cultural heritage of 'Lesachtal bread', which is part of the UNESCO Intangible Cultural Heritage list. How could this local intangible cultural heritage be saved and transferred from one generation to the next? Citizens can participate in narrative dialogue groups as an open communicative space to collect and discuss local knowledge, or write down and send their experiences to the collection of biographical records. Furthermore, students from secondary schools were instructed in the method of oral history and interviewed local elders about the traditional cultivation and its significance in their daily rural life. In this way, oral history interviews not only served as a tool of communication and mutual learning, but also as the empirical basis for several research outputs, such as a documentary of local narratives and local practices related to flax. Using this method of intergenerational communication could access and secure local narratives and traditional knowledge, and communicate interest and curiosity between the youngsters of the valley and elders of the community. This form of intergenerational oral history brings together people of different generations in a socially integrating way, with mutual interest and emotional bonds. As opposed to passive learning, oral history is very engaging and hands-on, not only collecting stories but also creating social bridges between generations.

www.lesachtalerbrot.wordpress.com

Storytelling and visualisation – tools touching the brain

Science information needs to be translated and this can be done by using tools like storytelling and visualisation. Visuals reach human brains many times faster than words and connect information with emotions (see box 30.6). Interesting citizen science stories translating policy issues to the public tend to be viewed as more relevant and of higher importance than those with a less developed narrative structure (Howlett & Ramesh 2009).

Storytelling is a powerful tool: it affects the brain and emotions in different ways with different effects. Information is likely to be transformed into personal ideas and experiences in the brain if it is told as a story (Gerrig 1993). This process is called neural coupling, which is mirrored by the same experience in listeners and the speaker (Stephens, Silbert & Hasson 2010). The neurotransmitter dopamine, released by the brain in emotionally charged events like an engaging story, helps people to remember and ensures that memories are relevant and accessible

---

Box 30.6. Creating a landscape memory

Can you still remember the landscape of your childhood? Landscapes undergo change; they become ever more homogenous. Yet we are generally unaware of this process. Changing Landscapes is a citizen science pilot project to foster public interest in landscape research and to jointly create a collective memory of the landscape. Citizens are asked to rummage through boxes of photos and family albums in order to find old snaps of landscapes. Then they go to the original location of the photo and, from the same perspective, take a new photo of the landscape. Furthermore, participants are encouraged to evaluate how they interpret the changes to the landscape, for example, positively or negatively, and which ecological effects they connect with these changes. From a scientific viewpoint, such as at the Leibniz Institute of Ecological Urban and Regional Development as well other partner institutes, the transformed landscapes are linked to data on biological diversity to investigate the relationship between changing landscapes and the impact on biological diversity. In some cases, landscapes are now used more intensively and the level of biodiversity is lower, but the opposite can also be found. In the study area of Saxon Switzerland we can say that 'everyone' is involved. Whether old or young, experienced photographers or absolute beginners, our project has managed to reach wide swathes of the public. In this way we can say that landscapes not only seem to be accessible to local people but also comprehensible, and indeed something tangible in their lives.

www.landschaft-im-wandel.de

Fig. 30.4 How would you evaluate this landscape change? Matched landscape photos from citizen science pilot project, Changing Landscapes. (Sources: Postcard top, 1908 collection Walz; photo bottom, Walz 2014)

for future adaptive behaviour, a concept called 'adaptive memory' (Shohamy & Adcock 2010). Many areas of the brain will also be activated by a good story (Barraza et al. 2015). All of this can happen if the story is about things that matter to people and stimulates listeners to care about the characters (see box 30.7).

Box 30.7. 'Blossoms' produced by the Flax project

The Framework Convention on the Value of Cultural Heritage for Society (Council of Europe 2005) highlights the need for local communities, citizens and civil society to take ownership of heritage to bring it alive and make it meaningful. Awareness-raising, identification, upkeep, development and knowledge and skills transmission are therefore essential and should be based on dialogue between the wider population and professionals with a view to mutual enrichment. But how could cultural heritage be conceived as a shared responsibility by citizens? And how could local knowledge be transmitted by contemporary media? 'Landscape and You-th – Tracing Flax' (2012–2015) is an oral history research project that explored the relationship between local knowledge, landscape and regional identity based on the cultivation and manufacturing of the plant flax. Several media outputs and performances, including an app, documentary film and rap song, were generated by students as part of this creative project. The project should enhance landscape awareness and sustainable tourism in the region and offer added value for all stakeholders.

The project runs in three steps: fieldwork on local knowledge (history re-enactment and oral history interviews), reflection and documentation of the results. In all steps, the residents of the rural region of Lesachtal, Austria, were the main researchers, accompanied by scientists from a university. The local young people decided which insights were important to show to the public and especially to tourists.

Connecting older people and students through historical re-enactment and its transfer into media outputs are increasingly regarded as factors in promoting local understanding and identity.

www.lesachtalerflachs.wordpress.com

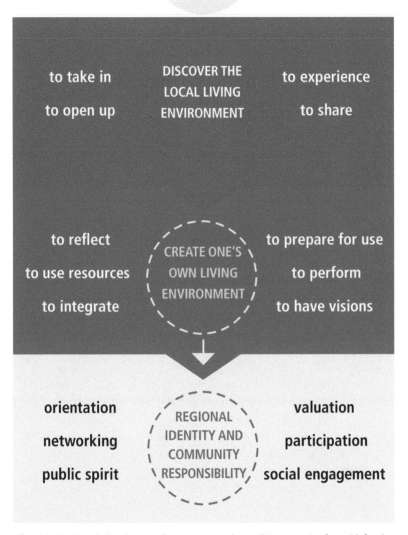

Fig. 30.5 Participative project process steps. (Source: Andrea Sieber)

Clash of two worlds – challenges in citizen science communication

One of the main challenges for science communication is bridging the values, expectations and needs of the research system, and the world of members of the public or policymakers (see box 30.8). Although anyone working in science is also part of the social system, being in a science role means living and acting within this system, which is defined by logics and aims that sometimes diverge from mainstream society (Weigold 2001). Science communication is therefore not a passive and linear process, but is characterised by complex transformative processes with the potential to influence both scientific discourse and societal debates (Bucchi 2008).

Box 30.8. City foxes – challenges in the co-operation between a research institute and a public broadcasting corporation

The co-operation between the Leibniz Institute for Zoo and Wildlife Research (IZW) and the Rundfunk Berlin-Brandenburg (rbb), an overall success, was also a clash of two worlds, with significant divergence in the partners' respective expectations and constraints. For the rbb, promoting a topic entails focusing on entertainment, storytelling and linking information with emotions. In contrast, the IZW's top priority is scientific quality, with the accuracy of the disseminated information taking precedence over emotionality. The two partners also work on very different time scales. The media cannot tell the same story twice, but need to keep the audience interested by continuously providing new angles. Scientific research, on the other hand, is often slow, visually unimpressive and does not produce novel results daily, making the constant requirement for new stories a challenge. Despite such conflicts, both partners consider this an exciting and worthwhile experience. The main lesson is that it is vital to make assumptions and expectations explicit from the start when working with partners in different sectors.

www.rbb-online.de/fuechse

Conclusions

Public engagement in citizen science provides unique opportunities to bring people together, learn from each other through multiway dialogue and make change possible. Science communication can benefit from these opportunities beyond more straightforward outreach. However, there is much more potential for societal outreach if scientists are better trained and thus able to explain their research in a narrative way. This applies not only to research results, but also to research background, problem formulation and research processes. Yet, these opportunities need appropriate skills and require openness from scientists.

The exposure of ideas and scientific assumptions to members of communities with unique knowledges, gaps, experiences and constructions of meaning is one innovation for science communication coming from citizen science. Here, innovation refers to the result of putting knowledge into a different frame and allowing it to be adopted, changed and filled

---

Box 30.9. Tips and helpful practices for communicating citizen science

1. Actively communicate your citizen science project outside the scientific community to increase visibility, raise awareness and stimulate participation.
2. Establish a good relationship with the media and take advantage of media attention.
3. Identify the aims of your communication, for example, motivate participation, inform about the project and provide educational information.
4. Identify what is relevant to people's life in your citizen science project and link to it when communicating, for example, stories to locality, issues of broader societal concern.
5. Plan and manage your communication accordingly, for example, understand your partner's needs, choose appropriate media and exchange information in adequate language.
6. Use visualisations and storytelling where possible to achieve people's understanding.
7. Allocate attention, time and resources to communication in citizen science.

with new meanings. This can result in unexpected developments, which might no longer be under the control of scientists. Citizen science, as with all science, can be an adventure because the results are unknown, and scientists should be open to a project changing in unexpected ways.

Communication in citizen science projects can also be challenging when there are different expectations, time frames or needs, which might not be explicit. Experience shows that focusing on the target audience and understanding and addressing their needs throughout the whole process can add to a project's success – both in a scientific way as well as for researchers as ambassadors for science (Druschke & Seltzer 2012). Careful project management and communication are key components in citizen science projects. Citizen science therefore demands advanced communication skills to master these interactive and innovative processes (Treise & Weigold 2002). It requires adequate flexibility on the part of scientists as dialogue and interaction with citizens and/or the media might develop in unforeseen ways.

At the same time, it offers the opportunity to contact other audiences, share knowledge, and create visions and emotional bonds; this is especially the case when communication is shaped as storytelling or other creative forms that are easy to grasp and to remember. Methods for external communication include newspapers, television and radio, social media and the internet, as well as classical forms of scientific knowledge dissemination such as journal articles and books. New technologies and the proliferation of information availability are drivers for this paradigm of exchange and active knowledge (see also Mazumdar et al. in this volume). These technologies make it easy to generate large-scale impact and outreach with relatively little effort when required in citizen science, for example, for a nationwide monitoring project.

To respond to the challenges of an open collaborative process, as well as maximise the effectiveness of projects, a concept of science communication is therefore integral to the practice of citizen science (see box 9 for communication recommendations).

Conclusions

31
Citizen science to foster innovation in open science, society and policy

Aletta Bonn[1,2,3], Susanne Hecker[1,3], Anne Bowser[4], Zen Makuch[5], Johannes Vogel[6] and Muki Haklay[3]

[1] Helmholtz Centre for Environmental Research – UFZ, Leipzig, Germany
[2] Friedrich Schiller University Jena, Germany
[3] German Centre for Integrative Biodiversity Research (iDiv) Halle-Jena-Leipzig, Germany
[4] Woodrow Wilson International Center for Scholars, Washington DC, US
[5] Imperial College London, UK
[6] Museum für Naturkunde Berlin, Germany
[7] University College London, UK

corresponding author email: aletta.bonn@idiv.de

In: Hecker, S., Haklay, M., Bowser, A., Makuch, Z., Vogel, J. & Bonn, A. 2018. *Citizen Science: Innovation in Open Science, Society and Policy*. UCL Press, London. https://doi.org/10.14324/111.9781787352339

Citizen science advances open science by operating at the interface of science, society and policy. As a result, its influence is growing in societal decision-making processes, including government policy. Citizen science has a long history and has recently entered a renaissance through expanded governance models, progressive education curricula, novel technologies and enhanced interest in open and participatory science. The chapters and extended case studies in this volume, contributed by authors from around the globe, provide new insight into the impacts, challenges, benefits and opportunities brought by citizen science.

The collection has identified the following key themes in citizen science (see figure 31.1):

1. *Innovation in science*, including methods, data and outcomes as well as advances in knowledge inclusivity and open science.
2. *Innovation with and for society*, including different formats of participation and its outcomes, in-depth two-way learning, social inclusivity and innovative science communication. In addition, the consideration of social, cultural and educational values advance diversity, inclusion and transparency in science, locally and globally.

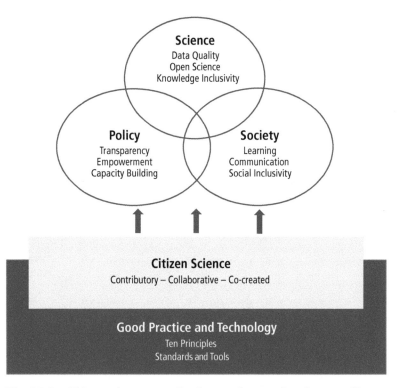

Fig. 31.1 Citizen science contribution to advances in science, policy and society

3. *Innovation and impact at the science-policy interface*, including public policy agendas, processes and outcomes. Citizen science may lead to empowerment and capacity building in advancing joint policy development and implementation.

Within these fields, citizen science provides:

4. *Innovation in technology and environmental monitoring*, including digital and technology literacy, dissemination, use and infrastructure promoting health and environmental well-being, ecosystem health and conservation.
5. *Innovation in science communication, education and higher education*, as citizen science provides interactive learning environments.
6. *Fitness for purpose* in organisational and institutional processes and standards, in connection to best practice.

Overall, citizen science can also contribute economic benefits for individuals and communities, as well as for science and policy, if through an enhanced evidence base and community engagement, environmental or social issues can be explored and subsequently addressed in policy and practice more effectively.

This concluding chapter turns to each key arena in which citizen science operates, to show how innovation and links to science, policy and society are allowing citizen science to enhance its impact. Each section also presents recommendations for different stakeholders. Overarching to this is the universal recommendation to apply the Ten Principles of Citizen Science developed by European Citizen Science Association (ECSA) members, the application of which will enhance science, policy and society by promoting best practice (see Robinson et al. in this volume). The application of traditional and innovative technology also enables citizen science in all its facets – from contributory to co-created formats – to reach its potential. As the chapters in this volume demonstrate, many of these principles are already supported through implementation, providing further evidence of best practice which can be drawn on in other contexts.

Innovation in science

Citizen science contributes to innovation in science itself, and indeed a genuine science outcome is a main principle of citizen science projects (Robinson et al. in this volume). For example, Shirk and Bonney (in this volume) demonstrate that citizen science is not only a distinct field of research, but also that it can make unique, novel and innovative contributions to scientific understanding. In addition, citizen science tends to provide a transdisciplinary paradigm, moving beyond segregated research disciplines towards matters of societal relevance across different fields (e.g., Parker & Owen; Ceccaroni & Piera, both in this volume). At the same time, it is independent on the subject and can address both basic and applied research. Basic knowledge or curiosity-driven endeavours such as citizen science investigations in astronomy or art are seen as equally relevant as more applied environmental studies in accordance with individual interests.

As citizen science projects can be designed to involve many volunteer contributors in large collaborative projects, such projects have the possibility to pursue research that could not be done otherwise. Examples of this come particularly from environmental projects, which cover a wide variety of topics, including the increasing threat of light pollution (Schroer et al. in this volume); air pollution (Volten et al. in this volume); the distribution of invasive mosquitoes (Palmer et al. in this volume); and, the characteristics of soundscapes (Li in this volume). Such projects combine scientific advancement while fulfilling practical monitoring needs in support of government agencies (Parker & Owen in this volume).

To analyse environmental changes worldwide, global citizen science project collaborations are required. Many such examples are related to the environmental and life sciences, in Europe (Hecker et al. 'European Citizen Science' in this volume) and in general (Kullenberg & Kasperowski 2016). Volunteers are often highly motivated to contribute to environmental measurements, when they are complementary to existing measurement networks, in order to contribute new, impactful findings and because of their high spatio-temporal resolution (Volten et al. in this volume). However, to enhance scientific advances from citizen science, there is an urgent need to properly archive data (Brenton et al. in this volume), to make databases interoperable with other systems and to assimilate data in models for analysis (Volten et al. in this volume).

In opening science to stakeholder communities, it is equally important to include indigenous and local knowledge (ILK, see, for example, Alessa et al. 2016; Mistry & Berardi 2016) as an added benefit to science, for example, in framing questions, designing projects, analysing results and understanding their possible impacts upon decision-making processes. Danielsen et al. (in this volume) demonstrate that both ILK and institutionally derived scientific understanding can be valuable in conservation planning activities. This knowledge inclusivity can bring specific expertise to citizen science projects and embed the results in the community affected.

Despite its potential, critics may question citizen science data. However, it is important to recognise, first, that citizen science participants are often experts in their respective fields, and sometimes have no parallel in academia, for example, for some animal or plant groups or some fields of history or cultural heritage. Indeed, estimates suggest that 80 to 90 per cent of baseline species data for biodiversity and ecosystem research and policy development in Europe is based on volunteer effort. In other words, citizen science often provides invaluable knowledge that cannot otherwise be ascertained, and it is questionable whether efforts from academic institutions would achieve other results.

Secondly, all data of known quality has value. Data with uncertain quality or non-conforming data can also be remediated or augmented with information from additional sources. It is therefore critical to contextualise data with metadata that describes the purpose and methods with which a dataset has been created (Williams et al. in this volume). When the results of a citizen science project are published, proper documentation helps readers to understand and evaluate the validity of the research findings (Brenton et al. in this volume). Documentation also enables other potential data users to evaluate fitness for reuse in other scientific research, and to understand how different data can inform management or other forms of decision-making. Many data collection activities could be standardised using protocols that encourage high quality and well documented citizen science contributions.

There is therefore a need to develop transparent mechanisms to produce high-quality data, including through the use of data standards, as discussed by Williams et al. (in this volume). Data quality should also be supported through the calibration and validation of low-cost sensors where possible (Volten et al.; Schroer et al., both in this volume). As Brenton et al. (in this volume) discuss, data generated by citizen science projects will have the greatest impact in both science and policy development when they can be supplied in a timely and accessible manner, and conform as much as possible to existing data and process standards. Here, the Atlas of Living Australia (ALA) and its global 'ALA family' (developed from adaptation of the ALA information technology [IT] infrastructure, knowledge and tools) is helping to provide a reliable infrastructure to enhance the scientific standards, visibility and application of citizen science project data.

Knowledge production through science and citizen science occurs within a particular social context. Here, the social sciences and humanities are critical to promoting ways of understanding and reflecting on citizen science (Mahr et al. in this volume). This includes evaluating the outcomes of projects and activities and reflecting on best practice while progressing the field (see also Kieslinger et al. in this volume). Social science-citizen science interactions can also inform the mechanisms for promoting sound decision-making through public policy. Here, transparency and representation are key components for legitimating public policy processes and their outcomes. To foster citizen science in academia, Wyler and Haklay (in this volume) call for university support to engage scientists and identify projects that fit university teaching or research needs. While citizen science can provide inspiration and innovation to science, this is not yet widely recognised and scientists engaging in citizen science are not always acknowledged for this. As such, there is a parallel need to support researchers both by valuing their work and by providing guidance and tools for those launching citizen science projects (Wyler & Haklay in this volume).

Recommendations for enhancing impact in science

- *Researchers, practitioners and their institutions* should recognise citizen science as an innovation tool for science, and demonstrate and share success. Knowledge inclusivity should also be valued as an asset in research design and analysis.
- *Researchers, practitioners and their institutions* should improve mechanisms for validating data from citizen science, and provide good interoperable infrastructures for archiving and analysing data.
- *Researchers, practitioners and their institutions* should develop and employ specific training and resources to conduct meaningful citizen science to achieve scientifically useful results. A good scientific study design and easy-to-use technical aids can facilitate and enhance standardised measurements and thereby foster the impact of citizen science projects in science and society.

- *Educators* in formal and informal education settings, such as schools, universities or museums should develop pilot citizen science for impact programmes, perhaps initially within their immediate communities, to train and engage potential and existing citizen scientists, their institutions and community groups.

Innovation with and for society

At the nexus of society, science and policy, a paradigm change is taking place, moving from a one-way transfer of information to a two-way exchange between stakeholders, including researchers and the public (Haklay; Hecker et al. 'Stories', both in this volume). Multiple information flows call for innovative interfaces between science, society and policy actors. Building capacity for citizen science is a prerequisite for supporting various forms of citizen science where different actors with various expectations and needs meet for collaboration.

The process of capacity building involves five main steps as discussed by Richter et al. (in this volume): (1) identifying and engaging different actors; (2) assessing capacities and needs for citizen science in the setting under focus; (3) developing a vision, missions and action plans; (4) developing resources such as websites and guidance; and (5) implementing and evaluating citizen science programmes. Capacity building is an iterative and adaptive process that requires sound engagement of all involved actors from society, science and policy (Richter et al. in this volume). As the field of citizen science grows, evaluation tools and criteria are needed to assess progress against the Ten Principles and to determine and secure impacts across society, science and policy – particularly at their interfaces. Kieslinger et al. (in this volume) suggest a framework for evaluation that allows project leaders and funders to assess projects proposed for funding and evaluate success after completion. This will also allow for reflexive adaptive project management during the course of a project. This framework, along with the Ten Principles themselves, is open to continuous development by the community to foster advances in citizen science.

The citizen science community has also moved on to distinguish between different forms of participation (Shirk et al. 2012) from 'contributory' to 'collaborative' and 'co-created' citizen science models. Haklay (in this volume) argues that there should be no assumption that greater participation is at all times required from or desired by participants. Building on Shirk et al. (2012), Ballard et al. (in this volume) show that no model is inherently better, but that different scientific and policy questions may suit different forms of participation. For example, contributory projects have great potential to generate large spatial and temporal-scale datasets tapping into widely dispersed (inter)national citizen science communities, such as large-scale environmental monitoring and observation projects. In this way, they can contribute to policy inputs indirectly via research. At the other end of the spectrum, collaborative and co-created projects allow for more intensive involvement of participants in local policy issues (Ballard et al. in this volume) and may support knowledge gains (Shirk et al. 2012). They can thereby facilitate in-depth civic participation that may contribute to more direct management advice and policy change, and more indirectly support education and capacity building in research. It is therefore more a question of appropriate fit that determines the choice of participation type. On this point, best practice should continue to be demonstrated, including using failures as learning experiences.

Observations of volunteers demonstrate that they have a range of motivations and reasons to be involved in citizen science, including both intrinsic and extrinsic motivations (Eveleigh et al. 2014; Geoghegan et al. 2016; Rotman et al. 2012). Motivations are important to the design of successful projects and generally vary from one project to another. For example, some participants may be highly motivated to contribute to air quality measurements that are complementary to existing measurement networks because of their high spatio-temporal resolution. They want to contribute to scientific research because of a special interest, for example, in aerosols and their impact on health and the environment (Volten et al. in this volume). They might also be motivated because they want to learn about their environment as a group, benefiting from group dynamics as well as individual learning and teaching concepts (Peltola et al. in this volume). At the same time, some motivations are common, such as personal enjoyment, interest, a stake in a topic of research, and/or community membership or socialisation through participation.

In addition to encompassing the diverse motivations of participants and responding to them, citizen science has the potential to allow for the broad diffusion of societal benefits. While citizen science does enhance participation beyond the usual actors, full inclusiveness is not an automatic outcome of citizen science, or one that is easily achieved. As Ceccaroni and Piera (in this volume) show, stakeholder mapping is one process that can help citizen science projects identify and engage interested parties. The degree of inclusiveness varies according to the techniques for involving citizens. Peltola et al. (in this volume) discuss how affective techniques like games for better recollection or visual stimulations and training material for species identification are crucial in involving less-experienced and less-privileged participants. As Novak et al. (in this volume) discuss, innovation in citizen science also increasingly builds on the use of digital technologies such as social networks, as well as open source hardware and software. These tools provide a means to empower participants in citizen science from various backgrounds and different expertise levels, regardless of their formal education, to get involved in the scientific research process.

This can help to harness opportunities for citizen science globally. While this volume aimed to include citizen science case studies from across the continents, including Africa and China (Danielsen et al.; Novak et al.; Li, all in this volume), the current distribution of citizen science projects reported in the literature is still skewed. Further expansion of citizen science to the global South has the potential to advance the local evidence base, including through the participation of illiterate volunteers (Stevens et al. 2014), contribute to closing spatial gaps in international environmental data (Amano, Lamming & Sutherland 2016; Chandler et al. 2017) and help at a very pertinent science-policy interface, for example, when responding to crises and contributing to early warning systems (Alessa et al. 2016; Cochran et al. 2009).

Recommendations for enhancing impact with and for society

- *Policymakers* should enable capacity building at the local, regional, national and international scale to move from knowledge transfer to knowledge exchange and joint knowledge creation at the science-society interface. Demonstrable citizen sciences outcomes and impacts for society should be rewarded with financial and other incentives that are proportionate to the benefits gained.

- *Researchers, practitioners and their institutions* should acknowledge participants as partners in the research process, respecting and considering their motivations, skills, abilities, values, expectations and knowledge.
- *Researchers, practitioners and their institutions* should consider that more participation is not necessarily better, as some volunteers prefer a limited level of engagement. In designing projects, they should also account for participant motivations that might change over time. Projects should therefore avoid simplistic framings and black-and-white approaches.
- *Researchers, practitioners and their institutions* should develop typologies for conducting citizen science projects according to their functionality, intended outcomes and impacts, as this could help optimise such projects.

- *Educators* and their institutions should familiarise themselves with the benefits and methods of citizen science as the basis for promoting education and training that mobilises students and trainees to be active in citizen science for science, policy and society.

- *Community members* should articulate and select levels of participation carefully, with awareness of how differences in education and participation levels shape projects, especially their direction and goals.

Innovation at the science-policy interface

Policy and regulation should be underpinned by sound science. However, the science (including citizen science) and policy communities do not have established ways of communicating with one another. This is a missed opportunity because at least some forms of scientific discovery and innovation require and therefore drive policy formation. Examples include stem cell research, cloning, biomedicine and other forms of genetic engineering. Other scientific endeavours are of foundational importance to policy formation. What would environmental policy formation look like without the underpinnings of environmental science?

Further to these points, there are critical junctures in the policy cycle where science can exercise fundamental information gathering, analysis and further supporting roles. Examples of these policy windows of opportunity include the following:

- Issue agenda-setting for policy development.
- Data and other evidence gathering as the basis for policy option formation and analysis.
- Consultation processes around option analysis and selection.
- Policy selection and related decision-making processes.
- Policy monitoring, evaluation and renewal.

Many chapters in this volume provide evidence for the increasing acceptance of citizen science within policy-making and implementation. One clear statement is by Nascimento et al. (in this volume), who acknowledge citizen science's potential to significantly impact local and national decision-making, empowering citizens and leading to a better and more transparent government. As noted by Parker and Owen (in this volume), Environmental Protection Agencies (EPAs) in Europe and the United States provide an example for this adoption. The EPAs have protection of the environment as their primary responsibility, as well as encouraging wise use of resources, protecting public health and enabling sustainable development. There is potential for citizen science activities to contribute to each of their activity areas. For EPAs, citizen science has an increasingly important role in providing evidence, raising environmental awareness and empowering the public. Citizen science has the potential to transform environmental protection by inviting the public to work with agencies to generate knowledge and find solutions to societal environmental challenges at any scale. Environmental protection agencies are increasingly turning to citizen science to assist in the achievement of environmental protection, and as Parker and Owen and Volten et al. (both in this volume) show, EPAs can both support and benefit from citizen science, for example with small sensor networks for monitoring environmental and related health indicators. As Smallman (in this volume) discusses, citizen science can foster advances in Responsible Research and Innovation (RRI), and the engagement of citizens can lead to concrete outcomes for conservation (Ballard et al.; Danielsen et al., both in this volume).

The organisational inertia and existing structures at the science-policy interface may need to change to facilitate contributions from citizen science. Concrete action is needed to achieve this, such as expanding the integration of citizen science data in policy monitoring and reporting, addressing organisational practices through co-ordination efforts across organisations, promoting legitimacy through support for pilots and experimentation, and establishing and supporting communities of practice (Nascimento et al. in this volume). The framing of, and links to, current political priorities also need to be identified to support active engagement with citizen science. This is all part of developing citizen engagement and empowerment in decision-making as addressed by both Parker et al. and Nascimento et al. (both in this volume). Examples of capacity-building activities are provided by Richter et al., with cases at the European and national scales that have led to the development of a Socientize *White Paper on Citizen Science for Europe* and the *Greenpaper Citizen Science Strategy 2020 for Germany* (Richter et al. in this volume).

The legal limitations and constraints of each political system also need to be considered when it comes to the integration of citizen science activities – a good demonstration of this is provided by Nascimento et al. (in this volume) with the US case of complying with the Paperwork Reduction Act.

The question of data quality, discussed above, is particularly important in the policy context. Data quality has many dimensions, including data reliability (i.e., whether data are complete and error free). All aspects of data quality are particularly important to consider when seeking alignment with environmental regulatory standards and monitoring requirements (Williams et al. in this volume). Some policymakers have advocated for a fitness-for-use approach, where key aspects like data quality, scale, cost, interoperability and data format must be taken into account when evaluating the value of citizen science for a particular policy question (Holdren 2015). Fitness-for-use encompasses data quality and also recognises that data must fit within existing indicators and government frameworks for decision-making. Naturally, a fitness-for-use test may not be required in circumstances where regulatory and policy frameworks are designed with the open data gathering, quality and formatting protocols that also accommodate non-government agency participation in these processes.

Recommendations for enhancing impact at the science-policy interface

- *Policymakers* should welcome and support participation from the citizen science community.
- *Policymakers* should jointly with practitioners, educators and researchers develop Green and White Papers with action plans to harness the potential of citizen science for innovation and participation in policy. The US Citizen Science & Crowdsourcing Act is a good example, and developments at national, international and agency level in Europe and elsewhere should progress to government implementation.
- *Policymakers* should learn from existing activities in EPAs and promote further uptake of citizen science by EPAs and other bodies.
- *Policymakers* should standardise open data management and other methodologies to facilitate citizen science participation in policy implementation, monitoring and renewal.

- *Researchers, practitioners and their institutions* should, where data protocols do not require inordinately high cost equipment or highly advanced technical knowledge, employ open data management processes as the norm rather than the exception.

- *Funders* should provide adequate funding to support citizen science. Funding needs to be available in a variety of formats to respond to the variety of citizen science activities and foster innovation. Consideration of long-term strategies is vital to embed citizen science in society, policy and practice.

- *Funders* should apply these recommendations at multiple spatial and temporal scales reflecting the range of jurisdictions (community, local government, sub-national, national, regional and international) at which citizen science makes a contribution to science, policy and society.

Innovation in technology and environmental monitoring

Recent and continuous improvements of information and communication technologies (ICT) offer new opportunities for social interactions that have changed communication patterns and networked people around the world (Novak et al. in this volume). In citizen science, ICT can broaden the opportunities for participation to new audiences and support collaborative, real-time data collection and dissemination (Newman et al. 2012). To realise this potential citizen science needs reliable, trusted, and high-quality technologies and infrastructures that are open, transparent and inclusive (Brenton et al. in this volume).

Technologies and infrastructures serve different needs. Project databases such as SciStarter and the Federal Catalog of Crowdsourcing and Citizen Science help volunteers find projects to contribute to (Mazumdar et al. in this volume). Some databases, like CitSci.org and Zooniverse, also support participation by facilitating observational data collection and management or analysis. Projects increasingly use social media such as Twitter, Facebook and Instagram to recruit and communicate with their volunteers. In addition to general-purpose social media websites, collective intelligence platforms for collaborative problem-solving can also help mobilise and connect people who have already demonstrated the desire to collaboratively address social problems (Mahr et al. in this volume).

Hardware, including sensors, smart devices or wearables, for example, smart watches, and drones can support data collection (Li; Ceccaroni & Piera; Mazumdar et al.; Volten et al.; Schroer et al., all in this volume). Data analysis is aided by platforms like Google Maps and toolkits, such as Mapbox, which allow for different types of data integration. Some technologies, like games or gamified applications, help improve the volunteer experience. Others, like open data portals and repositories, make it easier for researchers to find and work with citizen science data. While many pre-existing toolkits and platforms exist, innovations in hardware and software are also developed by hackers, makers, and DIY–citizen scientists working alone or together on spaces such as makerspaces or Fab Labs (Gold & Ochu; Mazumdar et al., both in this volume).

Tools and technologies must be carefully selected for a particular use. Brenton et al. (in this volume) argue that it is important to wisely consider the compliance of tools with applicable process and data standards, as well as their ability to connect with the information supply chain, for example, to established processes in science or government agencies. Open hardware and software communities such as Public Labs often democratise the research process by empowering communities to conduct their own investigations into issues of concern. In line with this goal, Public Labs is devoted to designing and producing low-cost, open source hardware that can be made available to as many people as possible. Other citizen science projects may have different goals, such as supporting reporting needs of authorities in relation to national and international targets or advancing scientific research. Achieving these goals often requires harvesting data collected in one context for use in another. In such cases, the importance of technologies like data standards, which promote interoperability and access to all data, may be more important, for example, than lowering barriers to participation through a low-cost technology, that produces data that does, however, not align with particular regulatory requirements. In all cases, good metadata documentation is required to help users understand the value of a citizen science tool, technology or dataset, and help other potential users assess fitness for purpose (Brenton et al.; Williams et al., both in this volume). Documentation of a project's results also ensures sustainability by allowing research activities to have impact even after a project formally ends.

There are therefore numerous challenges to developing citizen science technologies. Mahr et al. (in this volume) suggest that more work, potentially by Science and Technology Studies researchers, should be pursued to understand gaps in hardware and software as well as the motivations of communities like maker and hacker groups. Additional efforts to develop citizen science data and metadata standards are also required to expand early efforts from organisations like the Open Geospatial Consortium (OGC). The concept of fitness for purpose remains elusive. Helpful policy guidance could specify the technological requirements, in terms of hardware or particular data standards, that are required for citizen science activities to achieve impact in a range of research, policy-making and management contexts.

Recommendations for enhancing impact through harnessing advances in technology

- *Researchers, practitioners and their institutions* should employ technologies such as games or gamified applications to help improve the volunteer experience, and carefully prepare the scientific and technological design so that data standards can be met without impacting unduly on participant enjoyment.

- *Researchers, practitioners and their institutions* should aim to provide reliable information technology infrastructure and document metadata, so that citizen science data can be used in other contexts. Noting the new pressure to demonstrate relevance to their local communities, major universities should draw upon science, innovation, community and NGO actors to promote citizen science training and projects that benefit the local community. Innovation incubators and hack spaces aimed at community challenges are relevant in this regard, and also assist in building the community literacy of academic staff and students.

- *Community members* should develop and promote citizen science data and metadata standards, and choose technology tools wisely so that they align with regulatory requirements to optimise impact. This is especially relevant in environmental monitoring.

Innovation in science communication, education and higher education

Several chapters in this volume demonstrate the importance of citizen science in science communication and education – both formal and informal – and the way in which it increases scientific literacy and science capital (Edwards et al. in this volume). Collaborations between scientists and the media can benefit both partners, where scientists gain from an influx of participants to support research activities, and media channels find new stories to report. In addition, the use of the media improves the chances of scientific expertise and knowledge becoming effective in policy-making (Hecker et al. 'Stories' in this volume), offering the opportunity to disseminate knowledge about citizen science's impact and benefits.

It is clear that the benefits of citizen science can accrue throughout the education process. For example, as Makuch and Aczel (in this volume) demonstrate, children can both learn from and contribute to an evolving body of environmental knowledge and scientific enquiry in a meaningful way. This can make their learning more engaging as Harlin et al. (in this volume) point out. Makuch and Aczel (in this volume) emphasise that activities that take place outdoors, with direct access to the environment and the natural world, particularly enable children to develop environmental awareness and responsibility. In addition to the benefits to the children, their teachers gain from opportunities for professional development. To facilitate the inclusion of citizen science in the curriculum, teachers will need support and guidance in finding projects that suit their immediate classroom needs as discussed by Harlin et al. (in this volume). This can be achieved through curriculum development where teachers' unions can play an important leadership role alongside government. For universities, Wyler and Haklay (in this volume) recommend that specialist support is provided to researchers so they can maximise the benefits from citizen science in an ethical and a productive way. They point out that, just as in schools, citizen science provides opportunities to enrich the university curriculum and to enhance the ability of scientists, at an early career stage, to learn both how to build citizen science projects and how to engage the wider public.

More generally, across formal and informal education, there is a growing need for project designers to be aware of volunteers' background, education levels, and depth of engagement and learning (Edwards et al.; Haklay, both in this volume). Recognition is therefore required of both participants as individuals, including their diverging learning abilities and skills, and of the collective dynamics of learning. Peltola et al. (in this volume) point out that there is much potential in developing successful techniques that can broaden the role of participants by examining their particular concerns and supporting ownership of the learning process.

Recommendations for enhancing impact in science communication, education and higher education

- *Policymakers* should promote citizen science participation as a fundamental component of the education curriculum for schools, if not also as a fundamental component of good citizenship.

- *Researchers, practitioners and their institutions* should recognise volunteers' background, education levels, and depth of engagement and learning to develop successful projects in both formal and informal education.

- *Researchers, practitioners and their institutions* should help build and promote collaborations between scientists, the media and other societal stakeholders to promote participation in the design and delivery of citizen science projects for impact.

- *Educators* should develop supporting guidance for educators working at schools and universities and embed citizen science projects in curriculum activities to fully harness the benefit of citizen science in learning.

Innovation in organisational and institutional aspects and development of good practice

The authors of this volume have demonstrated the need to pay attention to organisational and institutional aspects that influence the implementation of citizen science projects. In particular, lessons can be learned from the efforts of supporting institutions including EPAs, and science or natural history museums. Here, innovations in technology and organisational practices are enabling a citizen-agency dialogue based on good data and feedback on the use of the evidence that the public provides (Parker et al. in this volume). A good example is provided by the US Federal Community of Practice for Crowdsourcing and Citizen Science (CCS), which played a large role in ensuring acceptance from federal authorities as discussed by Nascimento et al. (in this volume). Similarly, Sforzi et al. (in this volume) demonstrate the organisational transition in natural history museums. Museums play an increasing role at the interface between science and different publics, and in the promotion of citizen science. They are well-placed to support the key challenge of recording life in the natural world by actively engaging the public, since they are seen as trustworthy, visible portals to scientific research and information. Museums' large numbers of visitors thereby offer a powerful route to engaging the general public and by supporting citizen science they can boost the potential for meaningful participatory environmental research as a new source of public engagement. For museums, there is therefore a clear alignment between their social role and remit, and citizen science. Other organisations can learn from their experience and build on it to adapt their activities to also include citizen science.

As Brenton et al. (in this volume) point out, there is a growing community of people who share technical and organisational experience and tools, and who are working towards interoperability between citizen science projects and the data that they produce. Such a network of people and expertise can ensure the resilience of the system, as a whole, through the increased use of citizen science in different parts of government and research agencies. It is important for all projects to employ a combination of (digital) tools and to harness the benefits of different engagement opportunities using available technologies (e.g., Brenton et al.; Mazumdar et al., both in this volume), as well as to include structured events and activities, to promote active communication and to diversify participation. Gold and Ochu (in this volume) describe how ThinkCamps can create new spaces for sharing knowledge and foster an online and offline community. Overall, the success of communication in citizen science projects has implications for motivating and retaining participants and exchanging information, thus ensuring the main aim of most projects to collaboratively create knowledge (Hecker et al. 'Stories' in this volume).

Recommendations for enhancing impact through promoting good practice

- *Policymakers* should adopt citizen science charters for advancing their institutional missions, particularly where they cater to public interests and have an implicit responsibility to advance them through participation and policy implementation.

- *Researchers and practitioners* should assess and share case studies including unsuccessful projects and activities, so others can learn from them.

- *All community members* should share good practices, especially between diverse projects, communities and geographies, and continue to promote learning within the citizen science community. Good practice should be evaluated against the Ten Principles of Citizen Science and continue to evolve these.

- *All community members* should collect, document and share capacities, tools and other resources. Here, the citizen science associations (e.g., the Australian Citizen Science Association [ACSA], the Citizen Science Association [CSA] and the European Citizen Science Association [ECSA]) can help, and active participation by all can foster joint advances.

Outlook

The field of citizen science is growing and having an increasing ı
science, society and policy. In the face of significant challenges in ı
to environmental, economic, demographic and socio-political cha
these opportunities for participation in science should be harnessed a.
fostered to enhance knowledge generation and evidence provision. This
can in turn support societal goals, including progressive policy delivery
by civil institutions.

Citizen science can contribute to the scientific literacy, knowledge
and societal advancement that is needed to support societies at a time
where evidence-based policy-making, sound scientific expertise and certain foundational truths about democracy cannot be taken for granted in
even the most privileged societies. Engagement in citizen science can
provide in-depth learning opportunities through learning by doing (Bela
et al. 2016) and promote the public's ability to understand and deal with
variability and uncertainties in complex issues without the need to jump
to easy conclusions.

Participation at a range of levels and jurisdictional scales can foster
empowerment as an enlightened, effective citizen. Citizen science offers
opportunities to contribute evidence to better management of the environment, and to engage in developing policies. Building on current case
studies in citizen science, viewpoints can be broadened and experience,
tools and efforts shared to address both local and global policy challenges.

More research can help the local and global citizen science community understand and expand upon these benefits. The Ten Principles of
Citizen Science need to be tested in practice and evaluation criteria continuously developed and applied. Further understanding of the motivations and drivers for participation in citizen science can help to better
design projects. This can be supported through programmes that include
the provision of training, information technology infrastructure, reliable
and user-friendly small sensors, timely visualisation and inspiring communication of data, as well as complying with data standards and being
underpinned by appropriate levels of funding.

Citizen science thrives through bottom-up development and participation, and these characteristics should be supported and further legit-
imated by the powerful stakeholders that feature in the policy and
regulatory life of societies the world over. One of the most significant
impacts of citizen science in science, society and policy will be achieved
when data collected to solve a local problem can also feed into ongoing

downstream environmental, social and sustainable economic uses and benefits. Here, development is needed to combine the individuality of citizen science initiatives with robust scientific practices, to foster both social transformation and scientific achievements in citizen science.

Perhaps, above all, citizen science communities and formal institutions need to further develop and disseminate a participatory ethic – embedded throughout society – that pursues the relevance and impact of scientific citizenship as a key component of self-realisation in the contemporary world. In closing this volume, you are invited and encouraged to be part of this journey.

References

Abels, Gabriele. 2007. 'Citizen Involvement in Public Policy-Making: Does It Improve Democratic Legitimacy and Accountability? The Case of pTA', *Interdisciplinary Information Sciences* 13: 103–16.

Alabri, Abdulmonem, and Jane Hunter. 2010. 'Enhancing the Quality and Trust of Citizen Science Data'. In *e-Science, 2010 IEEE Sixth International Conference on e-Science*: 81–8.

Ala-Mutka, Kirsti. 2009. *Review of Learning in ICT-Enabled Networks and Communities*. Luxembourg: Publications Office of the European Union. http://ftp.jrc.es/EURdoc/JRC52394.pdf.

Alender, Bethany J. 2015. 'Understanding Volunteer Motivations to Participate in Citizen Science Projects: A Deeper Look at Water Quality Monitoring', *Journal of Science Communication* 15, no. 3: A04.

Alessa, Lilian, Andrew Kliskey, James Gamble, Maryann Fidel, Grace Beaujean, and James Gosz. 2016. 'The Role of Indigenous Science and Local Knowledge in Integrated Observing Systems: Moving Toward Adaptive Capacity Indices and Early Warning Systems', *Sustainability Science* 11: 91–102.

Allen, Liz, Ceri Jones, Kevin Dolby, David Lynn, and Mark Walport. 2009. 'Looking for Landmarks: The Role of Expert Review and Bibliometric Analysis in Evaluating Scientific Publication Outputs', *PLoS ONE* 4, no. 6: e5910. https://doi.org/10.1371/journal.pone.0005910

Alliance for Aquatic Resource Monitoring. Data Interpretation. 2010. https://www.dickinson.edu /download/downloads/id/7015/data_interpretation_manual_2017pdf.

Altbach, Philip G. 2011. 'The Past, Present, and Future of the Research University', *Economic and Political Weekly* 46, no. 16: 65–73.

Amano, Tatsuya, James D.L. Lamming, and William J. Sutherland. 2016. 'Spatial Gaps in Global Biodiversity Information and the Role of Citizen science', *Bioscience* 66: 393–400.

Anderson, Aston, Daniel Huttenlocher, Jon Kleinberg, and Jure Leskovec. 2012. 'Discovering Value from Community Activity on Focused Question Answering Sites: A Case Study of Stack Overflow.' In *KDD'12 Proceedings of the 18th ACM SIGKDD international conference on Knowledge discovery and data mining*: 850–8.

Anderson, Ben. 2014. *Encountering Affect. Capacities, Apparatuses, Conditions*. Surrey: Ashgate.

Andrzejaczek, Samantha, Jessica Meeuwig, Davud Rowat, Simon Pierce, Tim Davies, Rebecca Fisher, and Mark Meekan. 2016. 'The Ecological Connectivity of Whale Shark Aggregations in the Indian Ocean: A Photo-identification Approach', *Royal Society Open Science* 3, no. 11: 160455.

Antoniou, V., and A. Skopeliti. 2015. 'Measures and Indicators of VGI Quality: An Overview', *ISPRS Annals of the Photogrammetry, Remote Sensing and Spatial Information Sciences* 2: 345–51.

Antoniou, Vyron, Jeremy Morley, and Mordechai Haklay. 2010. 'Web 2.0 Geotagged Photos: Assessing the Spatial Dimension of the Phenomenon', *Geomatica* 64, no. 1: 99–110.

Archer, Louise, Emily Dawson, Jennifer DeWitt, Amy Seakins, and Billy Wong. 2015. '"Science capital": A Conceptual, Methodological, and Empirical Argument for Extending Bourdieusian

Notions of Capital Beyond the Arts', *Journal of Research in Science Teaching* 52: 922–48. https://doi.org/10.1002/tea.21227.

Arlinghaus, Robert, Ben Beardmore, Carsten Riepe, Jurgen Meyerhoff, and Thilo Pagel. 2014. 'Species-specific Preferences of German Recreational Anglers for Freshwater Fishing Experiences, with Emphasis on the Intrinsic Utilities of Fish Stocking and Wild Fishes', *Journal of Fish Biology* 85: 1843–67.

Árnason, Vilhjálmur. 2013. 'Scientific Citizenship in a Democratic Society', *Public Understanding of Science* 22, no. 8: 927–40. https://doi.org/10.1177/0963662512449598.

Arnstein, Sherry R. 1969. 'A Ladder of Citizen Participation', *Journal of the American Institute of Planners* 35: 216–24.

Arpin, Isabelle, Coralie Mounet, and David Geoffroy. 2015. 'Inventaires naturalistes et rééducation de l'attention. Le cas des jardiniers de Grenoble', *Etudes rurales* 195: 89–107.

Aubé, Martin, Johanne Roby, and Miroslav Kocifaj. 2013. 'Evaluating Potential Spectral Impacts of Various Artificial Lights on Melatonin Suppression, Photosynthesis, and Star Visibility', *PLoS ONE* 8, no. 7: e67789.

Aubrecht, Christoph, Malanding Jaiteh, Alexander De Sherbinin, and Travis Longcore. 2010. 'Monitoring Impact of Urban Settlements on Nearby Protected Areas from Space', *Geophysical Research Abstracts* 12: 12758.

August, Tom, Martin Harvey, Paula Lightfoot, David Kilbey, Timos Papadopoulos, and Paul Jepson. 2015. 'Emerging Technologies for Biological Recording', *Biological Journal of the Linnean Society* 115, no. 3: 731–49.

Australian Citizen Science Association – ACSA. n.d. Last accessed 24 January 2018. https://www.citizenscience.org.au/.

Azam, Clémentine, Isabelle Le Viol, Jean-François Julien, Yves Bas, and Christian Kerbiriou. 2016. 'Disentangling the Relative Effect of Light Pollution, Impervious Surfaces and Intensive Agriculture on Bat Activity with a National-Scale Monitoring Program', *Landscape Ecology* 31: 1–13.

Bailes, Helena J., and Robert J. Lucas. 2013. 'Human Melanopsin Forms a Pigment Maximally Sensitive to Blue Light (λmax ≈ 479 Nm) Supporting Activation of G(q/11) and G(i/o) Signalling Cascades', *Proceedings of the Royal Society B: Biological Sciences* 280: 20122987.

Bailey, Neil. 2016. 'Improving Participation in Citizen Science Using the Science Capital Concept'. Paper presented at the First International ECSA Conference, Berlin, 19–21 May.

Balestrini, Mara, Tomas Diez, Alexandre Pólvora, and Susana Nascimento. 2016. *Mapping Participatory Sensing and Community-led Environmental Monitoring Initiatives*. Report for Making Sense Project (Call H2020 Grant 688620). http://making-sense.eu/wp-content/uploads/2016/07/Making-Sense-D62-D41-Mapping-Participatory-Sensing.pdf

Ballard, Heidi L., Lucy D. Robinson, Alison N. Young, Gregory B. Pauly, Lila M. Higgins, Rebecca F. Johnson, and John C. Tweddle. 2017. 'Contributions to Conservation Outcomes by Natural History Museum-led Citizen Science: Examining Evidence and Next Steps', *Biological Conservation* 208: 87–97. https://doi.org/10.1016/j.biocon.2016.08.040

Banks, Sarah, Andrea Armstrong, Kathleen Carter, Helen Graham, Peter Hayward, Alex Henry, Tessa Holland et al. 2013. 'Everyday Ethics in Community-based Participatory Research', *Contemporary Social Science* 8, no. 3: 263–77. https://doi.org/10.1080/21582041.2013.769618.

Barber, Vanessa, Poppy Lakeman Fraser, Roger Fradera, David Slawson, and Laurence Evans. 2016. *OPAL – Exploring Nature Together. Findings and Lessons Learnt*. London: OPAL.

Barraza, J. A., V. Alexander, L. E. Beavin, E. T. Terris, and P. J. Zak. 2015. 'The Heart of the Story: Peripheral Physiology During Narrative Exposure Predicts Charitable Giving', *Biological Psychology* 105: 138–43.

Barstow, Daniel, and Cheick Diarra. 1997. 'Mars Exploration'. In *Internet Links for Science Education*, edited by Karen C. Cohen, 111–31. Boston: Springer.

Bastin, Lucy, Sven Schade, and Christian Schill. 2017. 'Data and Metadata Management for Better VGI Reusability'. In *Mapping and the Citizen Sensor*, edited by Giles Foody, Linda See, Steffen Fritz, Peter Mooney, Ana-Maria Olteanu-Raimond, Cidalia Costa Fonte, and Vyron Antoniou, 249–73. London: Ubiquity Press.

Bates, Adam J., Poppy Lakeman Fraser, Lucy Robinson, John C. Tweddle, Jon P. Sadler, Sarah E. West, Simon Norman, Martin Batson, and Linda Davies. 2015. 'The OPAL Bugs Count Survey: Exploring the Effects of Urbanisation and Habitat Characteristics Using Citizen Science', *Urban Ecosystems* 18: 1477–97.

Bauer, Martin W. 2016. 'Results of the Essay Competition on the "Deficit Concept"', *Public Understanding of Science* 25: 398–9.

Beck, Ulrich. 1991. *Politik in der Risikogesellschaft*. Frankfurt a. M.: Suhrkamp.

Becker, Mathias. 2015. 'Social Innovation and Ethical Guidelines'. CHEST project Deliverable D5.1, European Commission. http://www.chest-project.eu/wp-content/uploads/2015/10/RP2NEW_D5.1_Social_Innovation_and_Ethical_Guidelines.pdf.

Becker, Mathias, Ivan Ficano, James Hughes, Andries van Vugt, Fons Verhoef, Róbert Bjarnason, Mario Barile, et al. 2016. 'Report on Call 3 projects', CHEST project Deliverable D3.8, European Commission. http://www.chest-project.eu/wp-content/uploads/2015/10/D3.8_Report_on_Call3_projects_V1.1_PUBLIC.pdf.

Bela, Györgyi, Taru Peltola, Juliette C. Young, Isabelle Arpin, Balint Balázs, Jennifer Hauck, György Pataki, et al. 2016. 'Learning and the Transformative Potential of Citizen science', *Conservation Biology* 30: 990–9. https://doi.org.10.1111/cobi.12762.

Bell, Philip, Bruce Lewenstein, Andrew Shouse, and Michael Feder, eds. 2009. *Learning Science in Informal Environments: People, Places, and Pursuits*. Washington, DC: National Research Council of the National Academies.

Benkler, Yochai. 2006. *The Wealth of Networks: How Social Production Transforms Markets and Freedom*. New Haven: Yale University Press.

Bensaude-Vincent, Bernadette, and Christine Blondel, eds. 2008. *Science and Spectacle in the European Enlightenment*. Science, Technology and Culture, 1700–1945. Burlington: Ashgate.

Berkes, Fikret. 2012. *Sacred Ecology*, Third edition. New York: Routledge.

Bhatt, Samir, Peter W. Gething, Oliver J. Brady, Jane P. Messina, Andrew W. Farlow, Catherine L. Moyes, John M. Drake, et al. 2013. 'The Global Distribution and Burden of Dengue', *Nature* 496, no. 7446: 504–7. https://doi.org/10.1038/nature12060.

Bherer, Laurence, Pascale Dufour, and Françoise Montambeault. 2016. 'The Participatory Democracy Turn: An Introduction', *Journal of Civil Society* 12, no. 3: 225–30. https://doi.org/10.1080/17448689.2016.1216383.

Bi, Ke, Tyler Linderoth, Dan Vanderpool, Jeffery M. Good, Rasmum Nielsen, and Craig Moritz. 2013. 'Unlocking the Vault: Next-generation Museum Population Genomics', *Molecular Ecology* 22: 6018–32.

Bibby, Colin, Neil D. Burgess, David Hill, and Simon Mustoe. 2000. *Bird Census Techniques*, 2nd edition. London: Academic Press.

Bidwell, David. 2009. 'Is Community-Based Participatory Research Postnormal Science?' *Science, Technology & Human Values* 34, no. 6: 741–61.

Bishr, Mohamed and Werner Kuhn. 2013. 'Trust and Reputation Models for Quality Assessment of Human Sensor Observations', In *International Conference on Spatial Information Theory*, edited by Tenbrink T., J. Stell, A. Galton, and Z. Wood: 53–73. Scarborough, UK: Springer, Cham.

Blackwell Alan F., Lee Wilson, Alice Street, Charles Boulton, and John Nell. 2009. *Radical Innovation: Crossing Knowledge Boundaries with Interdisciplinary Teams*. Technical Report 760, University of Cambridge Computer Laboratory. https://www.cl.cam.ac.uk/techreports/UCAM-CL-TR-760.pdf

Blaney, Ralph J.P., Anne C.V. Philippe, Michael J.O. Pocock, and Glyn D. Jones. 2016. *Citizen Science and Environmental Monitoring: Towards a Methodology for Evaluating Opportunities, Costs and Benefits*. Report to the UK Environmental Observation Framework. Wallingford, UK: WRc, Fera Science, Centre for Ecology & Hydrology.

BMBF. 2017. Federal Ministry of Education and Research, Germany, 'Mitmachen und Forschen!' press release, 14 July 2017 https://www.bmbf.de/de/mitmachen-und-forschen-4503.html.

Bodmer, Walter. 1986. *The Public Understanding of Science*. London: The Royal Society. http://royalsociety.org/policy/publications/1985/public-understanding-science.

Bohensky, Erin L., and Yiheyis Maru. 2011. 'Indigenous Knowledge, Science, and Resilience: What Have We Learned from a Decade of International Literature on "Integration"?' *Ecology and Society* 16, no. 4: 6.

Bohensky, Erin L., James R. Butler, and Jocelyn Davies. 2013. 'Integrating Indigenous Ecological Knowledge and Science in Natural Resource Management: Perspectives from Australia', *Ecology and Society* 18, no. 3: 20.

Bond, Ross, and Lindsay Paterson. 2005. 'Coming Down from the Ivory Tower? Academics' Civic and Economic Engagement with the Community', *Oxford Review of Education* 31, no. 3: 331–51.

Bone, J., D. Barraclough, P. Eggleton, M. Head, D.T. Jones, and N. Voulvoulis. 2014. 'Prioritising Soil Quality Assessment through the Screening of Sites: The Use of Publicly Collected Data', *Land Degradation & Development* 25: 251–66.

Bonmati-Carrion, Maria Angeles, Raquel Arguelles-Prieto, Maria Jose Martinez-Madrid, Russel Reiter, Ruediger Hardeland, Maria Angeles Rol, and Juan Antonio Madrid. 2014. 'Protecting the Melatonin Rhythm through Circadian Healthy Light Exposure', *International Journal of Molecular Sciences* 15: 23448–500.

Bonn, Aletta, Anett Richter, Katrin Vohland, Lisa Pettibone, Miriam Brandt, Reinart Feldmann, Claudia Goebel, et al. 2016. *Greenpaper Citizen Science Strategy 2020 for Germany*. BürGEr schaffen WISSen / GEWISS, Berlin. http://www.buergerschaffenwissen.de/sites/default /files/assets/dokumente/gewiss_cs_strategy_englisch_0.pdf.

Bonney, Rick. n.d. Website Editorial: 'Citizen Science: Theory and Practice'. Last modified 24 October 2016. Last accessed 15 November 2017. http://theoryandpractice .citizenscienceassociation.org/about.

Bonney, Rick. 1996. 'Citizen Science: A Lab Tradition', *Living Birds* 15, no. 4: 7–15.

Bonney, Rick. 2004. 'Understanding the Process of Research'. In *Creating Connections: Museums and Public Understanding of Current Research*, edited by David Chittenden, Graham Farmelo, and Bruce Lewenstein, 199–210. Walnut Creek, CA: Altamira Press.

Bonney, Rick. 2007. 'Citizen science at the Cornell Lab of Ornithology'. In R. Yager & D. Falk (Eds.), *Exemplary science in informal education settings: Standards-based success stories*. Arlington, VA: NSTA Press.

Bonney, Rick, Heidi Ballard, Rebecca Jordan, Ellen McCallie, Tina Phillips, Jennifer Shirk, and Candie C. Wilderman. 2009. *Public Participation in Scientific Research: Defining the Field and Assessing its Potential for Informal Science Education*. A CAISE Inquiry Group Report. Washington, D.C.: Center for Advancement of Informal Science Education (CAISE).

Bonney, Rick, Caren B. Cooper, and Heidi Ballard. 2016. 'The Theory and Practice of Citizen Science: Launching a New Journal', *Citizen Science: Theory and Practice* 1, no.1.

Bonney, Rick, Caren B. Cooper, Janis Dickinson, Steve Kelling, Tina Phillips, Kenneth V. Rosenberg, and Jennifer Shirk. 2009. 'Citizen Science: A Developing Tool for Expanding Science Knowledge and Scientific Literacy', *BioScience* 59: 977–84.

Bonney, Rick and Andre Dhondt. 1997. 'Project FeederWatch'. In *Internet Links to Science Education: Student Scientist Partnerships*, edited by Karen Cohen, 31–54. New York: Plenum Press.

Bonney, Rick, Tina B. Phillips, Heidi L. Ballard, and Jody W. Enck. 2016. 'Can Citizen Science Enhance Public Understanding of Science?' *Public Understanding of Science* 25: 2–16.

Bonney, Rick, Jennifer L. Shirk, Tina B. Phillips, AndreaWiggins, Heidi L. Ballard, Abraham J. Miller-Rushing, and Julia K. Parrish. 2014. 'Next Steps for Citizen Science', *Science* 343: 1436–7. https://doi.org/10.1126/science.1251554.

Bonter, David N., and Caren B. Cooper. 2012. 'Data Validation in Citizen Science: A Case Study from Project FeederWatch', *Frontiers in Ecology and the Environment* 10: 305–7.

Bosak, Jon. 1998. 'The Cover Pages.' In *XML Developers' Conference, August 20–21*. Last accessed 28 August 2017. http://xml.coverpages.org/bosakXMLDayAnn9808.html.

Bower, Shannon D., Jacob W. Brownscombe, Kim Birnie-Gauvin, Matthew I. Ford, Andrew D. Moraga, Ryan J.P. Pusiak, Eric D. Turenne, et al. 2017. 'Making Tough Choices: Picking the Appropriate Conservation Decision-Making Tool', *Conservation Letters* 11: e12418. https://doi.org/10.1111/conl.12418.

Bowser, Anne, Derek Hansen, Yurong He, Carol Boston, Matthew Reid, Logan Gunnell, and Jennifer Preece. 2013. 'Using Gamification to Inspire New Citizen Science Volunteers'. In *The First International Conference on Gameful Design, Research, and Applications (Gamification '13)*, 19–25. Toronto.

Bowser, Anne, Derek Hansen, and Jenny Preece. 2013. 'Gamifying Citizen Science: Lessons and Future Directions'. In *CHI '13 Extended Abstracts on Human Factors in Computing Systems*, 3263–66. Proc. of CHI 2013 Workshop 'Designing gamification: Creating gameful and playful experiences'. Paris: ACM.

Bowser, Anne, Rachel McMonagle, and Elizabeth Tyson. 2015. 'Citizen Science and Crowdsourcing Metadata Workshop Summary'. Woodrow Wilson Center, Washington, DC: 9–10 July 2015. https://www.wilsoncenter.org/sites/default/files/2015_CL_EPA_metadataWS _summaryFINAL.pdf.

Bowser, Anne and Lea Shanley. 2013. *New Visions in Citizen Science*. Washington, DC: Woodrow Wilson International Center for Scholars. https://www.wilsoncenter.org/publication/new -visions-citizen-science.

Bowser, Anne, Katie Shilton, Jenny Preece, and Elizabeth Warrick. 2017. 'Accounting for Privacy in Citizen Science: Ethical Research in a Context of Openness'. In *Proceedings of the 2017 ACM Conference on Computer Supported Cooperative Work and Social Computing* (CSCW '17), 2124–36. New York: ACM.

Bowser, Anne, Andrea Wiggins, Lea Shanley, Jennifer Preece, and Sandra Henderson, 2014. 'Sharing Data while Protecting Privacy in Citizen Science', *Interactions* 21, no. 1: 70–3.

Bowser, Anne, Andrea Wiggins, and Robert D. Stevenson. 2013. *Data Policies for Public Participation in Scientific Research: A Primer*. Albuquerque, NM: DataONE.

Bowser, Anne and Andrea Wiggins. 2015. 'Privacy in Participatory Research: Advancing Policy to Support Human Computation', *Human Computation* 2: 19–44.

Bowser, Chris. 2016. 'The Hudson River Eel Project 2008–2016: Citizen Science Juvenile American Eel Surveys.' New York: New York State Department of Environmental Conservation. http://www.dec.ny.gov/docs/remediation_hudson_pdf/082415eelreport.pdf.

Boxshall, Geoff, and David Self. 2011. 'UK Taxonomy & Systematics Review – 2010.' *Results of survey undertaken by the Review Team at the Natural History Museum serving as contractors to the Natural Environment Research Council (NERC)*.

Brainard, George C., John P. Hanifin, Mark D. Rollag, Jeffrey Greeson, Brenda Byrne, Gena Glickman, Edward Gerner, and Britt Sanford. 2001. 'Human Melatonin Regulation Is Not Mediated by the Three Cone Photopic Visual System.' *The Journal of Clinical Endocrinology & Metabolism* 86: 433–6.

Brainard, George C., John P. Hanifin, Benjamin Warfield, Marielle K. Stone, Mary E. James, Melissa Ayers, Alan Kubey, Brenda Byrne, and Mark Rollag. 2015. 'Short-Wavelength Enrichment of Polychromatic Light Enhances Human Melatonin Suppression Potency', *Journal of Pineal Research* 58: 352–61.

Bramer, Rainer. 2010. *Natur: Vergessen? Erste Befunde des Jugendreports Natur 2010*. Bonn: Information Medien Agrar e.V.

Braschler, Brigitte, Kristen Mahood, Natasha Karenyi, Kevin Gaston, and Steven Chown. 2010. 'Realizing a Synergy between Research and Education: How Participation in Ant Monitoring Helps Raise Biodiversity Awareness in a Resource-poor Country', *Journal of Insect Conservation* 14:19–30.

Brewer, Carol. 2002. 'Outreach and Partnership Programs for Conservation Education. Where Endangered Species Conservation and Research Occur', *Conservation Biology* 16:4–6.

Bria, Francesca; Fabrizio Sestini, Mila Gascó, Peter Baeck, Harry Halpin, Esteve Almirall, and Frank Kresin. 2015. *Growing a Digital Social Innovation Ecosystem for Europe*. DSI Final Report, European Commission.

Brofeldt, Søren, Ida Theilade, Neil D. Burgess, Finn Danielsen, Michael K. Poulsen, Teis Adrian, Tran N. Bang et al. 2014. 'Community Monitoring of Carbon Stocks for REDD+: Does Accuracy and Cost Change over Time?' *Forests* 5, no. 8: 1834–54.

Brossard, Dominique, Bruce Lewenstein, and Rick Bonney. 2005. 'Scientific Knowledge and Attitude Change: The Impact of a Citizen Science Project', *International Journal of Science Education* 27, no 9: 1099–1121.

Bucchi, Massimiano. 2008. 'Of Deficits, Deviations and Dialogues: Theories of Public Communication of Science'. In *Handbook of Public Communication of Science and Technology*, edited by Massimiano Bucchi and Brian Trench, 57–76. London and New York: Routledge.

Buckingham-Hatfield, Susan, and Susan Percy, eds. 2005. *Constructing Local Environmental Agendas: People, Places and Participation*. London and New York: Routledge.

Budhathoki, Nama R. 2010. 'Participants' Motivations to Contribute Geographic Information in an Online Community'. PhD diss., University of Illinois at Urbana-Champaign.

Budhathoki, Nama R., and Caroline Haythornthwaite. 2013. 'Motivation for Open Collaboration: Crowd and Community Models and the Case of OpenStreetMap', *American Behavioral Scientist* 57, no. 5: 548–75.

Bull, Richard, Judith Petts, and James Evans. 2008. 'Social Learning from Public Engagement: Dreaming the Impossible?' *Journal of Environmental Planning and Management* 51: 701–16.

Bultitude, Karen. 2011. 'The Why and How of Science Communication'. In *Science Communication*, edited by P. Rosulek. Pilsen: European Commission. https://www.ucl.ac.uk/sts/staff /bultitude/KB_TB/Karen_Bultitude_-_Science_Communication_Why_and_How.pdf.

Burgess, Hillary K., Lauren B. DeBay, H.E. Froehlich, Natalaie Schmidt, Elli J. Theobald, Ailene K. Ettinger, Janneke Hille-RisLambers, Joshua Tewksbury, and Julia K. Parrish. 2017. 'The Science of Citizen Science: Exploring Barriers to Use as a Primary Research Tool', *Biological Conservation* 208: 113–20.

Burke, Jeffrey A., D. Estrin, Mark Hansen, Andrew Parker, Nithya Ramanathan, Sasank Reddy, and Mani B. Srivastava. 2006. 'Participatory Sensing', UCLA: Center for Embedded Network Sensing: 1–5. https://escholarship.org/uc/item/19h777qd.

Burri, Regula V. 2009. 'Coping with Uncertainty: Assessing Nanotechnologies in a Citizen Panel in Switzerland', *Public Understanding of Science* 18, no. 5: 498–511.

Busch, Julia A, Raul Bardaji, Luigi Ceccaroni, Anna Friedrichs, Jaume Piera, Carine Simon, Peter Thijsse, Marcel Wernand, Hendrik J. van der Woerd, and Oliver Zielinski. 2016. 'Citizen Bio-optical Observations from Coast- and Ocean and Their Compatibility with Ocean Colour Satellite Measurements', *Remote Sensing* 8, no. 8: 879.

BusinessDictionary.com. WebFinance, Inc. 19 November 2017. Last accessed 22 November 2017. http://www.businessdictionary.com/definition/best-practice.html.

Buytaert, Wouter, Zed Zulkafli, Sam Grainger, Luis Acosta, Tilashwork C. Alemie, Johan Bastiaensen, Bert De Bièvre, et al. 2014. 'Citizen Science in Hydrology and Water Resources: Opportunities for Knowledge Generation, Ecosystem Service Management, and Sustainable Development', *Frontiers in Earth Science* 2: 1–21.

Cartwright, Jon. 2016. 'Technology: Smartphone Science', *Nature* 531, no. 7596 : 669–71.

CASE – Centre for Ageing and Supportive Environments, n.d. 'Welcome to CASE!' Last accessed October 2017. http://www.med.lu.se/english/case.

Castro, Marcia C, Mary E. Wilson, and David E. Bloom. 2017. 'Disease and Economic Burdens of Dengue', *The Lancet Infectious Diseases* 17, no. 3: e70–78. https://doi.org/10.1016/S1473-3099(16)30545-X.

Causer, Tim, and Valerie Wallace. 2012. 'Building a Volunteer Community: Results and Findings from Transcribe Bentham', *Digital Humanities Quarterly* 6, no. 2. http://www.digitalhumanities.org/dhq/vol/6/2/000125/000125.html.

Cavalier, Darlene, and Eric B. Kennedy, eds. 2016. *The Rightful Place of Science: Citizen Science*. Tempe, AZ: Consortium for Science, Policy & Outcomes.

CCS. 2012. Web Archive. Last accessed 28 August 2017. https://web.archive.org/web/20120407102225/http://cybersciencesummit.org/challenges/.

CCS. 2014. Website. Last accessed 28 August 2017. http://cybersciencesummit.org/.

Ceccaroni, Luigi, Anne Bowser, and Peter Brenton. 2017. 'Civic Education and Citizen Science: Definitions, Categories, Knowledge Representation'. In *Analyzing the Role of Citizen Science in Modern Research*, edited by Luigi Ceccaroni and Jaume Piera, 1–23. Hershey, PA: IGI Global. https://doi/org/10.4018/978-1-5225-0962-2.ch001.

Ceccaroni, Luigi, and Jaume Piera, eds. 2017. *Analyzing the Role of Citizen Science in Modern Research*. Hershey, PA: IGI Global. https://doi/org/10.4018/978-1-5225-0962-2.ch001.

Ceci, Stephen V., and Wendy M. Williams. 2011. 'Understanding Current Causes of Women's Underrepresentation in Science', *Proceedings of the National Academy of Sciences* 108, no. 8: 3157–62.

Celino, Irene. 2013. 'Geospatial Dataset Curation through a Location-Based Game', *Semantic Web* 6: 121–30.

Center for Theory of Change. n.d. Last accessed 16 January 2018. http://www.Theoryofchange.org.

CFH. 2017. 'Chinese Field Herbarium'. Last accessed 6 June 2017. http://www.cfh.ac.cn/.

Chandler, Mark, Linda See, Kyle Copas, Astrid M.Z. Bonde, Bernat Claramunt López, Finn Danielsen, Jan Kristoffer Legind, et al. 2017. 'Contribution of Citizen Science Towards International Biodiversity Monitoring', *Biological Conservation* 213: 280–94.

Chari, Ramya, Luke J. Matthews, Marjory S. Blumenthal, Amanda F. Edelman, and Terese Jones. 2017. 'The Promise of Community Citizen Science', *Perspective: Expert Insights on a Timely Policy Issue*. RAND Corporation.

Cheng, J.C-H. and M.C. Monroe. 2012. 'Connection to Nature: Children's Attitude toward Nature', *Environment and Behavior* 44, no. 1: 31–49.

Cherry, L. and G. Braasch. 2008. 'Communicating Climate Science to Kids and Adults through Citizen Science, Hands-on Demonstrations, and a Personal Approach', *American Geological Union, Fall Meeting Abstract 2008:* abstract id. ED21B-0626. http://adsabs.harvard.edu/abs/2008AGUFMED21B0626C.

Chesbrough, Henry William. 2003. *Open Innovation: The New Imperative for Creating and Profiting from Technology*. Boston: Harvard Business School Publishing.

Chest. n.d. 'CHEST Project Partners'. Last modified 23 September 2016. http://www.chest-project .eu/partners/

Chilvers, Jason, and Matthew Kearnes. 2016. *Remaking Participation: Science, Environment and Emergent Publics*. Oxon & New York: Routledge.

Choi-Fitzpatrick, Austin. 2014. 'Drones for Good: Technological Innovations, Social Movements, and the State', *Journal of International Affairs* 68, no. 1: 19.

Christofferson, Rebecca C. 2016. 'Zika Virus Emergence and Expansion: Lessons Learned from Dengue and Chikungunya May Not Provide All the Answers', *American Journal of Tropical Medicine and Hygiene* 95, no. 1: 15–18. https://doi.org/10.4269/ajtmh.15-0866.

Citizen Cyberlab. n.d. Home page. Last accessed October 2016. http://www.citizencyberlab.org/.

Citizen Observatories. 2015. Last accessed 14 March 2018. swe4citizenscience GitHub. https://github.com/opengeospatial/swe4citizenscience.

Citizen Science Association. 2015. 'Introducing the Data and Metadata Working Group'. Last accessed 14 December 2017. http://citizenscience.org/2015/11/12/introducing-the-data -and-metadata-working-group/.

Citizen Science Association. n.d. Last accessed 24 January 2018. http://citizenscience.org.

Clark, Lauren, and William Ventres. 2016. 'Qualitative Methods in Community-Based Participatory Research Coming of Age', *Qualitative Health Research* 26, no. 1: 3–4.

CLRP. 2017. 'The Chintang Language Research Program.' Last accessed October 2017. http:// www.clrp.uzh.ch/index.html.

Cochran, Elizabeth S., Jesse F. Lawrence, Carl Christensen, and Ravi S. Jakka. 2009. 'The Quake-Catcher Network: Citizen Science Expanding Seismic Horizons', *Seismological Research Letters* 80: 26–30.

Cohen, Alison Klebanoff, Andrea Lopez, Nile Malloy, and Rachel Morello-Frosch. 2016. 'Surveying for Environmental Health Justice: Community Organizing Applications of Community-Based Participatory Research', *Environmental Justice* 9, no. 5: 129–36.

Cohn, Jeffrey P. 2008. 'Citizen Science: Can Volunteers do Real Research?' *Bioscience* 58: 192–7.

Colfer, Carol, J. Pierce, Marcus Colchester, Laxman Joshi, Rajindra Puri, Anja Nygren, and Citalli Lopez. 2005. 'Traditional Knowledge and Human Well-being in the 21st Century'. In *Forests in the Global Balance – Changing Paradigms, Vol. 17, IUFRO World Series,* edited by Mery, Gerardo, René I. Alfaro, Markku Kanninen, and Max Lobovikov, 173–82. Helsinki, Finland: International Union of Forest Research Organizations.

Collen, Ben, Nathalie Pettorelli, Jonathan E.M. Baillie, and Sarah M. Durant. 2013. *Biodiversity Monitoring and Conservation: Bridging the Gap between Global Commitment and Local Action*. Chichester, UK: John Wiley & Sons.

Collins, Sophia. n.d. '25 Things We Learned by Running a Radical Citizen Science Project about Nappies (as you do)'. Last modified 3 September 2016. https://nappysciencegang.wordpress .com/2016/09/03/25-things-we-learned-by-running-a-radical-citizen-science-project-about -nappies-as-you-do/.

Congress. 2016. Crowdsourcing and Citizen Science Act of 2016. *H.R.6414*. USA.

Conrad, Catherine T., and Tyson Daoust. 2008. 'Community-based Monitoring Frameworks: Increasing the Effectiveness of Environmental Stewardship', *Environmental Management* 41: 356–8.

Conrad, Cathy C., and Kirsta G. Hilchey. 2011. 'A Review of Citizen Science and Community-Based Environmental Monitoring: Issues and Opportunities', *Environmental Monitoring and Assessment* 176: 273–91.

Constant, Natasha and Liz Roberts. 2017. 'Narratives as a Mode of Research Evaluation in Citizen Science: Understanding Broader Science Communication Impacts', *Journal of Science Communication* 16, no 4: A03.

Convention on Biological Diversity. 2011. *The Tkarihwaié:ri – Code of Ethical Conduct to Ensure Respect for the Cultural and Intellectual Heritage of Indigenous and Local Communities Relevant to the Conservation and Sustainable Use of Biological Diversity*. Montreal, Canada: Secretariat of the Convention on Biological Diversity. http://www.cbd.int/traditional/code /ethicalconduct-brochure-en.pdf.

Cooper, Caren. 2016. *Citizen Science: How Ordinary People are Changing the Face of Discovery*. New York: The Overlook Press.

Cooper, Caren B., and Bruce V. Lewenstein. 2016. 'Two Meanings of Citizen Science'. In *The Rightful Place of Science: Citizen Science*, edited by D. Cavalier and E.B. Kennedy, 51–62. Tempe, AZ: Consortium for Science, Policy & Outcomes.

Cooper, Caren B., Jennifer Shirk, and Benjamin Zuckerberg. 2014. 'The Invisible Prevalence of Citizen Science in Global Research: Migratory Birds and Climate Change', *PLoS ONE* 9: e106508.

Coote, Anna, and Jo Lenaghan. 1997. *Citizens' Juries: Theory into Practice*. London: Institute of Public Policy Research.

Corburn, Jason. 2005. *Street Science: Community Knowledge and Environmental Health Justice*. Cambridge: MIT Press Books.

Corburn, Jason. 2007. 'Community Knowledge in Environmental Health Science: Co-producing Policy Expertise', *Environmental Science & Policy* 10: 150–61.

Cornell Lab of Ornithology. n.d. Last accessed October 2016. http://www.birds.cornell.edu.

Corrigan, Deborah. 2006. 'No Wonder the Kids are Confused: The Relevance of Science Education to Science'. In *2006 – Boosting Science Learning – What will it Take?* 1997-2008 Australian Council for Educational Research Conference Archive. Last accessed 29 May 2017. http://research.acer.edu.au/research_conference_2006/6.

Council of Europe. 2005. *Council of Europe Framework Convention on the Value of Cultural Heritage for Society*. Last accessed 2 July 2018. https://www.coe.int/en/web/conventions/full-list/-/conventions/treaty/199

Couvet, Denis, Frédéric Jiguet, Romain Julliard, Harold Levrel, and Anne Teyssedre. 2008. 'Enhancing Citizen Contributions to Biodiversity Science and Public Policy', *Interdisciplinary Science Reviews* 33: 95–103.

Cox, Joe, Eun Young Oh, Brooke Simmons, Chris Lintott, Karen Masters, Anita Greenhill, Gary Graham, and Kate Holmes. 2015. 'Defining and Measuring Success in Online Citizen Science: A Case Study of Zooniverse Projects', *Computing in Science & Engineering* 17, no. 4: 28–41.

Craglia, Max, and Carlos Granell. 2014. *Citizen Science and Smart Cities – Report of Summit, Ispra, 5–7th February 2014*. Luxembourg: Publications Office of the European Union. http://publications.jrc.ec.europa.eu/repository/bitstream/JRC90374/lbna26652enn.pdf.

Craglia, Max, and Lea Shanley. 2015. 'Data Democracy – Increased Supply of Geospatial Information and Expanded Participatory Processes in the Production of Data', *International Journal of Digital Earth* 8, no. 9: 679–93.

Crain, Rhiannon, Caren Cooper, and Janis L. Dickinson. 2014. 'Citizen Science: A Tool for Integrating Studies of Human and Natural Systems', *Annual Review of Environment and Resources* 39: 641–65.

Crall, Alycia W. 2010. 'Developing and Evaluating a National Citizen Science Program for Invasive Species', *Chemistry & Biodiversity* 1, no. 11: 1–185.

Crall, Alycia W., Rebecca Jordan, Kirsten Holfelder, Gregory J. Newman, Jim Graham, and Donald M. Waller. 2012. 'The Impacts of an Invasive Species Citizen Science Training Program on Participant Attitudes, Behavior, and Science Literacy', *Public Understanding of Science* 22: 745–64.

Crall, Alycia W., Gregory J. Newman, Thomas J. Stohlgren, Kirstin A. Holfelder, Jim Graham, and Donald M. Waller. 2011. 'Assessing Citizen Science Data Quality: An Invasive Species Case Study', *Conservation Letters* 4: 433–42.

Cunha, R. da, 2015. 'Are You Ready for Citizen Science?' Ecsites, *Digital Spokes*, Issue 11. http://www.ecsite.eu/activities-and-services/news-and-publications/digital-spokes/issue-11#section=section-indepth&href=/feature/depth/are-you-ready-citizen-science.

Curtis, Vickie. 2014. 'Online Citizen Science Games: Opportunities for the Biological Sciences.' *Applied & Translational Genomics* 3, no. 4: 90–4.

Curtis, Vickie. 2015. 'Online Citizen Science Projects: An Exploration of Motivation, Contribution and Participation'. PhD diss., The Open University.

Cutler, David, and Ruth Temple. 2010. 'What's in a Name? Taxonomy and Systematics under the microscope', *Biologist* 57: 192–8.

D'Auria, Convertito, and Luca D'Auria. 2016. 'Real-time Mapping of Earthquake Perception Areas in the Italian Region from Twitter Streams analysis'. In *Earthquakes and Their Impact on Society*, edited by Sebastiano D'Amico, 619–30. Switzerland: Springer.

Dallman, Winfried K., Vladislav Peskov, Olga A. Murashko, and Ekaterina Khmeleva. 2011. 'Reindeer Herders in the Timan-Pechora Oil Province of Northwest Russia: An Assessment of Interacting Environmental, Social, and Legal Challenges', *Polar Geography* 34: 229–47.

Danielsen, Finn. 2016. *Expanding the Scientific Basis for How the World Can Monitor and Manage Natural Resources*. DSc thesis, University of Copenhagen and NORDECO.

Danielsen, Finn, Neil Burgess, and Andrew Balmford. 2005. 'Monitoring Matters: Examining the Potential of Locally-based Approaches', *Biodiversity and Conservation* 14: 2507–42.

Danielsen, Finn, Neil Burgess, Andrew Balmford, Paul F. Donald, Mikkel Funder, Julia P.G. Jones, Philip Alviola, Danilo S. Balete, Tom Blomley, and Justin Brashares. 2009. 'Local Participation in Natural Resource Monitoring: A Characterization of Approaches', *Conservation Biology* 23, no. 1: 31–42.

Danielsen, Finn, Neil D. Burgess, Per M. Jensen, and Karin Pirhofer-Walzl. 2010. 'Environmental Monitoring: The Scale and Speed of Implementation Varies According to the Degree of Peoples Involvement', Journal of Applied Ecology 47, no. 6: 1166–68.

Danielsen, Finn, Martin Enghoff, Eyðfinn Magnussen, Tero Mustonen, Anna Degteva, Kia K. Hansen, Nette Levermann, Svein. D. Mathiesen, and Øystein Slettemark. 2017. 'Citizen Science Tools for Engaging Local Stakeholders and Promoting Local and Traditional Knowledge in Landscape Stewardship'. In *The Science and Practice of Landscape Stewardship*, edited by Claudia Bieling and Tobias Plieninger, 80–98. Cambridge, UK: Cambridge University Press.

Danielsen, Finn, Per M. Jensen, Neil D. Burgess, Ronald Altamirano, Philip A. Alviola, Heriao Andrianandrasana, Justin S. Brashares, A. Cole Burton, Indiana Coronado, and Nancy Corpuz. 2014. 'A Multicountry Assessment of Tropical Resource Monitoring by Local Communities', *BioScience* 64, no. 3: 236–51.

Danielsen, Finn, Per M. Jensen, Neil D. Burgess, Indiana Coronado, Sune Holt, Michael K. Poulsen, Ricardo M. Rueda et al. 2014. 'Testing Focus Groups as a Tool for Connecting Indigenous and Local Knowledge on Abundance of Natural Resources with Science-based Land Management Systems', *Conservation Letters* 7, no. 4: 380–9.

Danielsen, Finn, Marlynn M. Mendoza, Anson Tagtag, Phillip A. Alviola, Danilo S. Balete, Arne E. Jensen, Martin Enghoff, and Michael K. Poulsen. 2007. 'Increasing Conservation Management Action by Involving Local People in Natural Resource Monitoring', *AMBIO: A Journal of the Human Environment* 36, no. 7: 566–70.

Danielsen, Finn, Karin Pirhofer-Walzl, Teis P. Adrian, Daniel R. Kapijimpanga, Neil D. Burgess, Per M. Jensen, Rick Bonney et al. 2014. 'Linking Public Participation in Scientific Research to the Indicators and Needs of International Environmental Agreements', *Conservation Letters* 7, no. 1: 12–24.

Danielsen, Finn, Elmer Topp-Jørgensen, Nette Levermann, Piitaaraq Løvstrøm, Martin Schiøtz, Martin Enghoff, and PâviarâK Jakobsen. 2014. 'Counting What Counts: Using Local Knowledge to Improve Arctic Resource Management', *Polar Geography* 37: 69–91.

Daum, Andreas W. 2002. 'Science, Politics, and Religion: Humboldtian Thinking and the Transformations of Civil Society in Germany, 1830–1870', *Osiris* 17: 107–40.

Davies, Anna, and Julie Simon. 2012a. *Citizen Engagement in Social Innovation – A Case Study Report*. TEPSIE project deliverable. European Commission.

Davies, Anna, Julie Simon, Robert Patrick, and Will Norman. 2012b. *Mapping Citizen Engagement in the Process of Social Innovation*. TEPSIE project deliverable. European Commission.

Davies, Linda, Roger Fradera, Hauke Riesch, and Poppy Lakeman-Fraser. 2016. 'Surveying the Citizen Science Landscape: An Exploration of the Design, Delivery and Impact of Citizen Science through the Lens of the Open Air Laboratories (OPAL) Programme', *BMC Ecology* 16: 17.

Davies, Linda, Laura Gosling, Carolina Bachariou, Jennifer Eastwood, Roger Fradera, Nicola Manomaiudom, Nicola and Sallie Robins. 2013. *OPAL Community Environment Report – Exploring Nature Together*. London: OPAL Centre for Environmental Policy, Imperial College London. https://www.opalexplorenature.org/sites/default/files/7/file/Community-Environment-Report-low-res.pdf.

Davies, Sarah R., and Maja Horst. 2016. *Science Communication - Culture, Identity and Citizenship*. London: Palgrave Macmillan.

Davis, Clayton A., Julia Heiman, Erick Janssen, Stephanie Sanders, Justin Garcia, and Filippo Menczer. 2016. 'Kinsey Reporter: Citizen Science for Sex Research.' Paper presented at Let'sTalk About Sex (Apps) Workshop (CSCW 2015): Vancouver. arXiv:1602.04878.

Dayton, Paul K. 2003. 'The Importance of the Natural Sciences to Conservation: (An American Society of Naturalists Symposium Paper)', *The American Naturalist* 162: 1–13.

Declaration of the United Nations Conference on the Human Environment (Stockholm Declaration). 1972. http://www.globalhealthrights.org/wp-content/uploads/2014/06/Stockholm-Declaration1.pdf.

Delaney, David G., Corinne D. Sperling, Christiaan S. Adams, and Brian Leung. 2008. 'Marine Invasive Species: Validation of Citizen Science and Implications for National Monitoring Networks', *Biological Invasions* 10, no.1: 117–28.

Deterding, Sebastian. 2012. 'Gamification: Designing for Motivation', *Interactions* 19: 14–17.

Devictor, Vincent, Robert J. Whittaker, and Coralie Beltrame. 2010. 'Beyond Scarcity: Citizen Science Programmes as Useful Tools for Conservation Biogeography', *Diversity and distributions* 16, no.3: 354–62.

DeVilla, J. 2011. 'BarCamp Toronto, Anyone?' Global Nerdy. Last accessed 28 August 2017. http://www.globalnerdy.com/2011/06/09/barcamp-toronto-anyone/.

DeWitt, Jennifer, Louise Archer, and Ada Mau. 2016. 'Dimensions of Science Capital: Exploring its Potential for Understanding Students' Science Participation', *International Journal of Science Education* 38: 2431–49.

Diáz, Sandra, Sebsebe Demissew, Julia Carabias, Carlos Joly, Mark Lonsdale, Neville Ash, Anna Larigauderie et al. 2015. 'The IPBES Conceptual Framework – Connecting Nature and People', *Current Opinion in Environmental Sustainability* 14:1–16.

Dickerson, C. 2005. 'Hack Day at Yahoo!' *Chad Dickerson's Blog* (blog), 10 December 2005. Last accessed 28 August 2017. https://blog.chaddickerson.com/2005/12/10/hack-day-at-yahoo/.

Dickerson, C. 2006. 'Yahoo! Open Hack Day: How it All Came Together'. *Chad Dickerson's Blog* (blog), 3 October 2006. Last accessed 28 August 2017. https://blog.chaddickerson.com /2006/10/03/yahoo-open-hack-day-how-it-all-came-together/.

Dickinson, Janis L. and Rick Bonney, eds. 2015. *Citizen Science: Public Participation in Environmental Research*. Ithaca, NY: Comstock Publishing Associates.

Dickinson, Janis L., Jennifer L. Shirk, David N. Bonter, Rick Bonney, R.L. Crain, J. Martin, T. Phillips, and K. Purcell. 2012. 'The Current State of Citizen Science as a Tool for Ecological Research and Public Engagement', *Frontiers in Ecology and the Environment* 10, no. 6: 291–7.

Diez, Tomas and Alex Posada. 2013. 'The Fab and the Smart City: The Use of Machines and Technology for the City Production by its Citizens'. In *The 7th International Conference on Tangible, Embedded and Embodied Interaction,* 447–54. Toronto.

Dillon, Justin, Mark Rickinson, Kelly Teamey, Marian Morris, Mee Young Choi, Dawn Sanders, and Pauline Benfield. 2006. 'The Value of Outdoor Learning: Evidence from Research in the UK and Elsewhere', *School Science Review* 8: 320.

Disney, Jane E., Emma Fox, Anna Farrell, Carrie LeDuc, and Duncan Bailey. 2017. 'Engagement in Marine Conservation through Citizen Science: A Community-Based Approach to Eelgrass Restoration in Frenchman Bay, Maine, USA'. In *Citizen Science for Coastal and Marine Conservation* edited by J.A. Cigliano and H.L. Ballard, 153–77. New York: Routledge.

Dobreva, Milena. 2016. 'Collective Knowledge and Creativity: The future of citizen science in the humanities'. In *Knowledge, Information and Creativity Support Sytems,* edited by Susumu Kunifuii et al., 565–73. Munich: Springer International Publishing.

Donnelly, Alison, Olivia Crowe, Eugenie Regan, Sinead Begley, and Amelia Caffarra. 2014. 'The Role of Citizen Science in Monitoring Biodiversity in Ireland,', *International Journal of Biometeorology* 58, no.6: 1237.

DORA. 2012. 'Declaration on Research Assessment.' Last accessed October 2016. http://www .ascb.org/dora/.

Dosemagen, Shannon, and Gretchen Gehrke. 2016. 'Civic Technology and Community Science: A New Model for Public Participation in Environmental Decisions'. In *Confronting the Challenges of Public Participation: Issues in Environmental, Planning, and Health Decision-Making (Proceedings of the Iowa State University Summer Symposia on Science Communication)* edited by J. Goodwin. Ames, IA. Science Communication Project.

Dosemagen, Shannon, Jeff Warren, and Sara Wylie. 2011. 'Grassroots Mapping: Creating a Participatory Map-making Process Centered on Discourse', *The Journal of Aesthetics and Protest* 8: 1–12.

Douglas, Heather E. 2003. 'The Moral Responsibilities of Scientists (Tensions between Autonomy and Responsibility)', *American Philosophical Quarterly* 40, no. 1: 59–68.

Drollette, D. 2012. 'Citizen Science Enters a New Era.' BBC. Last accessed 6 June 2017. http://www.bbc.com/future/story/20120329-citizen-science-enters-a-new-era.

Druschke, Caroline Gottschalk, and Carrie E. Seltzer. 2012. 'Failures of Engagement: Lessons Learned from a Citizen Science Pilot Study', *Applied Environmental Education & Communication* 11: 178–88.

Dyson, Andrew, Jim Halpert, Diego Ramos, Richard van Schaik, Scott Thiel, Carol A.F. Umhoefer, and Patrick an Eecke. 2014. *Data Protection Laws of the World Handbook*. 3rd edition. DLA Piper: Online.

Eade, Deborah. 1997. 'Capacity-building: An Approach to People-centred Development'. Oxfam (UK and Ireland) Publisher. http://policy-practice.oxfam.org.uk/publications/capacity -building-an-approach-to-people-centred-development-122906.

Earle, Paul S., Daniel Bowden, and Michelle Guy. 2012. 'Twitter Earthquake Detection: Earthquake Monitoring in a Social World', *Annals of Geophysics* 54, no. 6: 708–15.

ECSA. 2015. 'Ten principles of Citizen Science.' The European Citizen Science Association. Last accessed 18 June 2018. http://ecsa.citizen-science.net/sites/default/files/ecsa_ten_principles _of_citizen_science.pdf.

ECSA. 2016a. 'Citizen Science ThinkCamp event'. Last accessed 28 August 2017. https://sites.google .com/a/gold-mobileinnovation.co.uk/ecsa2016---citsci-thinkcamp/About-the-Think-Camp/home.

ECSA. 2016b. *ECSA Policy Paper #3: Citizen Science as part of EU Policy Delivery - EU Directives*. http://ecsa.citizen-science.net/sites/default/files/ecsa_policy_paper_3.pdf.

ECSA. n.d. European Citizen Science Association. Last accessed 24 January 2018. http://ecsa .citizen-science.net.

Editorial. 2015. 'Rise of the Citizen Scientist', *Nature* 524: 265. http://www.nature.com/news /rise-of-the-citizen-scientist-1.18192.

Edwards, Richard, Diarmuid McDonnell, and Ian Simpson. 2016. 'Exploring the Relationship between Educational Background and Learning Outcomes in Citizen Science'. Paper presented at the First International ECSA Conference, Berlin.

Edwards, Richard, Tina Phillips, Rick Bonney, and Katherine Mathieson. 2015. *Citizen Science and Science Capital*. Stirling: University of Stirling.

Eitzel, Melissa, Jessica L Cappadonna, Chris Santos-Lang, Ruth E. Duerr, Arika Virapongse, Sarah E. West, Christopher C.M. Kyba, et al. 2017. 'Citizen Science Terminology Matters: Exploring Key Terms', *Citizen Science: Theory and Practice* 2, no. 1: 1. https://doi.org/10.5334 /cstp.96

Ellett, Kathleen and Alice Mayio. 1990. *Volunteer Water Monitoring: A Guide for State Managers*. Washington DC: US Environmental Protection Agency, EPA 440: 4–90.

Ellis, Rebecca, and Claire Waterton. 2004. 'Environmental Citizenship in the Making: The Participation of Volunteer Naturalists in UK Biological Recording and Biodiversity Policy', *Science and Public Policy* 31: 95–105.

Ellwood, Elizabeth, Henry Bart Jr, Michael Doosey, Dean Jue, Justin Mann, Gil Nelson, Nelson Rios, and Austin Mast. 2016. 'Mapping Life–Quality Assessment of Novice vs. Expert Georeferencers', *Citizen Science: Theory and Practice* 1, no. 1: 4.

Elmore, Andrew, Cathlyn Stylinski, and Kavya Pradhan. 2016. 'Synergistic Use of Citizen Science and Remote Sensing for Continental-Scale Measurements of Forest Tree Phenology', *Remote Sensing* 8, no. 6: 502.

Ernwein, Marion. 2015. 'Jardiner la ville néolibérale. La fabrique urbaine de la nature.' Thèse en Sciences économiques et sociales, mention géographie, l'Université de Genève.

Etzkowitz, Henry, and Loet Leydesdorff. 2000. 'The Dynamics of Innovation: From National Systems and "Mode 2" to a Triple Helix of University–Industry–Government Rela-tions', *Research Policy* 29, no. 2: 109–23.

European Commission. 2008. *Final Communication from the Commission to the Council, the European Parliament, the European Economic and Social Committee and the Committee of the Regions - Towards a Shared Environmental Information System (SEIS)*. Brussels: ENV. http://eur-lex.europa.eu/legal-content/EN/TXT/?uri=celex:52008DC0046.

European Commission. 2011. *COM/2011/0244 final: Our Life Insurance, Our Natural Capital: An EU Biodiversity Strategy to 2020*. Communication from the Commission to the

European Commission. 2013. *EU Shared Environmental Information System Implementation Outlook – Commission Staff Working Document*. Luxembourg: Publications Office of the European Union. Accessed 14 December 2017. http://ec.europa.eu/environment/archives /seis/pdf/seis_implementation_en.pdf.

European Commission. 2015. *Decision C (2015)2453: European Commission Decision of 17 April 2015 on the Horizon 2020 Work Programme 2014-2015, 16. Science with and for Society (revised)*. http://ec.europa.eu/research/participants/data/ref/h2020/wp/2014 _2015/main/h2020-wp1415-swfs_en.pdf.

European Commission. 2016a. *Commission Implementing Regulation (EU) 2016/1141 of 13 July 2016 Adopting a List of Invasive Alien Species of Union Concern Pursuant to Regulation (EU) No 1143/2014 of the European Parliament and of the Council.* http://eur-lex.europa.eu /legal-content/EN/TXT/?qid=1468477158043&uri=CELEX:32016R1141.

European Commission. 2016b. *Final Towards a Fitness Check of EU Environmental Monitoring and Reporting: To Ensure Effective Monitoring, More Transparency and Focused Reporting of EU Environment Policy – Commission Staff Working Document.* http://ec.europa.eu/environment /legal/reporting/pdf/SWD_2016_188_en.pdf.

European Commission. 2016c. *Open Innovation, Open Science, Open to the World – A vision for Europe.* Luxembourg: Publications Office of the European Union.

European Commission. 2016d. *Open Science.* Last accessed 29 June 2016. http://ec.europa.eu /research/openscience/index.cfm.

European Commission. 2017. INSPIRE Knowledge Base. Last accessed 14 March 2018. https://inspire.ec.europa.eu.

European Commission. n.d. 'Open Science Monitor'. Last accessed 23 October 2017. http://ec .europa.eu/research/openscience/index.cfm?pg=home&section=monitor.

European Council. 2005. *Decision 2005/370/EC: Council Decision of 17 February 2005 on the Conclusion, on Behalf of the European Community, of the Convention on Access to Information, Public Participation in Decision-making and Access to Justice in Environmental Matters.* http://eur-lex.europa.eu/legal-content/EN/TXT/?uri=CELEX%3A32005D0370.

European Foundation for the Improvement of Living and Working Conditions. 2011. *Second European Quality of Life Survey: Participation in Volunteering and Unpaid Work.* Luxembourg: Publications Office of the European Union.

European Parliament and Council. 2003. *Directive 2003/4/EC of the European Parliament and of the Council of 28 January 2003 on Public Access to Environmental Information and Repealing Council Directive 90/313/EEC.* http://eur-lex.europa.eu/legal-content/EN/TXT/?uri =CELEX%3A32003L0004.

European Parliament and Council. 2013. *Decision No 1386/2013/EU of the European Parliament and of the Council of 20 November 2013 on a General Union Environment Action Programme to 2020 'Living Well, within the Limits of Our Planet' Text with EEA Relevance.* http://eur-lex .europa.eu/legal-content/EN/TXT/?uri=CELEX%3A32013D1386.

European Parliament and Council. 2014. *Regulation (EU) No 1143/2014 of the European Parliament and of the Council of 22 October 2014 on the Prevention and Management of the Introduction and Spread of Invasive Alien Species.* http://eur-lex.europa.eu/legal-content/EN /TXT/?qid=1417443504720&uri=CELEX:32014R1143.

European Parliament, the Council, the Economic and Social Committee and the Commitee of the Regions. 2011. Brussels. http://eur-lex.europa.eu/legal-content/EN/TXT/?uri =CELEX:52011DC0244.

Eurostat. 2016. 'Educational Attainment Statistics'. Last updated 3 August 2016. http://ec.europa .eu/eurostat/statistics-explained/index.php/Educational_attainment_statistics.

Evans, Celia, Eleanor Abrams, Robert Reitsma, Karin Roux, Laura Salmonsen, and Peter Marra. 2005. 'The Neighborhood Nestwatch Program: Participant Outcomes of a Citizen-science Ecological Research Project', *Conservation Biology*, 19: 589–94.

Evans, Geoffrey, and John Durant. 1995. 'The Relationship between Knowledge and Attitudes in the Public Understanding of Science in Britain', *Public Understanding of Science* 4: 57–74.

Eveleigh, Alexandra, Charlene Jennett, Ann Blandford, Philip Brohan, and Anna L. Cox. 2014. 'Designing for Dabblers and Deterring Drop-outs in Citizen Science'. In *Proceedings of the 32nd Annual ACM Conference on Human Factors in Computing Systems*, 2985–94. New York: ACM.

Eveleigh, Alexandra, Charlene Jennett, Stuart Lynn, and Anna L. Cox. 2013. '"I Want to Be a Captain! I Want to Be a Captain!": Gamification in the Old Weather Citizen Science Project'. In *Gamification 2013 – First International Conference on Gameful Design, Research, and Applications*, 79–82. Ontario.

Evolution MegaLab. n.d. 'Evolution MegaLab'. Last accessed 13 March 2018. http://www .evolutionmegalab.org/.

Extreme Citizen Science. n.d. Last accessed October 2016. https://www.ucl.ac.uk/excites.

Falchi, Fabio, Pierantonio Cinzano, Dan Duriscoe, Christopher C.M. Kyba, Christopher D. Elvidge, Kimberly Baugh, Boris A. Portnov, Nataliya A. Rybnikova, and Riccardo Furgoni. 2016. 'The New World Atlas of Artificial Night Sky Brightness', *Science Advances* 2: e1600377.

Falk, John, Jonathan Osborne, Lynn Dierking, Emily Dawson, Matthew Wenger, and Billy Wong. 2012. *Analysing the UK Science Education Community: The Contribution of Informal Providers*. London: Wellcome Trust.

Fenwick, Tara, Richard Edwards, and Peter Sawchuk. 2011. *Emerging Approaches to Educational Research: Tracing the Socio-material*. Oxon and New York: Routledge.

Fernandez-Gimenez, Maria, Heidi Ballard, and Victoria Sturtevant. 2008. 'Adaptive Management and Social Learning in Collaborative and Community-based Monitoring: A Study of Five Community-based Forestry Organizations in the Western USA', *Ecology and Society* 13, no. 2: 4.

Ficano, Ivan. 2014. *Open Call report*. CHEST project Deliverable D2.2. European Commission. http://www.chest-project.eu/wp-content/uploads/2015/10/d2.2_chest_open_call_report _v1.pdf.

FIDRA. n.d. 'Nurdle Free Oceans: Reducing Plastic Pollution in Our Seas'. Last accessed 13 March 2018. https://www.nurdlehunt.org.uk/whats-the-solution.html.

Fischer, Frank. 2000. *Citizens, Experts and the Environment: The Politics of Local Knowledge*. Durham: Duke University Press.

Flanagin, Andrew J. and Miriam J. Metzger. 2008. 'The Credibility of Volunteered Geographic Information', *GeoJournal* 72, nos. 3 and 4: 137–48.

Flückiger, Yves, and Nikhil Seth. 2016. 'SDG Indicators Need Crowdsourcing', *Nature* 531: 448.

Foldit. n.d. Accessed October 2016. http://fold.it/portal/.

Follett, Ria and Vladimir Strezov. 2015. 'An Analysis of Citizen Science Based Research: Usage and Publication Patterns', *PLoS ONE* 10: e0143687.

Fontaine, Benoît, and Mathilde Renard. 2010. *PROPAGE, protocole de suivi des papillons par les gestionnaires*. Muséum d'histoire naturelle et Noé conservation.

Fontaine, Benoît, Kees van Achterberg, Miguel Angel Alonso-Zarazaga, Rafael Araujo, Manfred Asche, Host Aspöck, Ulrike Aspöck, Paolo Audisio, Berend Aukema, and Nicolas Bailly. 2012. 'New Species in the Old World: Europe as a Frontier in Biodiversity Exploration, a Test Bed for 21st Century Taxonomy', *PloS ONE* 7: e36881.

Fowler, Amy, J. Duncan Whyatt, Gemma Davies, and Rebecca Ellis. 2013. 'How Reliable are Citizen-derived Scientific Data? Assessing the Quality of Contrail Observations Made by the General Public', *Transactions in GIS* 17, no. 4: 488–506.

Franzoni, Chiara, and Henry Sauermann. 2014. 'Crowd Science: The Organization of Scientific Research in Open Collaborative Projects', *Research Policy* 43, no. 1: 1–20.

Freitag, Amy, Ryan Meyer, and Liz Whiteman. 2016. 'Strategies Employed by Citizen Science Programs to Increase the Credibility of Their Data', *Citizen Science: Theory and Practice* 1: 2.

Freitag, Amy, and Max J. Pfeffer. 2013. 'Process, Not Product: Investigating Recommendations for Improving Citizen Science 'Success', *PLoS ONE* 8, no. 5: e64079. https://doi.org/10.1371 /journal.pone.0064079

Frewer, Lynn, Steve Hunt, Mary Brennan, Sharron Kuznesof, Mitchell Ness, and Chris Ritson. 2003. 'The Views of Scientific Experts on How the Public Conceptualize Uncertainty', *Journal of Risk Research* 6: 75–85.

Frickel, Scott, Sarah Gibbon, Jeff Howard, Joanna Kempner, Gwen Ottinger, and David J. Hess. 2010. 'Undone Science: Charting Social Movement and Civil Society Challenges to Research Agenda Setting', *Science, Technology & Human Values* 35, no. 4: 444–73.

Friedland, Gerald and Jaeyoung Choi. 2011. 'Semantic Computing and Privacy: A Case Study Using Inferred Geo-location', *International Journal of Semantic Computing* 5, no. 1: 79–93.

Fritz, Steffen and Cidália Costa Fonte. 2016. 'Citizen Science and Earth Observation', *Remote Sensing* (Special Issue) 8, no. 11. http://www.mdpi.com/journal/remotesensing/special _issues/earth_observation

Fritz, Steffen, Ian McCallum, Christian Schill, Christoph Perger, Linda See, Dmitry Schepas-chenko, Marijn Van der Velde, Florian Kraxner, and Michael Obersteiner. 2012. 'Geo-Wiki: An Online Platform for Improving Global Land Cover', *Environmental Modelling & Software* 31: 110–23.

Fritz, Steffen, Linda See, Ian McCallum, Christian Schill, Michael Obersteiner, Marijn Van der Velde, Hannes Boettcher, Petr Havlík, and Frédéric Achard. 2011. 'Highlighting Continued Uncertainty in Global Land Cover Maps for the User Community', *Environmental Research Letters* 6, no. 4: 044005.

Fritz, Steffen, Linda See, Christoph Perger, Ian McCallum, Christian Schill, Dmitry Schepas-chenko, Martina Duerauer et al. 2017. 'A Global Dataset of Crowdsourced Land Cover and Land Use Reference Data,' *Scientific Data* 4: 170075.

Frobel, Von Kai, and Helmut Schlumprecht. 2016. 'Erosion der Artenkenner. Ergebnisse einer Befragung und notwendige Reaktionen', *Naturschutz und Landschaftsplanung* 48: 105–13.

Frost-Nerbonne, Julia F., and Kristen C. Nelson. 2004. 'Volunteer Macroinvertebrate Monitoring in the United States: Resource Mobilization and Comparative State Structures', *Society and Natural Resources* 17, no. 9: 817–39.

Fujitani, Marie, Andrew McFall, Christoph Randler, and Robert Arlinghaus. 2016. 'Efficacy of Lecture-based Environmental Education for Biodiversity Conservation: A Robust Controlled Field Experiment with Recreational Anglers Engaged in Self-organized Fish Stocking', *Journal of Applied Ecology* 53: 25–33.

Funder, Mikkel, Finn Danielsen, Yonika Ngaga, Martin R. Nielsen, and Michael K. Poulsen. 2013. 'Reshaping Conservation: The Social Dynamics of Participatory Monitoring in Tanzania's Community-managed Forests', *Conservation and Society* 11: 218–32.

Funtowicz, Silvio O., and Jerome R. Ravetz. 1992. 'Three Types of Risk Assessment and the Emergence of Postnormal Science'. In *Social Theories of Risk*, edited by Sheldon Krimsky and Dominic Golding, 251–73. Westport: Greenwood.

Funtowicz, Silvio O., and Jerome R. Ravetz. 1993. 'Science for the Post-normal Age', *Futures* 25: 735–55.

Galaxy Zoo. n.d. 'Galaxy Zoo'. Accessed October 2017. https://www.galaxyzoo.org/.

Gamborg, Christian, Rag Parsons, Rajindra K. Puri, and Peter Sandøe. 2012. 'Ethics and Research Methodologies for the Study of Traditional Forest-Related Knowledge'. In *Traditional Forest-Related Knowledge: Sustaining Communities, Ecosystems and Biocultural Diversity*, World Forests Volume 12, edited by John A. Parrotta, and Ronald L. Trosper, 535–60. London and New York: Springer.

GAO. 2016. *Open Innovation: Practices to Engage Citizens and Effectively Implement Federal Initiatives*. GAO-17-14. Washington DC: Government Accountability Office. Accessed 14 December 2017. http://www.gao.gov/products/GAO-17-14.

Gao, Huiji, Geoffrey Barbier, and Rebecca Goolsby. 2011. 'Harnessing the Crowdsourcing Power of Social Media for Disaster Relief', *IEEE Intelligent Systems* 26: 10–14.

Garcia-Martí, Irene, Raul Zurita-Milla, Arno Swart, Kees C. van den Wijngaard, Arnold J.H. van Vliet, Sita Bennema, and Margriet Harms. 2016. 'Identifying Environmental and Human Factors Associated with Tick Bites using Volunteered Reports and Frequent Pattern Mining', *Transactions in GIS* 21, no. 2: 277–99.

Garibay Group. 2015. *Driven to Discover: Summative Evaluation Report*. Minnesota: University of Minnesota Extension.

Gaskell, George. 2004. 'Science Policy and Society: The British Debate over GM Agriculture', *Current Opinion in Biotechnology* 15: 241–45.

Gaston, Kevin J., James P. Duffy, and Jonathan Bennie. 2015. 'Quantifying the Erosion of Natural Darkness in the Global Protected Area System', *Conservation Biology* 29: 1132–41.

Gates, Bill. 2018. Speech. Malaria Summit London, April 18, 2018. https://www.gatesfoundation.org/Media-Center/Speeches/2018/04/Malaria-Summit.

Gauntlett, David. 2011. *Making is Connecting, the Social Meaning of Creativity, from DIY and Knitting to YouTube and Web 2.0.* Cambridge: Polity Press.

GBIF. 2017. 'What is GBIF?' Last accessed 14 March 2018. https://www.gbif.org/what-is-gbif.

Gedney, Melissa, and Lea Shanley. 2014. 'Barriers and Accelerators to Crowdsourcing and Citizen Science in Federal Agencies: An Exploratory Study'. Wilsons Commons Lab. Accessed 14 November 2016. https://wilsoncommonslab.org/2014/09/07/an-exploratory-study-on-barriers/.

Geiger, Matthias F., Jonas J. Astrin, Thomas Borsch, Ulrich Burkhardt, Peter Grobe, Ralf Hand, Axel Hausmann, Karin Hohberg, Lars Krogmann, and Matthias Lutz. 2016. 'How to Tackle the Molecular Species Inventory for an Industrialized Nation – Lessons from the First Phase of the German Barcode of Life initiative GBOL (2012–2015)', *Genome* 59: 661–70.

Gellman, Robert. 2015. *Crowdsourcing, Citizen Science, and the Law: Legal Issues Affecting Federal Agencies*. Washington DC: Woodrow Wilson International Center for Scholars. Accessed 14 December 2017. https://www.wilsoncenter.org/sites/default/files/CS_Legal_Barriers_Gellman.pdf.

GEO BON. 2017. Vision & Goals. Last accessed 14 March 2018. http://geobon.org/about/vision-goals./

Geoghegan, Hilary, Alison Dyke, Rachel Pateman, Sarah West, and Glyn Everett. 2016. *Understanding Motivations for Citizen Science*. Swindon, UK: UK Environmental Observation Framework.

http://www.ukeof.org.uk/resources/citizen-science-resources/MotivationsforCSREPORTFIN
ALMay2016.pdf.

Gerrig, Richard J. 1993. *Experiencing Narrative Worlds: On the Psychological Activities of Reading*.
New Haven: Yale University Press.

Gershenfeld, Neil. 2008. *Fab: The Coming Revolution on Your Desktop – from Personal Computers to
Personal Fabrication*. New York: Basic Books.

Gewin, Virginia. 2016. 'Turning Point: Aerial Archaeologist', *Nature* 534, no. 7607: 427.

Gibbons, Michael. 1999. 'Science's New Social Contract with Society', *Nature* 402, no. 6761:
C81-C84. https://doi.org/10.1038/35011576.

Gibbs, Anita. 1997. 'Focus Groups', *Social Research Update* 19 (Winter). http://sru.soc.surrey.ac.uk
/SRU19.html.

Gibbs, Samuel. 2016. 'Mobile Web Browsing Overtakes Desktop for the First Time'. *The Guardian*,
2 November 2016. https://www.theguardian.com/technology/2016/nov/02/mobile-web
-browsing-desktop-smartphones-tablets.

Gilbert, Sarah. 2015. 'Participants in the Crowd: Deliberations on the Ethical Use of Crowdsourcing in Research'. Workshop Ethics for Studying Online Sociotechnical Systems in a Big Data
World at ACM CSCW 2015, Vancouver.

Göbel, Claudia. 2017. 'Following up with the ECSA Inclusiveness Challenge'. ECSA. Last
accessed 28 August 2017. https://ecsa.citizen-science.net/blog/following-ecsa
-inclusiveness-challenge.

Göbel, Claudia, Victoria Y. Martin, and Mónica Ramirez-Andreotta. 2016. *Stakeholder Analysis:
International Citizen Science Stakeholder Analysis on Data Interoperability Final Report*.
Washington, DC: Woodrow Wilson International Center for Scholars.

Göbel, Claudia, Jessica L. Cappadonna, Gregory J. Newman, Jian Zhang, and Katrin Vohland.
2016. 'More Than Just Networking for Citizen Science: Examining Core Roles of Practitioner
Organizations'. In *Analyzing the Role of Citizen Science in Modern Research*, edited by Luigi
Ceccaroni and Jaume Piera, 24–49. Hershey, PA: IGI Global.

Gold, Margaret. 2011. 'MC ThinkCamp M-Health'. Last accessed 28 August 2017. https://sites
.google.com/a/gold-mobileinnovation.co.uk/mc-thinkcamp---mhealth/.

Gold, Margaret. 2012. 'Community Building for Collaborative Problem Solving'. Last accessed 28
August 2017. http://mobilecollective.co.uk/community-building-for-collaborative-problem
-solving/.

Golinski, Jan. *Science as Public Culture: Chemistry and Enlightenment in Britain, 1760–1820*.
Cambridge: Cambridge University Press, 1999.

Gollan, John, Lisa L. de Bruyn, Nick Reid, and Lance Wilkie. 2012. 'Can Volunteers Collect Data
that are Comparable to Professional Scientists? A Study of Variables Used in Monitoring the
Outcomes of Ecosystem Rehabilitation', *Environmental Management* 50, no. 5: 969–78.

Gommerman, Luke, and Martha C. Monroe. 2017. 'Lessons Learned from Evaluations of Citizen
Science Programs'. EDIS publication FOR291. Gainesville, Florida: University of Florida,
School of Forest Resources and Conservation Department, Electronic Data Information
Source of UF/IFAS Extension. http://edis.ifas.ufl.edu.

Goodchild, Michael F., and Linna Li. 2012. 'Assuring the Quality of Volunteered Geographic
Information', *Spatial statistics* 1: 110–20.

Gratani, Monica, James Butler, Frank Royee, Peter Valentine, Damien Burrows, Warren Canendo,
and Alexander Anderson. 2011. 'Is Validation of Indigenous Ecological Knowledge a
Disrespectful Process? A Case Study of Traditional Fishing Poisons and Invasive Fish
Management from the Wet Tropics, Australia', *Ecology and Society* 16, no. 3: 1–14.

Gray, Stephen, Johanna Hilsberg, Andrew McFall, and Robert Arlinghaus. 2015. 'The Structure
and Function of Angler Mental Models about Fish Population Ecology: The Influence of
Specialization and Target Species', *Journal of Outdoor Recreation Tourism* 12: 1–13.

Greenwave Project. n.d. Accessed 16 November 2016. http://www.primaryscience.ie/greenwave
_introduction.php.

Greenwood, Jeremy J. D. 2007. 'Citizens, Science and Bird Conservation', *Journal of Ornithology* 148, no. 1: 77–124.

Griessler, Erich, Peter Biegelbauer, and Janus Hansen. 2011. 'Citizens' Impact on Knowledge-
Intensive Policy: Introduction to a Special Issue', *Science and Public Policy*, 38: 583–8.

Groom, Quentin, Lauren Weatherdon, and Ilse R. Geijzendorffer. 2017. 'Is Citizen Science an
Open Science in the Case of Biodiversity Observations?' *Journal of Applied Ecology* 54:
612–17.

Grove-White, Robin, Claire Waterton, Rebecca Ellis, Johannes Vogel, Gill Stevens, and Bridget Peacock. 2007. 'Amateurs as Experts: Harnessing New Networks for Biodiversity'. End of Award Report: Lancaster University and Natural History Museum. Accessed 14 December 2017. http://csec.lancs.ac.uk/docs/%20Amateurs%20as%20Experts%20Final%20Report.pdf.

Grunig, James E., and Larissa A. Grunig. 2001. *Guidelines for Formative and Evaluative Research in Public Affairs – A Report for the Department of Energy Office of Science*. Gainesville: Institute for Public Relations.

Gustetic, Jenn, Kristen Honey, and Lea Shanley. 2015. 'Open Science and Innovation: Of the People, By the People, For the People'. Obama Whitehouse Archives. Last modified 9 September 2015. https://obamawhitehouse.archives.gov/blog/2015/09/09/open-science-and-innovation -people-people-people.

Hadler, James L., Dhara Patel, Roger S. Nasci, Lyle R. Petersen, James M. Hughes, Kristy Bradley, Paul Etkind, Lilly Kan, and Jeffrey Engel. 2015. 'Assessment of Arbovirus Surveillance 13 Years after Introduction of West Nile Virus, United States', *Emerging Infectious Diseases* 21, no. 7: 1159–66. https://doi.org/10.3201/eid2107.140858.

Haklay, Mordechai (Muki). 2010. 'How Good is Volunteered Geographical Information? A Comparative Study of OpenStreetMap and Ordnance Survey Datasets', *Environment and planning B: Planning and design* 37, no. 4: 682–703.

Haklay, Mordechai (Muki). 2012. 'London Citizen Cyberscience Summit – New Collaborations and Ideas'.*Po Ve Sham* (blog), 3 March 2012. Last accessed 28 August 2017. https:// povesham.wordpress.com/2012/03/03/london-citizen-cyberscience-summit-new -collaborations-and-ideas/.

Haklay, Mordechai (Muki). 2013. 'Citizen Science and Volunteered Geographic Information: Overview and Typology of Participation'. In *Crowdsourcing Geographic Knowledge: Volunteered Geographic Information (VGI) in Theory and Practice*, edited by Daniel Sui, Sarah Elwood, and Michael Goodchild, 105–22. Dordrecht: Springer.

Haklay, Mordechai (Muki). 2014. 'Citizen Cyberscience Summit – Day 3'. *Po Ve Sham* (blog), 23 February 2014. Last accessed 28 August 2017. https://povesham.wordpress.com/2014/02 /23/citizen-cyberscience-summit-day-3/.

Haklay, Mordechai (Muki). 2015. 'Citizen Science Policy: A European Perspective'. Case Study Series Vol. 4. Washington DC: Woodrow Wilson International Center for Scholars. https://www .wilsoncenter.org/sites/default/files/Citizen_Science_Policy_European_Perspective_Haklay.pdf.

Haklay, Mordechai (Muki). 2016a. 'Why is Participation Inequality Important?' In *European Handbook on Crowdsourced Geographic Information*, edited by Cristina Capineri et al., 35–45. London: Ubiquity Press.

Haklay, Mordechai (Muki). 2016b. 'ECSA2016 ThinkCamp Challenge: How Can Overleaf Support Collaborative Writing between Academics and Citizen Scientists?' *Po Ve Sham* (blog), 24 May 2016. https://povesham.wordpress.com/2016/05/24/ecsa2016-thinkcamp-challenge -how-can-overleaf-support-collaborative-writing-between-academics-and-citizen-scientists/.

Haklay, Mordechai (Muki). 2016c. 'Making Participatory Sensing Meaningful'. In *The Participatory City*, edited by Yasminah Beebeejaun, 152–9. Berlin: Jovis.

Haklay, Mordechai (Muki). 2017. 'Volunteered Geographic Information, Quality Assurance'. In *The International Encyclopedia of Geography: People, the Earth, Environment, and Technology*, edited by Douglas Richardson, Noel Castree, Michael F. Goodchild, Audrey Kobayashi, Weidong Liu, and Richard A. Marston, 1–6. Hoboken: Wiley/AAG.

Haklay, Mordechai (Muki), Vyron Antoniou, Sofia Basiouka, Robert Soden, and Peter Mooney. 2014. *Crowdsourced Geographic Information Use in Government*. Global Facility for Disaster Reduction & Recovery (GFDRR). London: World Bank. https://www.gfdrr.org/sites/gfdrr /files/publication/Crowdsourced%20Geographic%20Information%20Use%20in%20 Government.pdf.

Hampton, Stephanie E., Carly A. Strasser, Joshua J. Tewksbury, Wendy K. Gram, Amber E. Budden, Archer L. Batcheller, Clifford S. Duke and John H. Porter. 2013. 'Big Data and the Future of Ecology', *Frontiers in Ecology and the Environment* 11: 156–62.

Hansen, Kirsten, Gillian Nowlan, and Christina Winter. 2012. 'Pinterest as a Tool: Applications in Academic Libraries and Higher Education', *The Canadian Journal of Library and Information Practice and Research* 7: 1–11.

Hara, Noriko, Paul Solomon, Seung-Lye Kim, and Diane H. Sonnenwald. 2001. *An Emerging View of Scientific Collaboration: Scientists' Perspectives on Collaboration and Factors that Impact*

Collaboration. Chapel Hill, NC: School of Information and Library Science, University of North Carolina. 27599-3360 SILS Technical Report TR-2001-08 (Draft 12/18/01).

Hargrove, Robert. 1997. *Mastering the Art of Creative Collaboration.* Columbus, OH: McGraw-Hill.

Harkavy, Ira. 2006. 'The Role of Universities in Advancing Citizenship and Social Justice in the 21st Century', *Education, Citizenship and Social Justice* 1, no. 1: 5–37.

Harnik, Paul G. and Robert M. Ross. 2003. 'Assessing Data Accuracy When Involving Students in Authentic Paleontological Research', *Journal of Geoscience Education* 51, no. 1: 76–84.

Harris, Christopher G., and Padmini Srinivasan. 2012. 'Crowdsourcing and Ethics: The Employment of Crowdsourcing Workers for Tasks that Violate Privacy and Ethics'. In *Security and Privacy in Social Networks,* edited by Y. Altshuler, Y. Elovici, A.B. Cremers, N. Aharony, and A. Pentland, 67–83. New York: Springer.

Hart, Robert A. 1997. *Children's Participation: The Theory and Practice of Involving Young Citizens in Community Development and Environmental Care.* London: Earthscan.

Haywood, Benjamin K. 2014. 'A "Sense of Place" in Public Participation in Scientific Research', *Science Education* 98: 64–83.

Haywood, Benjamin K., and John C. Besley. 2014. 'Education, Outreach, and Inclusive Engagement: Towards Integrated Indicators of Successful Program Outcomes in Participatory Science', *Public Understanding of Science* 23: 92–106.

Hecker, Susanne, Rick Bonney, Muki Haklay, Franz Hölker, Heribert Hofer, Claudia Goebel, Margaret Gold, et al. 2018. 'Innovation in Citizen Science – Perspectives on Science-Policy Advances.' *Citizen Science: Theory and Practice* 3, no. 1. doi: http://doi.org/10.5334/cstp.114.

Hemment, Drew, Rebecca Ellis, and James Wynne. 2011. 'Participatory Mass Observation and Citizen Science', *Leonardo* 44, no 1: 62–3. https://doi.org/10.1162/LEON_a_00096.

Henriquez, Laurence. 2016. *Amsterdam Smart Citizen Lab. Towards Community Driven Data Collection.* Edited by Frank Kresin and Natasha de Sena. AMS/Waag Society: Amsterdam. https://waag.org/sites/waag/files/public/media/publicaties/amsterdam-smart-citizen-lab -publicatie.pdf.

Herlo, Bianca, Florian Sametinger, Jennifer Schubert, and Andreas Unteidig. 2015. 'Participatory Design and the Hybrid City. The Living Lab Mehringplatz, Berlin, and the Project "Community Now? Conflicts, Interventions, New Publics"'. In *Proceedings of Hybrid City 2015: Data to the People,* 151-160. Athens: University Research Institute of Applied Communication (U.R.I.A.C.).

Herring, Scarlett R., Brett R. Jones, and Brain P. Bailey. 2009. 'Idea Generation Techniques among Creative Professionals'. In *Proceedings of the 42nd Hawaii International Conference on System Sciences.* Washington DC: IEEE Computer Society.

Hess, David J. 2011. 'To Tell the Truth: On Scientific Counterpublics', *Public Understanding of Science* 20, no. 5: 627–41.

Hess, David J. 2015. 'Undone Science and Social Movements: A Review and Typology'. In *The Routledge International Handbook of Ignorance Studies,* edited by Matthias Gross and Linsey McGoey, 141–54. New York: Routledge.

Hidalgo-Ruz, Valeria, and Martin Thiel. 2013. 'Distribution and Abundance of Small Plastic Debris on Beaches in the SE Pacific (Chile): A Study Supported by a Citizen Science Project', *Marine Environmental Research* 87-88: 12–18.

Higgins, Christopher I., Jamie Williams, Didier G. Leibovici, Ingo Simonis, Mason J. Davis, Conor Muldoon, Paul van Genuchten, Gregory O'Hare, and Stefan Wiemann. 2016. 'Citizen OBservatory WEB (COBWEB): A Generic Infrastructure Platform to Facilitate the Collection of Citizen Science Data for Environmental Monitoring', *International Journal of Spatial Data Infrastructures Research* 11, no. 1: 20–48.

Hill, William C., James D. Hollan, Dave Wroblewski, and Tim McCandless. 1992. 'Edit Wear and Read Wear'. In *Proceedings of the SIGCHI Conference on Human Factors in Computing Systems,* 3–9. New York: ACM.

Hillman, Thomas, and Åsa Mäkitalo. 2016. 'Considering External Resource Use in Forum Discussions as an Indicator of Citizen Scientist Learning'. Paper presented at the First International ECSA Conference, Berlin, May 19–21.

Hinchliffe, Steve, Les Levidow, and Sue Oreszczyn. 2014. 'Engaging Cooperative Research', *Environment and Planning A* 46: 2080–94.

Hindin, David, Ben Grumbles, George Wyeth, Kristen Benedict, Tim Watkins, George (Tad) Aburn, Jr., Megan Ulrich, Steve Lang, Kelly Poole, and Alexandra Dapolito Dunn. 2016. 'Advanced Monitoring Technology: Opportunities and Challenges. A Path Forward for EPA, States, and

Tribes', *EM, The Magazine for Environmental Managers,* November 2016. https://www.epa
.gov/sites/production/files/2016-11/documents/article-adv-mon-technology.pdf.

Hindson, James, and Lucy Carter. 2009. *Natural History Societies and Recording Schemes in the UK: A Consultation into the Factors that Limit their Functioning and Development.* London: OPAL.

HKBWS. 2015. *China Coastal Waterbird Census Report 1.2010-12.2011.* Hong Kong: 4M Studio.

Hochachka, Wesley M., Daniel Fink, Rebecca A. Hutchinson, Daniel Sheldon, Weng-Keen Wong, and Steve Kelling. 2012. 'Data-intensive Science Applied to Broad-scale Citizen Science', *Trends in Ecology & Evolution* 27, no. 2: 130–7.

Hoffman, Sharona. 2014. 'Citizen Science: The Law and Ethics of Public Access to Medical Big Data', *Berkeley Technology Law Journal.* Case Legal Studies Research Paper No. 2014-21.

Hogg, Michael A., and Deborah Terry. 2000. 'Social Identity and Self-Categorisation Processes in Organizational Contexts', *Academy of Management Review* 25, no, 1: 121–40.

Holdren, John P. 2015. *Addressing Societal and Scientific Challenges through Citizen Science and Crowdsourcing.* Memorandum to the Heads of Executive Departments and Agencies. Washington DC: White House Office of Science and Technology Policy. https:// obamawhitehouse.archives.gov/sites/default/files/microsites/ostp/holdren_citizen_science _memo_092915_0.pdf.

Hölker, Franz, Timothy Moss, Barbara Griefahn, Werner Kloas, and Christian C. Voigt. 2010. 'The Dark Side of Light: A Transdisciplinary Research Agenda for Light Pollution Policy', *Ecology and Society* 15: 13.

Hölker, Franz, Christian Wolter, Elizabeth K. Perkin, and Klement Tockner. 2010. 'Light Pollution as a Biodiversity Threat', *Trends in Ecology and Evolution* 25: 681–2.

Hölker, Franz, Christain Wurzbacher, Carsten Weißenborn, Michael T. Monaghan, Stephanie I.J. Holzhauer, and Katrin Premke. 2015. 'Microbial Diversity and Community Respiration in Freshwater Sediments Influenced by Artificial Light at Night', *Philosophical Transactions of the Royal Society of London. Series B, Biological Sciences* 370: 20140130.

Holocher-Ertl Teresa, and Barbara Kieslinger. 2015. 'Citizen Science: BürgerInnen schaffen Innovationen'. In *Wissenschaft und Gesellschaft im Dialog: Responsible Science.* Bundesministe-rium für Wissenschaft, Forschung und Wirtschaft (BMWFW).

Holøs, Sverre. 2016. 'High School Students as Citizen Scientists in Air Quality Research – Lessons Learned'. Paper presented at the First International ECSA Conference, Berlin.

Hopkins, G.W. and R.P. Freckleton. 2002. 'Declines in the Numbers of Amateur and Professional Taxonomists: Implications for Conservation', *Animal Conservation* 5, no. 3: 245–9.

Horlick-Jones, Tom, John Walls, Gene Rowe, Nick Pidgeon, Wouter Poortinga, and Tim O'Riordan. 2006. 'On Evaluating the GM Nation? Public Debate about the Commercialisation of Transgenic Crops in Britain', *New Genetics and Society* 25: 265–88

House of Lords Science and Technology Select Committee. 2000. *Science in Society.* London: Science and Technology Committee Publications. http://www.publications.parliament.uk/pa /ld199900/ldselect/ldsctech/38/3801.htm

Howlett, Michael, and M. Ramesh. 2009. *Studying Public Policy: Policy Cycles and Policy Subsystems.* Oxford: Oxford University Press.

Hsing, Pen-Yuan. 2016. ECSA ThinkCamp Etherpad. https://pad.okfn.org/p/ECSAThinkCamp _CameraTrap. n.d. Last accessed 11 June 2018.

Hunter, Jane, Abdulmonem Alabri, and Catherine Ingen. 2013. 'Assessing the Quality and Trustworthiness of Citizen Science Data', *Concurrency and Computation: Practice and Experience* 25, no. 4: 454–66.

Huntington, Henry P. 1998. 'Observations on the Utility of the Semi-directive Interview for Documenting Traditional Ecological Knowledge', *Arctic* 51: 237–42.

Huntington, Henry P. 2011. 'The Local Perspective', *Nature* 478: 182–3.

Hyder, Kieran, Bryony Townhill, Lucy G. Anderson, Jane Delany, and John K. Pinnegar. 2015. 'Can Citizen Science Contribute to the Evidence-base that Underpins Marine Policy?' *Marine Policy* 59: 112–20.

Iacovides, Ioanna, Janet Charlene, Cassandra Cornish-Trestrail, and Anna L. Cox. 2013. 'Do Games Attract or Sustain Engagement in Citizen Science? A Study of Volunteer Motivations'. In *Proceedings of CHI Extended Abstracts,* 1101–6. Paris: ACM Press.

Imperial College London. 2016. OPAL. Last accessed 14 July 2016. https://www .opalexplorenature.org/.

Indicia. n.d. 'Indicia'. Last accessed 27 July 2017. http://www.indicia.org.uk.

Ingold, Tim. 2001. 'From the Transmission of Representations to the Education of Attention'. In *The Debated Mind: Evolutionary Psychology Versus Ethnography,* edited by Harvey Whitehouse, 113–53. Oxford: Berg.

Irish Environmental Protection Agency. n.d. 'See It Say It'. Last modified 2017. http://www.epa.ie /enforcement/report/seeit/.

Irwin, Alan. 1995. *Citizen Science: A Study of People, Expertise, and Sustainable Development.* London: Routledge.

Irwin, Alan. 2001. 'Constructing the Scientific Citizen: Science and Democracy in the Biosciences', *Public Understanding of Science* 10: 1–18.

Irwin, Alan. 2006. 'The Politics of Talk: Coming to Terms with the "New" Scientific Governance', *Social Studies of Science* 36, no. 2: 299–320.

Irwin, Alan, and Mike Michael. 2003. *Science, Social Theory and Public Knowledge.* Berkshire: Open University Press.

Israel, Barbara A., Amy J. Schulz, Edith A. Parker, and Adam B. Becker. 2001. 'Community-based Participatory Research: Policy Recommendations for Promoting a Partnership Approach in Health Research', *Education for Health* 14, no. 2: 182–97.

James, Alexander, Kevin J. Gaston, and Andrew Balmford. 2001. 'Can We Afford to Conserve Biodiversity?' *AIBS Bulletin* 51, no. 1: 43–52.

James-Creedon, Jackie. n.d. 'Jackie James-Creedon: The Whole Truth'. *The Whole Truth* (blog), Last modified 2 March 2016. https://jackiejamescreedon.wordpress.com/.

Jasanoff, Sheila. 2003. 'Technologies of Humility: Citizen Participation in Governing Science', *Minerva* 41, no. 3: 223–44.

Jasanoff, Sheila. 2004. *States of Knowledge: The Co-Production of Science and the Social Order.* New York and London: Routledge.

Jenkins, Lynda L. 2011. 'Using Citizen Science beyond Teaching Science Content: A Strategy for Making Science Relevant to Students' Lives', *Cultural Studies of Science Education* 6: 501–8.

Jennett, Charlene, Laure Kloetzer, Anna L. Cox, Daniel Schneider, Emily Collins, Mattia Fritz, Michael J. Bland, et al. 2017. 'Creativity in Citizen Cyberscience', *Human Computation* 3: 181–204.

Jennett, Charlene, Laure Kloetzer, Daniel Schneider, Ioanna Iacovides, Anna Cox, Margaret Gold, Brian Fuchs et al. 2016. 'Motivations, Learning and Creativity in Online Citizen Science', *Journal of Science Communication* 15, no. 3: A05.

Jiang, Qijun, Frank Kresin, Arnold Bregt, Lammert Kooistra, Emma Pareschi, Edith van Putten, Hester Volten, and Joost Wesseling. 2016. 'Citizen Sensing for Improved Urban Environmental Monitoring', *Journal of Sensors*: 1–9.

Johnson, Katherine. 2016. 'Real Life Science with Dandelions and Project BudBurst', *Journal of Microbiology & Biology Education* 17: 115.

Johnson, Kenneth G., Stephen J. Brooks, Phillip B. Fenberg, Adrain G. Glover, Karen E. James, Adrian M. Lister, Ellinor Michel, Mark Spencer, Jonathan A. Todd, and Eugenia Valsami-Jones. 2011. 'Climate Change and Biosphere Response: Unlocking the Collections Vault', *Bioscience* 61: 147–53.

Johnson, Noor, Carolina Behe, Finn Danielsen, Eva-M. Krümmel, Scott Nickels, and Peter L. Pulsifer. 2016. *Community-Based Monitoring and Indigenous Knowledge in a Changing Arctic: A Review for the Sustaining Arctic Observing Networks.* Sustain. Arctic Observing Network Task # 9. Ottawa: Inuit Circumpolar Council.

Johnson, Vicky, Roger Hart, and Jennifer Colwell. 2014a. *Steps to Engaging Young Children in Research – Vol. 1: The Guide.* Education Research Centre, University of Brighton. https://bernardvanleer.org/publications-reports/steps-engaging-young-children-research -volume-1-guide/.

Johnson, Vicky, Roger Hart, and Jennifer Colwell. 2014b. *Steps to Engaging Young Children in Research—Vol. 2: The Researcher Toolkit.* Education Research Centre, University of Brighton. https://bernardvanleer.org/publications-reports/steps-engaging-young-children -research-volume-2-researcher-toolkit/.

Jones, Damon E., Mark Greenberg, and Max Crowley. 2015. 'Early Social-Emotional Functioning and Public Health: The Relationship between Kindergarten Social Competence and Future Wellness', *American Journal of Public Health* 105, no. 11: 2283–90.

Jong de, Maaike. 2016. 'Illuminating a Bird's World – Effects of Artificial Light at Night.'. PhD diss., Wageningen University.

Jordan, Rebecca, Heidi Ballard, and Tina Phillips. 2012. 'Key Issues and New Approaches for Evaluating Citizen-science Learning Outcomes', *Frontiers in Ecology and the Environment* 10, no. 6: 307–9. https://doi.org/10.1890/110280.

Jordan Rebecca, Alycia Crall, Steven Gray, Tina Phillips, and David Mellor. 2015. 'Citizen Science as a Distinct Field of Inquiry', *BioScience* 65, no. 2: 208–11. https://doi.org/10.1093/biosci/biu217.

Jordan, Rebecca, Steven Gray, David Howe, Wesley Brooks, and Joan Ehrenfeld. 2011. 'Knowledge Gain and Behavioral Change in Citizen Science Programs', *Conservation Biology* 25: 1148–54.

Joss, Simon, and John Durant. 1995a. 'The UK National Consensus Conference on Plant Biotechnology', *Public Understanding of Science* 4: 195–204.

Joss, Simon, and John Durant, eds. 1995b. *Public Participation in Science: the Role of Consensus Conferences in Europe*. London: Science Museum.

Julliard, Romain, Frédéric Jiguet, and Denis Couvet. 2004. 'Evidence for the Impact of Global Warming on the Long-term Population Dynamics of Common Birds'. In *Proceedings of the Royal Society of London B: Biological Sciences* 271: S490–2.

Jupp, Eleanor. 2008. 'The Feeling of Participation: Everyday Spaces and Urban Change', *Geoforum* 39: 331–43.

Kaartinen, Riikka, Bess Hardwick, and Tomas Roslin. 2013. 'Using Citizen Scientists to Measure an Ecosystem Service Nationwide', *Ecology* 94, no. 11: 2645–52.

Kahn, Peter H., Jr., Rachel L. Severson, and Jolina H. Ruckert. 2009. 'The Human Relation with Nature and Technological Nature', *Current Directions in Psychological Science* 18, no. 1: 37–42.

Kambouri, Maria. 2015a. 'Investigating Early Years Teachers' Understanding and Response to Children's Preconceptions', *European Early Childhood Education Research Journal* 24, no. 6: 907–27.

Kambouri, Maria. 2015b. 'Children's Preconceptions of Science: How these Can be Used in Teaching', *Early Years Educator (EYE)* 16, no. 11: 38–44.

Kampen, Helge, Jolyon M. Medlock, Alexander G.C. Vaux, Constantianus J.M. Koenraadt, Arnold J.H. van Vliet, Frederic Bartumeus, Aitana Oltra, Carla A. Sousa, Sébastien Chouin, and Doreen Werner. 2015. 'Approaches to Passive Mosquito Surveillance in the EU', *Parasites & Vectors* 8, no. 11: 1–13. https://doi.org/10.1186/s13071-014-0604-5.

Kapos, Valerie, Andrew Balmford, Rosalind Aveling, Philip Bubb, Peter Carey, Abigail Entwistle, John Hopkins, Teresa Mulliken, Roger Safford, and Alison Stattersfield. 2008. 'Calibrating Conservation: New Tools for Measuring Success', *Conservation Letters* 1, 4: 155–64.

Kawrykow, Alexander, Gary Roumanis, Alfred Kam, Daniel Kwak, Clarence Leung, Chu Wu, Eleyine Zarour et al. 2012. 'Phylo: A Citizen Science Approach for Improving Multiple Sequence Alignment', *PLoS ONE* 7, no. 3: e31362.

Kellert, Stephen R. 2002. 'Experiencing Nature: Affective, Cognitive, and Evaluative Development'. In *Children and Nature: Psychological, Sociocultural, and Evolutionary Investigations*, edited by P.H. Kahn and S.R. Kellert, 117–52. MIT Press: Cambridge, MA: MIT Press.

Kellett, Mary. 2005. *Children as Active Researchers: A New Research Paradigm for the 21st Century?* ESRC National Centre for Research Methods, NCRM/003. http://eprints.ncrm.ac.uk/87/1/MethodsReviewPaperNCRM-003.pdf.

Kelling, Steve, Daniel Fink, Frank A. La Sorte, Alison Johnston, Nicholas E. Bruns, and Wesley M. Hochachka. 2015. 'Taking a "Big Data" Approach to Data Quality in a Citizen Science Project', *Ambio* 44: S601–11.

Kelling, Steve, Jun Yu, Jeff Gerbracht, and Weng-Keen Wong. 2011. 'Emergent Filters: Automated Data Verification in a Large-scale Citizen Science Project', In *e-Science Workshops (eScienceW), 2011 IEEE Seventh International Conference on*, 20–7.

Kennett, Rod, Finn Danielsen, and Kirsten M. Silvius. 2015. 'Conservation management: Citizen science is not enough on its own', *Nature* 521: 161.

Khatib, Firas, Seth Cooper, Michael D. Tyka, Kefan Xu, Ilya Makedon, Zoran Popović, and David Baker. 2011. 'Algorithm discovery by protein folding game players', *Proceedings of the National Academy of Sciences* 108, no. 47: 18949–53.

Kieslinger Barbara, Teresa Schäfer, and Claudia M. Fabian. 2015. *Kriterienkatalog zur Bewertung von Citizen Science Projekten und Projektanträgen*. Studie im Auftrag des Bundesministerium für Wissenschaft, Forschung und Wirtschaft (BMWFW).

Kieslinger, Barbara, Teresa Schäfer, Florian Heigl, Daniel Dörler, Anett Richter, and Aletta Bonn. 2017. 'The Challenge of Evaluation: An Open Framework for Evaluating Citizen Science Activities', *SocArXiv*. https://osf.io/preprints/socarxiv/enzc9.

Kim, Jinseop S., Matthew J. Greene, Aleksandar Zlateski, Kisuk Lee, Mark Richardson, Srinivas C. Turaga, Michael Purcaro et al. 2014. 'Space-time wiring specificity supports direction selectivity in the retina', *Nature* 509: 331–6.

Kirn, Sarah. 2016. 'Designed for Learning: Impacts and Future Direction of the Vital Signs Program'. Paper presented at the First International ECSA Conference, Berlin.

Kitzinger, Jenny. 1995. 'Qualitative Research-introducing Focus Groups', *British Medical Journal* 311: 299–302.

Klein, Julie Thompson. 2004. 'Prospects for Transdisciplinarity', *Futures* 36, no. 4: 515–26.

Klemmer, Scott R., Bjorn Hartmann, and Leila Takayama. 2006. 'How Bodies Matter: Five Themes for Interaction Design'. In *The 6th Conference on Designing Interactive Systems*, 140–9. San Antonio.

Kloetzer, Laure, Daniel Schneider, and Charlene Jennett. 2016. 'Evaluating Learning in Online Citizen Science: Reflections on a Mixed Methods Approach'. Paper presented at the First International ECSA Conference, Berlin.

Kloetzer, Laure, Daniel Schneider, Charlene Jennett, Ioanna Iacovides, Alexandra Eveleigh, Anna L. Cox, and Margaret Gold. 2013. 'Learning by Volunteer Computing, Thinking and Gaming: What and How are Volunteers Learning by Participating in Virtual Citizen Science?' In *ESREA 2013: Changing Configurations of Adult Education in Transitional Times Proceedings*, 73–92. European Society for Research on the Education of Adults (ESREA). Berlin.

Knol, Anne and Joost Wesseling. 2014. 'Burgermetingen luchtkwaliteit in Nederland: milieude-fensie meet en RIVM rekent. De resultaten van en ervaringen met een bijzondere samenwerking', *Tijdschrift Lucht* 6: 12–15.

Köhler, Denis. 2015. 'The Lighting Master Plan as an Instrument for Municipalities?' In *Urban Lighting, Light Pollution and Society*, edited by J. Meier, U. Hasenöhrl, K. Krause, and M. Pottharst, 141–58. New York: Routledge.

Kohler, Robert E. 2002. *Landscapes & Labscapes: Exploring the Lab-Field Border in Biology*. Chicago: University of Chicago Press.

Kohler, Robert E. 2006. *All Creatures: Naturalists, Collectors, and Biodiversity, 1850–1950*. Princeton, NJ: Princeton University Press.

Kolman, Jacob M. 2016. 'The Ethics of Biomedical Big Data: Busting Myths', *Journal of Science and Law* 2, no. 3.

Kosmala, Margaret, Andrea Wiggins, Alexandra Swanson, and Brooke Simmons. 2016. 'Assessing Data Quality in Citizen Science', *Frontiers in Ecology and the Environment* 14: 551–60.

Koster, Jeremy M. 2007. 'Hunting and Subsistence among the Mayangna and Miskito of Nicaragua's Bosawas Biosphere Reserve'. PhD diss., Pennsylvania State University.

Kountoupes, Dina, and Karen Oberhauser. 2008. 'Citizen Science and Youth Audiences: Educational Outcomes of the Monarch Larva Monitoring Project', *Journal of Community Engagement and Scholarship* 1: 10–20.

Kraemer, Moritz U.G., Marianne E. Sinka, Kirsten A. Duda, Adrian Q.N. Mylne, Freya M. Shearer, Christopher M. Barker, Chester G. Moore, et al. 2015. 'The Global Distribution of the Arbovirus Vectors Aedes Aegypti and Ae. Albopictus', *eLife* 4, no. e08347: 1–18. https://doi.org/10.7554/eLife.08347.

Krasny, Marianne E., and Rick Bonney. 2005. Scientific research and education collaboration. In E. Johnson & M. Mappin (Eds.), Environmental education and advocacy: Changing perspectives of ecology and education. Cambridge, England: Cambridge University Press.

Krasny, Marianne E., and Keith G. Tidball. 2009. 'Community Gardens as Contexts for Science, Stewardship, and Civic Action Learning', *Cities and the Environment* 2, no. 1: 1-18. http://digitalcommons.lmu.edu/cgi/viewcontent.cgi?article=1037&context=cate.

Krathwohl, David R., Benjamin S. Bloom, and Bertram B. Masia. 1964. *Taxonomy of Educational Objectives: Handbook II: Affective Domain*. New York: David McKay Co.

Kuhn, Thomas. 2012. *The Structure of Scientific Revolutions*. Chicago: University of Chicago Press.

Kullenberg, Christopher, and Dick Kasperowski. 2016. 'What Is Citizen Science? – A Scientometric Meta-Analysis', *PLoS ONE* 11, no. 1: e0147152. https://doi.org/10.1371/journal.pone.0147152.

Kuznetsov, Stacey, George N. Davis, Eric Paulos, Mark D. Gross, and Jian C. Cheung. 2011. 'Red Balloon Green Balloon Sensors in the Sky', In *Proceedings of the 13th International Conference on Ubiquitous Computing*, 237–46.

Kyba, Christopher C.M. 2018. 'How Can We Tell if Light Pollution is Getting Better or Worse?' *Nature Astronomy* 2: 267–269.

Kyba, Christopher C.M., Stefanie Garz, Helga U. Kuechly, Alejandro Sánchez de Miguel, Jamie Zamorano, Jürgen Fischer, and Franz Hölker. 2015. 'High-Resolution Imagery of Earth at Night: New Sources, Opportunities and Challenges', *Remote Sensing* 7: 1–23.

Kyba, Christopher C.M., Andreas Hänel, and Franz Hölker. 2014. 'Redefining Efficiency for Outdoor Lighting', *Energy & Environmental Science* 7: 1806–9.

Kyba, Christopher C.M., Theres Kuester, Alejandro Sánchez de Miguel, Kimberley Baugh, Andreas Jechow, Franz Hölker, Jonathan Bennie et al. 2017. 'Artificially Lit Surface of Earth at Night Increasing in Radiance and Extent', *Science Advances* 3: 1–9.

Kyba, Christopher C.M., Kai Pong Tong, Jonathan Bennie, Ingnacio Birriel, Jennifer J. Birriel, Andrew Cool, Arne Danielsen et al. 2015. 'Worldwide Variations in Artificial Skyglow', *Scientific Reports* 5: 8409.

Kyba, Christopher C.M., Janna M. Wagner, Helga U. Kuechly, Constance E. Walker, Christopher D. Elvidge, Rubio Falchi, Thomas Ruhtz, Jürgen Fischer and Franz Hölker. 2013. 'Citizen Science Provides Valuable Data for Monitoring Global Night Sky Luminance', *Scientific Reports* 3: 1835.

Lakeman-Fraser, Poppy, Laura Gosling, Andy J. Moffat, Sarah E. West, Roger Fradera, Linda Davies, Maxwell A. Ayamba, and René Wal. 2016. 'To Have Your Citizen Science Cake and Eat It? Delivering research and outreach through Open Air Laboratories (OPAL)', *BMC Ecology* 16: 16.

Land-Zandstra, Anne M., Jeroen L.A. Devilee, Frans Snik, Franka Buurmeijer, Jos M. van den Broek. 2016. 'Citizen Science on a Smartphone: Participants' Motivations and Learning', *Public Understanding of Science* 25: 45–60.

Lanza, E. L., C. V. A. L. Gascón, and F. Sanz. 2014. 'SOCIENTIZE Participatory Experiments, Dissemination and Networking Activities in Perspective', *Human Computation* 1: 119-135. http://dx.doi.org/10.15346/hc.v1i2.4.

Laso Bayas, Juan Carlos, Myroslava Lesiv, François Waldner, Anne Schucknecht, Martina Duerauer, Linda See, Steffan Fritz, Steffen, Dilek Fraisl, Inian Moorthy, and Ian McCallum. 2017. 'A Global Reference Database of Crowdsourced Cropland Data Collected Using the Geo-Wiki Platform', *Scientific Data* 4: 170136.

Laso Bayas, Juan Carlos, Linda See, Steffan Fritz, Tobias Sturn, Christoph Perger, Martina Dürauer, Mathias Karner et al. 2016. 'Crowdsourcing In-situ Data on Land Cover and Land Use Using Gamification and Mobile technology', *Remote Sensing* 8, no. 11: 905.

Latour, Bruno, and Peter Weibel, eds. 2005. *Making Things Public: Atmospheres of Democracy*. Cambridge: The MIT Press.

Laut, Jeffrey, Francesco Cappa, Oded Nov, and Maurizio Porfiri. 2015. 'Increasing Patient Engagement in Rehabilitation Exercises Using Computer-based Citizen Science', *PloS ONE* 10: e0117013.

Lawless, James G., and Barrett N. Rock. 1998. 'Student-Scientist Partnerships and Data Quality', *Journal of Science Education and Technology* 7, no. 1: 5–13.

Lawrence, Anna. 2006. '"No Personal Motive?" Volunteers, Biodiversity, and the False Dichotomies of Participation', *Ethics, Place and Environment* 9: 279–98.

Lawrence, Anna. 2010. *Taking Stock of Nature: Participatory Biodiversity Assessment for Policy, Planning and Practice*. Cambridge: Cambridge University Press.

Le Coz, Jérôme, Antoine Patalano, Daniel Collins, Nicolás Federico Guillén, Carlos Marcelo García, Graeme M. Smart, Jochen Bind, et al. 2016. 'Crowdsourced Data for Flood Hydrology: Feedback from Recent Citizen Science Projects in Argentina, France and New Zealand', *Journal of Hydrology* 541: 766–77.

Leach, Melissa, Ian Scoones, and Brian Wynne, eds. 2005. *Science and Citizens: Globalization and the Challenge of Engagement*. London: Zed Books.

League of European Research Universities. 2016. *Citizen Science at Universities: Trends, Guidelines and Recommendations*. Advice paper No. 20. https://www.leru.org/files/Citizen-Science-at -Universities-Trends-Guidelines-and-Recommendations-Full-paper.pdf.

Lee, Tracy, Michael S. Quinn, and Danah Duke. 2006. 'Citizen, Science, Highways, and Wildlife: Using a Web-based GIS to Engage Citizens in Wildlife Information', *Ecology and Society* 2006 11:11. http://www.scopus.com/inward/record.url?eid=2-s2.0-33745887483&partnerID =40&md5=b997b0e387d1486bf8e1fe730ea1aed1.

Lee, Jeehyung, Wipapat Kladwang, Minjae Lee, Daniel Cantu, Matin Azizyan, Hanjoo Kim, Alex Limpaecher, Snehal Gaikwad, Sungroh Yoon, and Adrien Treuille. 2014. 'RNA Design Rules from a Massive Open Laboratory', *Proceedings of the National Academy of Sciences* 111: 2122–7.

Leibovici, Didier G., Julian F. Rosser, Crona Hodges, Barry Evans, Mike J. Jackson, and Chris I. Higgins. 2017a. 'On Data Quality Assurance and Conflation Entanglement in Crowdsourcing for Environmental Studies', *ISPRS International Journal of Geo-Information* 6, no. 3: 78.

Leibovici, Didier G., Jamie Williams, Julain F. Rosser, Crona Hodges, Colin Chapman, Chris Higgins, and Mike J. Jackson. 2017b. 'Earth Observation for Citizen Science Validation, or Citizen Science for Earth Observation Validation? The Role of Quality Assurance of Volunteered Observations', *Data* 2, no. 4: 35.

Leino, Helena, and Juha Peltomaa. 2012. 'Situated Knowledge-Situated Legitimacy: Consequences of Citizen Participation in Local Environmental Governance', *Policy and Society* 31, no. 2: 159–68.

Levidow, Les. 2008. 'Democratising Technology Choices? European Public Participation in Agbiotech Assessments', *IIED Gatekeeper Series* 135. London: International Institute for Environment and Development.

Levrel, Harold, Benoît Fontaine, Pierre-Yves Henry, Frédéric Jiguet, Romain Julilard, Christian Kerbiriou, and Denis Couvet. 2010. 'Balancing State and Volunteer Investment in Biodiversity Monitoring for the Implementation of CBD Indicators: A French Example', *Ecological Economics* 69, no. 7: 1580–6.

Levy, Steven. 2012. 'Tim O'Reilly's Key to Creating the Next Big Thing', Wired Magazine. 21 December 2012. https://www.wired.com/2012/12/mf-tim-oreilly-qa/.

Lewandowski, Eva, Wendy Caldwell, Dane Elmquist, and Karen Oberhauser. 2017. 'Public Perceptions of Citizen Science', *Citizen Science: Theory and Practice*. 2, no. 1: 3. https://doi.org/10.5334/cstp.77.

Lewandowski, Eva, and Hannah Specht. 2015. 'Influence of Volunteer and Project Characteristics on Data Quality of Biological Surveys', *Conservation Biology* 29: 713–23.

Li, Chunming, Dong Wei, Jonathan Vause, and Jianping Liu. 2013. 'Towards a Societal Scale Environmental Sensing Network with Public Participation', *International Journal of Sustainable Development & World Ecology* 20, no. 3: 261–6.

Li, XueYan, Lu Lian, Peng Gong, Yang Liu, and FeiFei Liang. 2013. 'Bird Watching in China Reveals Bird Distribution Changes', *Chinese Science Bulletin* 58, no. 6: 649–56.

Liebenberg, Louis. 2015. 'Citizen Science: Creating an Inclusive, Global Network for Conservation', *The Guardian*, 7 January 2015. https://www.theguardian.com/science/2015/jan/07/citizen-science-creating-an-inclusive-global-network-for-conservation.

Liegl, Michael, Rachel Oliphant, and Monika Büscher. 2015. 'Ethically Aware IT Design for Emergency Response: From Co-Design to ELSI Co-Design.'. In *Proceedings of the ISCRAM 2015 12th International Conference on Information Systems for Crisis Response and Management,* edited by Leysia Palen, Monika Buscher, Tina Comes, and Amanda Hughes. http://iscram2015.uia.no/wp-content/uploads/2015/05/4-6.pdf.

Lisjak, Josip, Sven Schade, and Alexander Kotsev. 2017. 'Closing Data Gaps with Citizen Science? Findings from the Danube region', *ISPRS International Journal of Geo-Information* 6: 277.

Littledyke, Michael. 2004. 'Primary Children's Views on Science and Environmental Issues: Examples of Environmental Cognitive and Moral Development', *Environmental Education Research* 10, no. 2: 217–35.

Lolkema, Dorien E., H. Noordijk, A.P. Stolk, R. Hoogerbrugge, M.C. van Zanten, and W. A. Van Pul. 2015. 'The Measuring Ammonia in Nature (MAN) Network in the Netherlands', *Biogeosciences* 12: 5133–42.

Longueville, Bertrand de, Gianluca Luraschi, Paul Smits, Stephen Peedell, and Tom De Groeve. 2010. 'Citizens as Sensors for Natural Hazards: A VGI Integration Workflow', *Geomatica* 64, no. 1: 41–59.

Lorimer, Jamie. 2008. 'Counting Corncrakes: The Affective Science of the UK Corncrake Census.' *Social Studies of Science* 38: 377–405.

Lóscio, Bernadette Farias, Caroline Burle, Newton Calegari, Annette Greiner, Antoine Isaac, Carlos Iglesias, and Carlos Laufer. 2016. *'Data on the Web Best Practices.'*. W3C Working Draft 19.

Losey, John, Leslie Allee, and Rebecca Smyth. 2012. 'The Lost Ladybug Project: Citizen Spotting Surpasses Scientist's Surveys.', *American Entomologist* 58, no. 1(1): 22–4.

Louv, Richard. 2005. *Last Child in the Woods: Saving Our Children from Nature-Deficit Disorder.* Chapel Hill, NC: Algonquin Books.

Lund, Kate, Paul Coulton, and Andrew Wilson. 2011. 'Participation Inequality in Mobile Location Games'. In *Proceedings of the 8th International Conference on Advances in Computer Entertainment Technology*, 27. New York: ACM.

Luzar, Jeffrey B., Kirsten M. Silvius, Han Overman, Sean T. Giery, Jane M. Read, and José M. V. Fragoso. 2011. 'Large-scale Environmental Monitoring by Indigenous People', *BioScience* 61: 771–81.

Ma, Ming D., Luming Shen, John Sheridan, Jefferson Zhe Liu, Chao Chen, and Quanshui Zheng. 2011. 'Friction of Water Slipping in Carbon Nanotubes', *Physical Review E* 83, no. 3: 036316.

Maasen, Sabine, and Sascha Dickel. 2016. 'Partizipation, Responsivität, Nachhaltigkeit. Zur Realfiktion eines neuen Gesellschaftsvertrags', In *Handbuch Wissenschaftspolitik*, edited by Dagmar Simon, Andreas Knie, Stefan Hornbostel and Karin Zimmermann, 225–42. Wiesbaden: Springer VS.

Macey, Gregg P., Ruth Breech, Mark Chernaik, Caroline Cox, Denny Larson, Deb Thomas, and David O. Carpenter. 2014. 'Air Concentrations of Volatile Compounds near Oil and Gas Production: A Community-based Exploratory Study', *Environmental Health* 13, no. 1: 82.

MacLennan, Gregor. n.d. 'We Built a Drone'. Digital Democracy. Last accessed 1 September 2011. https://www.digital-democracy.org/blog/we-built-a-drone/.

Macnaghten, Phil, Matthew Kearnes, and Brian Wynne. 2005. 'Nanotechnology, Governance, and Public Deliberation: What Role for the Social Sciences?' *Science Communication* 27: 268–91.

Magnussen, Rikke. 2017. 'Involving Lay People in Research and Professional Development through Gaming: A Systematic Mapping Review', In *ECGBL 2017 11th European Conference on Game-Based Learning*, 394–401.

Mahr, Dana (geb. Dominik). 2014. *Citizen Science: partizipative Wissenschaft im späten 19. und frühen 20. Jahrhundert.* Wissenschafts- und Technikforschung, Bd.12. Baden-Baden: Nomos.

Mahr, Dana. 2016. 'Entfremdung, Entzauberung und eine ungenügende Öffentlichkeitskonzep-tion. Historische Hintergründe der partizipativen Wende in den Wissenschaften', *oead.news* 101: 18–20.

Mahr, Dana, and Sascha Dickel. Forthcoming. Citizen Science beyond Invited Participation, submitted to *History of Science*.

Makechnie, Colin, Lindsay Maskell, Lisa Norton, and David Roy. 2011 'The Role of "Big Society" in Monitoring the State of the Natural Environment', *Journal of Environmental Monitoring* 13, no. 10: 2687–91.

Maly, Tim. 2012. 'Citizen Smart-Kites Check China's Air', Wired. Last accessed 6 June 2017. https://www.wired.com/2012/12/chinese-air-quality-kites/.

Margoluis, Richard, Carolyn Stem, Vinaya Swaminathan, Marcia Brown, Arlyne Johnson, Guillermo Placci, Nick Salafsky, and Ilke Tilders. 2013. 'Results Chains: A Tool for Conservation Action Design, Management, and Evaluation', *Ecology and Society* 18, no. 3: 22.

Marine Conservation Society. n.d. 'Beach Cleans'. Last accessed 13 March 2018. https://www.mcsuk.org/get-active/beachcleans.

Marres, Noortje. 2007. 'The Issues Deserve More Credit: Pragmatist Contributions to the Study of Public Involvement in Controversy', *Social Studies of Science* 37: 759–80.

Masters, Karen L., Eun Y. Oh, Joe Cox, Brooke Simmons, Chris Lintott, Gary Graham, Anita Greenhill, and Kate Holmes. 2016. 'Science Learning via Participation in Online Citizen Science', *Journal of Science Communication* 15, no. 3: 1–33.

Mattessich, Paul W., and Barbara R. Monsey. 1992. *Collaboration: What Makes it Work.* St. Paul, Minnesota: Amherst H. Wilder Foundation.

Mazumdar, Suvodeep, Vita Lanfranchi, Neil Ireson, Stuart Wrigley, Clara Bagnasco, Uta When, Rosalind McDonagh et al. 2016. 'Citizens Observatories for Effective Earth Observations: The WeSenseIt Approach', *Environmental Scientist* 25: 56–61.

Mazumdar, Suvodeep, Stuart Wrigley, and Fabio Ciravegna. 2017. 'Citizen Science and Crowdsourcing for Earth Observations: An Analysis of Stakeholder Opinions on the Present and Future', *Remote Sensing*, 9, no. 1: 87.

McElfish, James, John Pendergrass, and Talia Fox. 2016. *Clearing the Path: Citizen Science and Public Decision-Making in the United States.* Washington DC: Woodrow Wilson International Center for Scholars.

McKinley, Duncan C., Russell D. Briggs, and Ann M. Bartuska. 2013. 'When peer-reviewed publications are not enough! Delivering science for natural resource management, *Forest Policy and Economics* 37: 9–19.

McKinley, Duncan C., Abe J. Miller-Rushing, Heidi L. Ballard, Rick Bonney, Hutch Brown, Susan C. Cook-Patton, Daniel M. Evans, et al. 2015. 'Investing in Citizen Science Can Improve Natural Resource Management and Environmental Protection', *Issues in Ecology* 19. https://www.esa.org/esa/wp-content/uploads/2015/09/Issue19.pdf.

McKinley, Duncan C., Abe J. Miller-Rushing, Heidi L Ballard, Rick Bonney, Hutch Brown, Susan C. Cook-Patton, Daniel M. Evans et al. 2017. 'Citizen Science Can Improve Conservation Science, Natural Resource Management, and Environmental Protection', *Biological Conservation* 208: 15–28.

Meek, Sam, Mike Jackson, and Didier G. Leibovici. 2014. 'A Flexible Framework for Assessing the Quality of Crowdsourced Data'. In *Connecting a Digital Europe Through Location and Place AGILE'2014 International Conference on Geographic Information Science.*

Meek, Sam, Mike Jackson, and Didier G. Leibovici. 2016. 'A BPMN Solution for Chaining OGC Services to Quality Assure Location-based Crowdsourced Data', *Computers & Geosciences* 87: 76–83.

Melchior, Alan, and Lawrence Bailis. 2003. '2001 – 2002 Earth Force Evaluation: Program Implementation and Impacts.' Center for Youth and Communities. Heller Graduate School, Brandeis University, Waltham, MA.

Menozzi, Marie-Jo. 2007. '"Mauvaises herbes", qualité de l'eau et entretien des espaces', *Natures Sciences Sociétés* 15: 144–53.

Merriam-Webster, Incorporated. 2017. s.v. 'Infrastructure'. Last modified 14 November 2017. Last accessed 22 November 2017. https://www.merriam-webster.com/dictionary/infrastructure.

Miah, Andy. 2017. 'Nanoethics, Science Communication, and a Fourth Model for Public Engagement', *Nanoethics* 11: 139. https://doi.org/10.1007/s11569-017-0302-9.

Micheel, Isabel, Paraskevi Lazaridou, Jasminko Novak, Giorgia Baroffio, Andrea Caminola, Andrea Castelletti, Piero Fraternali et al. 2014. *Use Cases And Early Requirements.* SmartH2O Project Deliverable D2.1. European Commission.

Miczajka, Victoria L., Alexandra-Maria Klein, and Gesine Pufal. 2015. 'Elementary School Children Contribute to Environmental Research as Citizen Scientists', *PLoS ONE* 10, no. 11: e0143229.

Miller, Brian, William Conway, Richard P. Reading, Chris Wemmer, David Wildt, Devra Kleiman, Steven Monfort, Alan Rabinowitz, Beth Armstrong, and Michael Hutchins. 2004. 'Evaluating the Conservation Mission of Zoos, Aquariums, Botanical Gardens, and Natural History Museums', *Conservation Biology* 18, no. 1: 86–93.

Miller, James R. 2005. 'Biodiversity Conservation and the Extinction of Experience', *TRENDS in Ecology and Evolution* 20, no. 8:430–4.

Miller, Kenton, Elsa Chang, and Nels Johnson. 2001. *Defining Common Ground for the Mesoamerican Biological Corridor.* Washington, DC: World Resources Institute.

Miller-Rushing, Abraham, Richard Primack, and Rick Bonney. 2012. 'The History of Public Participation in Ecological Research', *Frontiers in Ecology and the Environment* 10: 285–90. https://doi.org/10.1890/110278.

Miller-Rushing, Abraham J. and Richard B. Primack. 2008. 'Global Warming and Flowering Times in Thoreau's Concord: A Community Perspective', *Ecology* 89, no. 2: 332–41.

Minson, Sarah E., Benjamin A. Brooks, Craig L. Glennie, Jessica R. Murray, John O. Langbein, Susan E. Owen, Thomas H. Heaton, Robert A. Iannucci, and Darren L. Hauser. 2015. 'Crowdsourced Earthquake Early Warning', *Science advances* 1, no. 3: e1500036.

Mistry, Jayalaxshmi, and Andrea Berardi. 2016. 'Bridging Indigenous and Scientific Knowledge', *Science* 352: 1274–5.

Mobasheri, Amin, Alexander Zipf, Mohamed Bakillah, and Steve H.L. Liang. 2013. 'QualEvS4Geo: A Peer-to-Peer System Architecture for Semi-Automated Quality Evaluation of Geo-data in SDI', In *Innovative Computing Technology (INTECH), 2013 Third International Conference on*, 7–11.

Moedas, Carlos. 2016. 'Open Science: Share and Succeed'. Oral intervention at the *Open Science Conference, Amsterdam, April 4 2016.* Last accessed 14 December 2017. http://europa.eu/rapid/press-release_SPEECH-16-1225_en.htm.

Molinier, Matthieu, Carlos A. López-Sánchez, Timo Toivanen, Timo, Ilkka Korpela, José J. Corral-Rivas, Renne Tergujeff, and Tuomas Häme. 2016. 'Relasphone – Mobile and Participative In Situ Forest Biomass Measurements Supporting Satellite Image Mapping', *Remote Sensing* 8, no. 10: 869.

Monitoring matters. n.d. Last accessed 11 June 2018. www.monitoringmatters.org

Mooney, Peter, and Padraig Corcoran. 2012. 'Who are the Contributors to OpenStreetMap and What Do They Do?' In *Proceedings of the GIS Research UK 20th Annual Conference, Lancaster (GBR)*, 355–60.

Mooney, Peter, Padraig Corcoran, and Blazej Ciepluch. 2013. 'The Potential for Using Volunteered Geographic Information in Pervasive Health Computing Applications', *Journal of Ambient Intelligence and Humanized Computing* 4: 731–45.

Mooney, Peter, Marco Minghini, Mari Laakso, Vyron Antoniou, Ana-Maria Olteanu-Raimond, and Andriani Skopeliti. 2016. 'Towards a Protocol for the Collection of VGI Vector Data', *ISPRS International Journal of Geo-Information* 5, no. 11: 217.

Moran, Lisa, and Henrike Rau. 2016. 'Mapping Divergent Concepts of Sustainability: Lay Knowledge, Local Practices and Environmental Governance.', *Local Environment* 21: 344–60.

Mori, E., A. Sforzi, and M. Di Febbraro. 2013. 'From the Apennines to the Alps: Recent Range Expansion of the Crested Porcupine Hystrix cristata L., 1758 (Mammalia: Rodentia: Hystricidae) in Italy', *Italian Journal of Zoology* 80: 469–80.

Mroz, Ann. 2011. Leader: 'Citizens, to the Lab and front Bench'. Times Higher Education. 17 November. https://www.timeshighereducation.com/comment/leader/leader-citizens-to-the -lab-and-front-bench/417842.article.

Mueller, Jocelyn G., Issoufou H. B. Assanou, Iro D. Guimbo, and Astier M. Almedom. 2010. 'Evaluating Rapid Participatory Rural Appraisal as an Assessment of Ethnoecological Knowledge and Local Biodiversity Patterns', *Conservation Biology* 24, no. 1: 140–50.

Mueller, Michael P., Deborah Tippins, and Lynn A. Bryan. 2012. 'The Future of Citizen Science', *Democracy & Education* 20, no. 1: article 2. http://democracyeducationjournal.org/home /vol20/iss1/2/.

Muenich, Rebecca Logsdon, Sara Peel, Laura C. Bowling, Megan Heller Haas, Ronald F. Turco, Jane R. Frankenberger, and Indrajeet Chaubey. 2016. 'The Wabash Sampling Blitz: A Study on the Effectiveness of Citizen Science', *Citizen Science: Theory and Practice*. 1, no. 1: 3. https://doi.org/10.5334/cstp.1.

Mukundarajan, Haripriya, Felix Jan Hein Hol, Erica Araceli Castillo, Cooper Newby, and Manu Prakash. 2017. 'Using Mobile Phones as Acoustic Sensors for High-Throughput Mosquito Surveillance', *eLife* 6, no. e27854: 1–26. https://doi.org/10.7554/eLife.27854.

Müller, Michael J. 2002. 'Participatory Design: The Third Space in HCI'. In *The human-computer interaction handbook*, edited by Julie A. Jacko and Andrew Sears, 1051–68. Hillsdale, NJ, USA: L. Erlbaum Associates Inc.

Nadasdy, Paul. 1999. 'The Politics of TEK: Power and the "Integration" of Knowledge', *Arctic Anthropolgy* 36: 1–18.

Nagy, Christopher, Kyle Bardwell, Robert F. Rockwell, Rod Christie, and Mark Weckel. 2012. 'Validation of a Citizen Science-Based Model of Site Occupancy for Eastern Screech Owls with Systematic Data in Suburban New York and Connecticut', *Northeastern Naturalist* 19, no. 6: 143–58.

Nascimento, Susana, Ângela Guimarães Pereira, and Alessia Ghezzi. 2014. *From Citizen Science to Do It Yourself Science: An Annotated Account of an Ongoing Movement.* Luxembourg: Publications Office of the European Union. http://publications.jrc.ec.europa.eu/repository /bitstream/JRC93942/ldna27095enn.pdf.

Nascimento, Susana, and Alexandre Pólvora. 2015. 'Social Sciences in the Transdisciplinary Making of Sustainable Artifacts', *Social Science Information* 55, no. 1: 28–42.

Nascimento, Susana, Alexandre Pólvora, Alexandra Paio, Sancho Oliveira, Vasco Rato, Maria João Oliveira, Bárbara Varela, and João Pedro Sousa. 2016. 'Sustainable Technologies and Transdisciplinary Futures: From Collaborative Design to Digital Fabrication', *Science as Culture* 25, no. 4: 520–37.

National Advisory Council for Environmental Policy and Technology (NACEPT). 2016. *Environmental Protection Belongs to the Public: A Vision for Citizen Science at EPA.* NACEPT Report.

National Audubon Society. 2017. 'Christmas Bird Count'. Last accessed 27 July 2017. http://www .audubon.org/conservation/science/christmas-bird-count.

National Research Council. 2009. *Learning Science in Informal Environments: People, Places, and Pursuits.* Washington, DC: The National Academies Press.

Newman, G., M. Chandler, M. Clyde, B. McGreavy, H. Haklay, H. Ballard, S. Gray et al. 2017. 'Leveraging the Power of Place in Citizen Science for Effective Conservation Decision Making', *Biological Conservation* 208: 55–64.

Newman, Greg, Jim Graham, Alycia W. Crall, and M. Laituri. 2011. 'The art and science of multi-scale citizen science support.', *Ecological Informatics* 6, nos. 3 and 4(3–4): 217–27.

Newman, Greg, Philip Roetman, and Johannes Vogel. 2015. 'Open Letter to Nature'. *Citizen Science.* http://citizenscience.org/wp-content/uploads/2015/09/response_nature.pdf.

Newman, Greg, Andrea Wiggins, Alycia Crall, Eric Graham, Sarah Newman, and Kevin Crowston. 2012. 'The Future of Citizen Science: Emerging Technologies and Shifting Paradigms',

Frontiers in Ecology and the Environment 10, no. 6: 298–304. https://doi.org/0.1890/110294.

Newman, Greg, Don Zimmerman, Alycia Crall, Melinda Laituri, John Graham, and Linda Stapel. 2010. 'User-friendly Web Mapping: Lessons from a Citizen Science Website', *International Journal of Geographical Information Science* 24: 1851–69.

Nielsen, Jakob. 2006. 'Participation Inequality: Encouraging More Users to Contribute'. Jakob Nielsen's Alertbox 9.

Nielsen, Jakob. n.d. 'The 90-9-1 Rule for Participation Inequality in Social Media and Online Communities'. Nielsen Norman Group. Last modified 10 December 2016. https://www.nngroup.com/articles/participation-inequality/.

Nieuwenhuijsen, Mark J., David Donaire-Gonzalez, Ioar Rivas, Montserrat De Castro, Marta Cirach, Gerard Hoek, Edmund Seto, Michael Jerrett, and Jordi Sunyer. 2015. 'Variability in and Agreement between Modeled and Personal Continuously Measured Black Carbon Levels Using Novel Smartphone and Sensor Technologies', *Environmental Science & Technology* 49: 2977–82.

Niewöhner, Jörg. 2016. 'Co-laborative Anthropology: Crafting Reflexivities Experimentally'. In *Analysis and Interpretation*, edited by Jukka Jouhki and Tytti Steel, 81-124. Helsinki: Ethnos.

Nold, Christian, and Louise Francis. 2017. Participatory Sensing: Recruiting Bipedal Platforms or Building Issue-centred Projects?' In *Participatory Sensing, Opinions and Collective Awareness*, 23–35. Berlin: Springer International Publishing.

Notley, Tanya, and Camellia Webb-Gannon. 2016. 'FCJ-201 Visual Evidence from Above: Assessing the Value of Earth Observation Satellites for Supporting Human Rights', *The Fibreculture Journal* 27: Networked War/Conflict. https://doi.org/10.15307/fcj.27.201.2016.

Nov, Oded, Ofer Arazy, and David Anderson. 2011a. 'Technology-Mediated Citizen Science Participation: A Motivational Model'. In *Fifth International AAAI Conference on Weblogs and Social Media*, 249–56.

Nov, Oded, Ofer Arazy, and David Anderson. 2011b. 'Dusting for science: motivation and participation of digital citizen science volunteers', In *Proceedings of the 2011 iConference* (iConference '11), 68–74.

Nov, Oded, Mor Naaman, and Chen Ye. 2009. 'Analysis of Participation in an Online Photo-Sharing Community: A Multidimensional Perspective', *Journal of the American Society for Information Science and Technology* 61: 555–6.

Novacek, Michael J. 2008. 'Engaging the Public in Biodiversity Issues'. In *Proceedings of the National Academy of Sciences* 105: 11571–8.

Novak, Jasminko. 2009. 'Mine, Yours . . . Ours? Designing for Principal-Agent Collaboration in Interactive Value Reation'. In *Proceedings of Wirtschaftsinformatik 2009*, 305–14. Vienna: Österreichische Computer Gesellschaft.

Nowotny, Helga, Peter Scott, and Michael Gibbons. 2001. *Re-Thinking Science: Knowledge and the Public in an Age of Uncertainty*. Cambridge, UK: Polity Press.

Oberhauser, Karen, and Gretchen LeBuhn. 2012. 'Insects and Plants: Engaging Undergraduates in Authentic Research through Citizen Science', *Frontiers in Ecology and the Environment* 10, no. 6: 318–20.

Oberhauser, Karen, and Michelle D. Prysby. 2008. 'Citizen Science: Creating a Research Army for Conservation', *American Entomologist* 54, no. 2: 103–4.

Office of Science and Technology Policy. 2015. 'Memorandum to the Heads of Executive Departments and Agencies: Addressing Societal and Scientific Challenges through Citizen Science and Crowdsourcing'. Obama Whitehouse Archives. Last modified 30 September 2015. https://obamawhitehouse.archives.gov/sites/default/files/microsites/ostp/holdren_citizen_science_memo_092915_0.pdf.

OGC. 2016. OGC/CS DWG. Last accessed 20 October 2017. http://external.opengeospatial.org/twiki_public/CitizenScienceDWG/DraftCharter.

Oltra, Aitana, John R.B. Palmer, and Frederic Bartumeus. 2016. 'AtrapaelTigre.com: Enlisting Citizen-Scientists in the War on Tiger Mosquitoes'. In *European Handbook of Crowdsourced Geographic Information*, edited by Cristina Capineri, Muki Haklay, Haosheng Huang, Vyron Antoniou, Juhani Kettunen, Frank Ostermann, and Ross Purves, 295–308. London: Ubiquity Press. http://dx.doi.org/10.5334/bax.

Open Air Laboratories (OPAL). n.d. Citizen Science and Engagement Project, Imperial College, London, UK (D. Slawson, Director). Last accessed 15 February 2016. http://www.imperial.ac.uk/opal/public-engagement/.

Open Geospatial Consortium. 2016. 'OGC requests participation in its Citizen Science DWG'. Last accessed 14 December 2017. http://www.opengeospatial.org/pressroom/pressreleases/2453.

OpenSystem. n.d. Last accessed October 2017. http://www.ub.edu/opensystems/.

Ottinger, Gwen. 2010. 'Buckets of Resistance: Standards and the Effectiveness of Citizen Science', *Science, Technology, & Human Values* 35: 244–70.

Ottinger, Gwen. 2016. 'Social Movement-Based Citizen Science'. In *The Rightful Place of Science: Citizen Science*, edited by Darlene Cavalier and Eric B. Kennedy, 89–103. Tempe, AZ and Washington, DC: Consortium for Science, Policy and Outcomes.

Overdevest, Christine, Cailin Orr, and Kristine Stepenuck. 2004. 'Volunteer stream monitoring and local participation in natural resource issues', *Human Ecology Review* 11: 177–85.

Owen, Harrison. 1993. *Open Space Technology: A User's Guide*. San Francisco: Berrett-Koehler Publishers.

Owen, Harrison. 2008. *Open Space Technology: A User's Guide*. Third edition. San Francisco: Berrett-Koehler Publishers.

Owen, Richard, John Bessant, and Maggy Heintz, eds. 2013. *Responsible Innovation*. Chichester, UK: John Wiley & Sons, Ltd.

Owen, Richard, Phil Macnaghten, and Jack Stilgoe. 2012. 'Responsible Research and Innovation: From Science in Society to Science for Society, with Society', *Science and Public Policy* 39: 751–60

Paige, Kathryn, Robert Hattam, and Christopher B. Daniels. 2015. 'Two Models for Implementing Citizen Science Projects in Middle School', *The Journal of Educational Enquiry* 14, no. 2: 4–17.

Palmer, John R.B., Aitana Oltra, Francisco Collantes, Juan Antonio Delgado, Javier Lucientes, Sarah Delacour, Mikel Bengoa, et al. 2017. 'Citizen Science Provides a Reliable and Scalable Tool to Track Disease-Carrying Mosquitoes', *Nature Communications* 8, no. 1: 916. https://doi.org/10.1038/s41467-017-00914-9.

Pandya, Rajul E. 2012. 'A Framework for Engaging Diverse Communities in Citizen Science in the US', *Frontiers in Ecology and the Environment* 10: 314–17.

Parcak, Sarah. 2015. 'Remote Sensing with Satellite Technology', In *Emerging Trends in the Social and Behavioral Sciences: An Interdisciplinary, Searchable, and Linkable Resource*, edited by Robert A. Scott and Marlis Buchmann, 1–13. https://doi.org/10.1002/9781118900772.etrds0281.

Parker, Alison. 2016. 'Paperwork Reduction Act and Development of a Generic Clearance'. Presentation to the US Geological Survey Community for Data Integration, Washington, DC, September 14, 2016. Last accessed 14 November 2016. https://my.usgs.gov/confluence/display/cdi/CDI+Monthly+Meeting+20160914.

Parliamentary Office of Science and Technology, The. 2014. *Environmental Citizen Science*. POSTnote 476. London: House of Parliament. http://researchbriefings.files.parliament.uk/documents/POST-PN-476/POST-PN-476.pdf.

Participatory Monitoring and Management Partnership (PMMP). 2015. 'Manaus Letter: Recommendations for the Participatory Monitoring of Biodiversity'. In *International Seminar on Participatory Monitoring of Biodiversity for the Management of Natural Resources 2014*, edited by Pedro A. L. Constantino, Kirsten, M. Silvius, Jan Kleine Büening, Paulina Arroyo, Finn Danielsen, Carlos C. Durigan et al. Manaus, Brazil. http://docs.wixstatic.com/ugd/8d7574_869904b775da441896aa91d49d28daad.pdf

Patterson, Lucy. 2016. 'Citizen Science Disco'. Storify. Last accessed 28 August 2017. https://storify.com/lucypatterson/citizen-science-disco.

Patterson, Lucy. 2017. 'DIY, Science Policy, Brussels, etc.' Medium. Last accessed 28 August 2017. https://medium.com/@lu_cyP/diy-science-policy-brussels-etc-58f8d663a916.

Peltola, Taru, and Johanna Tuomisaari. 2015. 'Making a Difference: Forest Biodiversity, Affective Capacities and the Micro-politics of Expert Fieldwork', *Geoforum* 64: 1–11.

Peplow, Mark. 2016. 'Citizen Science Lures Garners into Sweden's Human Protein Atlas', *Nature Biotechnology* 34, no. 5: 452–3.

Pereira, H.M., S. Ferrier, M. Walters, G.N. Geller, R.H.G. Jongman, R.J. Scholes, M.W. Bruford, et al. 2013. 'Essential Biodiversity Variables', *Science* 339, no. 6117: 277–8.

Peres, Carlos A. 1999. 'General Guidelines for Standardizing Line-transect Surveys of Tropical Forest Primates', *Neotropical Primates* 7, no. 1:11–16.

Peters, Hans Peter, Harald Heinrichs, Arlena Jung, Monika, Kallfass, and Imme Petersen. 2008. 'Medialization of Science as a Prerequisite of Its Legitimazation and Political Relevance'. In *Communicating Science in Social Contexts – New Models, New Practices,* edited by Donghong Cheng, Michel Claessens, Toss Gascoigne, Jenni Metcalfe, Bernard Schiele, and Shunke Shi, 71–92. New York: Springer.

Peters, Monica A., Chris Eames, and David Hamilton. 2015. 'The Use and Value of Citizen Science Data in New Zealand', *Journal of the Royal Society of New Zealand* 45, no. 3: 151–60.

Peters, Monica A., David Hamilton, Chris Eames, John Innes, and Norman W.H. Mason. 2016. 'The Current State of Community-based Environmental Monitoring in New Zealand', *New Zealand Journal of Ecology* 40, no. 3: 279–88.

Petersen, Lyle R., and Ann M. Powers. 2016. 'Chikungunya: Epidemiology [version 1; referees: 2 approved]', *F1000Research* 5: 82. https://doi.org/10.12688/f1000research.7171.1.

Pettibone, Lisa, Katrin Vohland, Aletta Bonn, Anett Richter, Wilhelm Bauhus, Birgit Behrisch, Rainer Borcherding et al. 2016. *Citizen Science for All. A Guide for Citizen Science Practitioners.* Deutsches Zentrum für Integrative Biodiversitätsforschung (iDiv) Halle-Jena-Leipzig, Helmholtz-Zentrum für Umweltforschung - UFZ, Leipzig; Berlin-Brandenburgisches Institut für Biodiversitätsforschung (BBIB), Museum für Naturkunde (MfN) - Leibniz-Institut für Evolutions- und Biodiversitätsforschung, Berlin. http://www.buergerschaffenwissen.de/sites /default/files/assets/dokumente/handreichunga5_engl_web.pdf.

Pettibone, Lisa, Katrin Vohland, and David Ziegler. 2017. 'Understanding the (Inter) Disciplinary and Institutional Diversity of Citizen Science: A Survey of Current Practice in Germany and Austria', *PloS ONE* 12, no. 6: e0178778. https://doi.org/10.1371/journal.pone.0178778.

Phillips, Tina. 2016. 'Examining how Participation and Engagement in Citizen Science Influence Learning: A Mixed Methods, Collaborative Research Project'. Paper presented at the First International ECSA Conference, Berlin.

Phillips, Tina, Rick Bonney, and Jennifer Shirk. 2012. 'What is Our Impact? Toward a Unified Framework for Evaluating Impacts of Citizen Science'. In *Citizen Science: Public Participation in Environmental Research,* edited by J.L. Dickinson and R. Bonney, 82–95. Ithaca, NY: Cornell University Press.

Phillips, Tina, Marion Ferguson, Matthew Minarchek, Norman Porticella, and Rick Bonney. 2014. *User's Guide for Evaluating Learning Outcomes in Citizen Science.* Ithaca, NY: Cornell Lab of Orinthology. http://www.birds.cornell.edu/citscitoolkit/evaluation.

Phills Jr., James A., Kriss Deiglmeier, and Dale T. Miller. 2008. 'Rediscovering Social Innovation'. Stanford Social Innovation Review. Last modified 4 October 2016. http://www.ssireview.org /articles/entry/rediscovering_social_innovation/.

Pike, Angela G., and Máiréad Dunne. 2011. 'Student Reflections on Choosing to Study Science Post-16', *Cultural Studies of Science Education* 6, no. 2: 485-500.

Plant Tracker. 2017. 'Plant Tracker.' Last accessed 13 March 2018. http://planttracker.org.uk/.

Pocock, Michael J.O., Daniel S. Chapman, Lucy J. Sheppard, and Helen E. Roy. 2013. *Developing a Strategic Framework to Support Citizen Science Implementation in SEPA. Final Report on behalf of SEPA.* Wallingford: NERC Centre for Ecology & Hydrology.

Pocock, Michael J.O., Daniel S. Chapman, Lucy J. Sheppard, and Helen E. Roy. 2014a. *A Strategic Framework to Support the Implementation of Citizen Science for Environmental Monitoring. Final Report to SEPA.* Wallingford: NERC Centre for Ecology & Hydrology. http://www.ceh.ac .uk/sites/default/files/hp1114final_5_complete.pdf.

Pocock, Michael J.O., Daniel S. Chapman, Lucy J. Sheppard, and Helen E. Roy. 2014b. *Choosing and Using Citizen Science: A Guide to When and How to Use Citizen Science to Monitor Biodiversity and the Environment.* Wallingford: NERC Centre for Ecology & Hydrology. https://www.ceh.ac.uk/sites/default/files/sepa_choosingandusingcitizenscience_interactive _4web_final_amended-blue1.pdf.

Pocock, Michael J.O., Helen E. Roy, Chris D. Preston, and David B. Roy. 2015. 'The Biological Records Centre: A Pioneer of Citizen Science', *Biological Journal of the Linnean Society* 115: 475–93.

Pocock, Michael J.O., John C. Tweddle, Joanna Savage, Lucy D. Robinson, and Helen E. Roy. 2017. 'The Diversity and Evolution of Ecological and Environmental Citizen Science'. *PloS ONE* 12: e0172579.

Polanyi, Michael. 1962. 'The Republic of Science: Its Political and Economic Theory', *Minerva* 1: 54–74.

Postles, Matt, and Madeleine Bartlett. 2014. *The Rise and Rise of BioBlitz: Public Engagement and Wildlife Recording Events in the UK*. Bristol Natural History Consortium.

PPSR-Core project. n.d. Last accessed 13 March 2008. https://www.wilsoncenter.org/article/ppsr-core-metadata-standards.

Prange, Erica, Micah Lande, and Darlene Cavalier. 2018. *Citizen Science Maker Summit Reports: Learning Outcomes and Next Steps*. Arizona State University White Paper. Tempe, AZ.

Preece, Jennifer, and Ben Shneiderman. 2009. 'The Reader-to-Leader Framework: Motivating Technology-mediated Social Participation', *AIS Transactions on Human-Computer Interaction* 1, no. 1: 13–32.

Prestopnik, Nathan R., and Kevin Crowston. 2012. 'Citizen Science System Assemblages: Understanding the Technologies that Support Crowdsourced Science.'. In *Proceedings of the 2012 iConference*, Toronto, Canada, 168–76. New York: ACM.

Project Budburst. n.d. 'Project Budburst'. Last accessed 13 April 2017. http://budburst.org.

Purcell, Karen, Cecilia Garibay, and Janis L. Dickinson. 2012. 'A Gateway to Science for All: Celebrate Urban Birds'. In *Citizen Science: Public Participation in Environmental Research*, edited by J.L. Dickinson and R. Bonney, 191–200. Ithaca, US: Comstock Pub. Associates.

Raddick, M. Jordan., Georgia Bracey, Pamela L. Gay, Chris J. Lintott, Carie Cardamone, Phil Murray, Kevin Schawinski, Alexander S. Szalay, and Jan Vandenberg. 2013. 'Galaxy Zoo: Motivations of Citizen Scientists', *Astronomy Education Review* 12, no. 1. https://doi.org/10.3847/AER2011021.

Ramanauskaite, Egle. 2016. 'Back in Berlin! DIY & Citizen Science Stakeholder Roundtable'. *Seplute* (blog), 8 November 2016. Last accessed 14 March 2018. https://seplute.tumblr.com/post/152953730485/back-in-berlin-diy-citizen-science-stakeholder.

Ramanauskaite, Egle. 2017a. 'Camera Traps for Citizen Science'. Hackaday.Io Last accessed 14 March 2018. https://hackaday.io/project/11088-camera-traps-for-citizen-science.

Ramanauskaite, Egle. 2017b. The We Cure ALZ Challenge. Last accessed 14 March 2018. https://sites.google.com/a/gold-mobileinnovation.co.uk/ecsa2016---citsci-thinkcamp/About-the-Think-Camp/the-challenges/2-the-wecurealz-challenge.

Ramasubramanian, Laxmi. 2008. *Geographic Information Science and Public Participation*. CUNY Academic Works. http://academicworks.cuny.edu/hc_pubs/16.

Ramirez-Andreotta, Monica D., Mark L. Brusseau, Janick Artiola, Raina M. Maier, and A. Jay Gandolfi. 2015. 'Building a Co-created Citizen Science Program with Gardeners Neighboring a Superfund Site: The Gardenroots Case Study', *International Public Health Journal* 7, no. 1: 139–53.

Ranard, Benjamin L., Yoonhee P. Ha, Zachary F. Meisel, David A. Asch, Shawndra S. Hill, Lance B. Becker, Anne K. Seymour, and Raina M. Merchant. 2014. 'Crowdsourcing – Harnessing the Masses to Advance Health and Medicine, a Systematic Review', *Journal of General Internal Medicine* 29, no. 1: 187–203.

Randin, Christophe F., Robin Engler, Signe Normand, Massimiliano Zappa, Niklaus E. Zimmermann, Peter B. Pearman, Pascal Vittoz, Wilfried Thuiller, and Antoine Guisan. 2009. 'Climate Change and Plant Distribution: Local Models Predict High-elevation Persistence', *Global Change Biology* 15: 1557–69.

Reiter, Russel J., Emilio Sanchez-Barcelo, Maria Mediavilla, Eloisa Gitto, and Ahmet Korkmaz. 2011. 'Circadian Mechanisms in the Regulation of Melatonin Synthesis: Disruption with Light at Night and the Pathophysiological Consequences', *Journal of Experimental and Integrative Medicine* 1: 13–22.

Research Triangle Environmental Health Collaborative. 2016. 'Community Engaged Research and Citizen Science: Advancing Environmental Public Health to Meet the Needs of Our Communities'. Research Triangle Environmental Health Collaborative Summit, Raleigh: North Carolina State University.

Resnik, David, Kevin Elliott, and Aubrey A. Miller. 2015. 'A Framework for Addressing Ethical Issues in Citizen Science', *Environmental Science & Policy* 54: 475–81.

Reznik, Tomas, Karel Charvat, Sarka Horakova, and Zbyněk Křivánek. 2016. 'VGI Profile for Precision Farming: Unified Data Model and Applications'. Paper presented at INSPIRE 2016, Barcelona.

Richter, Anett, and Lisa Pettibone. 2014. 'BürGER schaffen WISSen (GEWISS) – ein Projekt zur Entwicklung von Citizen Science Kapazitäten in Deutschland stellt sich vor', In *Tagfalter-Monitoring Deutschland – Jahresbericht 2013, Oedippus Band,* edited by E. Kühn, M. Musche,

A. Harpke, R. Feldmann, M. Wiemers, N. Hirneisen, B. Metzler, P. Wiemers, and J. Settele, 30, 45–57.

Richter, Anett, Lisa Pettibone, David Ziegler, Katrin Vohland, and Aletta Bonn. 2017. *Entwicklung von Citizen Science-Kapazitäten in Deutschland.*, BürGEr schaffen WISSen – Wissen schafft Bürger (GEWISS). Endbericht. http://www.buergerschaffenwissen.de.

Richter, Anett, Tabe Turrini, Karin Ulbrich, Anika Mahla, and Aletta Bonn. 2016. 'Citizen Science – Möglichkeiten in der Umweltbildung'. In *Nachhaltigkeit erfahren. Engagement als Schlüssel einer Bildung für nachhaltige Entwicklung*, edited by A. Bittner, T. Pyhel, and V. Bischoff, 95–115. München: Oekom Verlag.

Riesch, Hauke, and Clive Potter. 2014. 'Citizen Science as Seen by Scientists: Methodological, Epistemological and Ethical Dimension', *Public Understanding of Science* 23, no. 1: 107–20.

Rio Declaration on Environment and Development. 1992. http://www.un.org/documents/ga /conf151/aconf15126-1annex1.htm.

River Obstacles. 2017. 'River Obstacles'. Last accessed 13 March 2018. https://www.river -obstacles.org.uk/.

Riverfly Partnership, The. 2007. Accessed 14 December 2017. http://www.riverflies.org/press -releases.

Riverfly Partnership, The. 2017. 'ARMI Data: riverflies.org'. http://www.riverflies.org/riverflies -gis-home.

Robinson, Lucy, Jade Lauren Cawthray, Dai Lee, and John Tweddle. 2016. 'Citizen Science: Authentic Science Research at the Natural History Museum'. In *Museum Participation: New Directions for Audience Collaboration,* edited by K. McSweeney and J, Kavanagh. Edinburgh, UK: MuseumsEtc.

Robinson, Lucy, John Tweddle, Matt Postles, Sarah West, and Jack Sewell. 2013. 'Guide to Running a BioBlitz'. Natural History Museum, Bristol Natural History Consortium, University of York and Marine Biological Association, UK.

Rochford, Linda. 1991. 'Generating and Screening New Product Ideas', *Industrial Marketing Management* 20, no. 4: 2870296.

Rodríguez, Eduardo. 2015. 'Together We Look for Answers'. In *EcoJustice, Citizen Science and Youth Activism Situated Tensions for Science Education,* Vol. 1 of Environmental Discources in Science Education, edited by M.P. Mueller and Deborah J. Tippins, 11–19. Switzerland: Springer.

Rosas, Lisa G., Deborah Salvo, Sandra J. Winter, David Cortes, Juan Rivera, Nicole M. Rodriguez, and Abby C. King. 2016. 'Harnessing Technology and Citizen Science to Support Neighborhoods that Promote Active Living in Mexico', *Journal of Urban Health* 93, no. 6: 953–73.

Roth, Wolff-Michael, and Stuart Lee 2002. 'Scientific Literacy as Collective Praxis', *Public Understanding of Science* 11: 33–56.

Rotman, Dana, Jenny Preece, Jen Hammock, Kezee Procita, Derek Hansen, Cynthia Parr, Darcy Lewis, and David Jacobs. 2012. 'Dynamic Changes in Motivation in Collaborative Citizen-Science Projects'. In *Proceedings of the ACM 2012 Conference on Computer Supported Cooperative Work,* 217–26. New York: ACM.

Rowland, Katherine. 2012. 'Citizen Science Goes "Extreme"', Nature, 17 February, 2012. https://doi.org/10.1038/nature.2012.10054.

Roy, Helen, Michael J.O. Pocock, Christopher D. Preston, D.B. Roy, Joanna Savage, John C. Tweddle, and Lucy D. Robinson. 2012. *Understanding Citizen Science and Environmental Monitoring.* Final report on behalf of UK Environmental Observation Framework. NERC Centre for Ecology & Hydrology and Natural History Museum. https://www.ceh.ac.uk/sites /default/files/citizensciencereview.pdf.

RRI Tools. 2016. 'RRI Tools'. Last accessed 13 March 2018. https://www.rri-tools.eu.

Rubinstein, Albert. 1994. 'At the Front End of the R&D / Innovation Process – Idea Development and Entrepreneurship', *International Journal of Technology Management* 9, nos. 5, 6, and 7: 652–77.

Russell, Sharman Apt. 2014. *Diary of a Citizen Scientist: Chasing Tiger Beetles and Other New Ways of Engaging the World.* Oregon, US: Oregon State University Press.

Ruzic, Roxanne, Lindsay Goodwin, Rochelle Mothokakobo, and Theresa S. Talley. 2016. *Developing a Citizen Science Model to Engage Members of Underrepresented Minority Groups.* Exploratory study final report. San Diego, CA: Ruzic Consulting, Inc., Ocean Discovery Institute and Scripps Institution of Oceanography.

Salathé, Marcel. 2016. 'Digital Pharmacovigilance and Disease Surveillance: Combining Traditional and Big-Data Systems for Better Public Health', *The Journal of Infectious Diseases* 214, suppl 4: S399–403.

Salk, Carl, Tobias Sturn, Linda See, and Steffen Fritz. 2016a. 'Local Knowledge and Professional Background Have a Minimal Impact on Volunteer Citizen Science Performance in a Land-cover Classification Task', *Remote Sensing* 8, no. 9: 774.

Salk, Carl F., Tobia Sturn, Linda See, and Steffen Fritz. 2017. 'Limitations of Majority Agreement in Crowdsourced Image Interpretation', *Transactions in GIS* 21, no 2: 207–23.

Salk, Carl F., Tobia Sturn, Linda See, Steffen Fritz, and Christoph Perger. 2016b. 'Assessing Quality of Volunteer Crowdsourcing Contributions: Lessons from the Cropland Capture Game', *International Journal of Digital Earth* 9, no. 4: 410–26.

Sánchez de Miguel, Alejandro, José Gómez Castaño, Jaime Zamorano, Sergio Pascual, M. Ángeles, L. Cayuela, Peter Martinez, Guillermo Martín Challupner, and Christopher C.M. Kyba. 2014. 'Atlas of Astronaut Photos of Earth at Night', *Astronomy and Geophysics* 55: 4–36.

Sanders, Elizabeth, and Pieter Stappers. 2008. 'Co-creation and the New Landscapes of Design', *CoDesign* 4, no. 1: 5–18. https://doi.org/10.1080/15710880701875068.

Saunders, Tom, and Geoff Mulgan. 2017. *Governing with Collective Intelligence.* London: Nesta. http://www.nesta.org.uk/sites/default/files/governing_with_collective_intelligence.pdf.

SAŽP Banská Bystrica. n.d. Verejnost' – Informačný systém environmentálnych zát'aží. http://envirozataze.enviroportal.sk/.

Savio, Lorenzo del, Alena Buyx, and Barbara Prainsack. 2015. 'The Ethics of Citizen Science in Biomedicine', *Das Gesundheitswesen* 77, nos. 8/9: A231.

Savio, Lorenzo del, Barbara Prainsack, and Alena Buyx. 2016. 'Crowdsourcing the Human Gut. Is Crowdsourcing also "Citizen Science"?' *Journal of Science Communication* 15, no.3: A03.

Scassa, Teresa, and Haewon Chung. 2015a. *Typology of Citizen Science Projects from an Intellectual Property Perspective.* Washington, DC.: Wilson Center – Commons Lab. https://www .wilsoncenter.org/sites/default/files/Typology_of_Citizen_Science_IP_Rights_Scassa.pdf.

Scassa, Teresa, and Haewon Chung. 2015b. *Managing Intellectual Property Rights in Citizen Science: A Guide for Researchers and Citizen Scientists.* Washington DC, Woodrow Wilson International Center for Scholars. https://www.wilsoncenter.org/sites/default/files /managing_intellectual_property_rights_citizen_science_scassa_chung.pdf.

Schade, Sven, and Chrysi Tsinaraki. 2016. *Survey Report: Data Management in Citizen Science Projects.* Luxembourg: Publications Office of the European Union. http://publications.jrc.ec .europa.eu/repository/bitstream/JRC101077/lb-na-27920-en-n%20.pdf.

Schäfer, Teresa, and Kieslinger Barbara. 2016. 'Supporting Emerging Forms of Citizen Science: A Plea for Diversity, Creativity and Social Innovation', *Journal of Science Communication (JCOM)* 15, no. 2: Y02.

Schiele, Bernard. 2008. 'On and About the Deficit Model in an Age of Free Flow'. In *Communicating Science in Social Contexts – New Models, New Practices,* edited by Donghong Cheng, Michel Claessens, Toss Gascoigne, Jenni Metcalfe, Bernard Schiele and Shunke Shi, 91–117. New York: Springer.

Schierenberg, A., A. Richter, M. Kremer, P. Karrasch, and A. Bonn. 2016. *Anleitung zur Entwicklung von Bürgerwissenschafts-Projekten - Citizen Science in den Nationalen Naturlandschaften.* EUROPARC Deutschland, Berlin; Helmholtz-Zentrum für Umweltforschung - UFZ, Deutsches Zentrum für Integrative Biodiversitätsforschung (iDiv) Halle-Jena-Leipzig, Leipzig. http://www.buergerschaffenwissen.de/citizen-science/ressourcen.

Schmucki, Reto, Guy Pe'Er, David B. Roy, Constantí Stefanescu, Chris A.M. Van Swaay, Tom H. Oliver, Mikko Kuussaari, et al. 2016. 'A Regionally Informed Abundance Index for Supporting Integrative Analyses across Butterfly Monitoring Schemes', *Journal of Applied Ecology* 53, no. 2: 501–10.

Schroer, Sibylle, Katja Felsmann, Franz Hölker, Stephan, Mummert, Michael T. Monaghan, Christian Wurzbacher, and Katrin Premke. 2016. 'The Impact of Outdoor Lighting on Ecosystem Function – Gaining Information with a Citizen Science Approach Using a Questionnaire'. In *Front. Environ. Sci. Conference Abstract: Austrian Citizen Science Conference 2016.*

Schroer, Sibylle, and Franz Holker. 2016. 'Impact of Lighting on Flora and Fauna'. In *Handbook of Advanced Lighting Technology,* edited by Robert Karlicek, Ching-Chern Sun, Georgis Zissis, and Ruiqing Ma, 1–33. Switzerland: Springer International Publishing.

Schroer, Sibylle, and Franz Holker. 2017. 'Light Pollution Reduction'. In *Handbook of Advanced Lighting Technology*, edited by Robert Karlicek, Ching-Chern Sun, Georgis Zissis, and Ruiqing Ma, 991–1010. Switzerland: Springer International Publishing.

Schroer, Sibylle, Franz Hölker, and Oscar Corcho. 2016. 'The Impact of Citizen Science on Research about Artificial Light at Night', *Environmental Scientist* 25: 18–24.

Schultz, Wesley. 2002. 'Inclusion with Nature: The Psychology of Human–Nature Relations'. In *Psychology of sustainable development*, edited by Peter Schmuck and Wesley Schultz, 61–78. Boston: Kluwer Academic Publisher.

SciStarter. n.d. 'What is Citizen Science?' Last modified 3 October 2017. https://scistarter.com /citizenscience.html.

Sclove, Richard E. 2010. 'Reinventing Technology Assessment', *Issues in Science and Technology* 27, no. 1: 34–8.

Scottish Environment Protection Agency. 2017a. 'Rainfall'. Last accessed 13 March 2018. http://www.sepa.org.uk/environment/water/rainfall/.

Scottish Environment Protection Agency. 2017b. 'Learning about Air Quality'. Last accessed 13 March 2018. https://www.environment.gov.scot/educational-resources/learning-about-air -quality/.

Scottish Environment Protection Agency. 2017c. 'Scotland's Environment – Citizen Science Portal'. Last accessed 13 March 2018. https://envscot-csportal.org.uk/.

Secretariat of the Convention on Biological Diversity. 2014. *Global Biodiversity Outlook 4*. Montréal.

See, Linda, Alexis Comber, Carl Salk, Steffen Fritz, Marjin van der Velde, Christoph Perger, Christian Schill, Ian McCallum, Florian Kraxner, and Michael Obersteiner. 2013. 'Comparing the Quality of Crowdsourced Data Contributed by Expert and Non-experts', *PloS ONE* 8, no. 7: e69958.

See, Linda, Steffen Fritz, Christoph Perger, Christian Schill, Ian McCallum, Dimitry Schepas-chenko, Martina Duerauer, et al. 2015. 'Harnessing the Power of Volunteers, the Internet and Google Earth to Collect and Validate Global Spatial Information Using Geo-Wiki', *Technological Forecasting and Social Change* 98: 324–35.

Seed, Lindsay, Pat Wolseley, Laura Gosling, Linda Davies, and Sally A. Power. 2013. 'Modelling Relationships between Lichen Bioindicators, Air Quality and Climate on a National Scale: Results from the UK OPAL Air Survey', *Environmental Pollution* 182: 437–47.

Segal, Avi, Ya'akov (Kobi) Gal, Robert J. Simpson, Victoria Homsy, Mark Hartswood, Kevin R. Page, and Marina Jirotka. 2015. 'Improving Productivity in Citizen Science through Controlled Intervention'. In *Proceedings of the 24th International Conference on World Wide Web*, 331-337. Geneva: ACM.

Seifert, Franz. 2006. 'Local Steps in an International Career: A Danish-style Consensus Conference in Austria', *Public Understanding of Science* 15, no. 1: 73–88.

Sequeira, Ana M.M., Philip E.J. Roetman, Christopher B. Daniels, Andrew K. Baker, and Corey J.A. Bradshaw. 2014. 'Distribution Models for Koalas in South Australia Using Citizen Science-collected Data', *Ecology and Evolution* 4, no. 11: 2103–14.

Serrano Sanz, Fermin, Teresa Holocher-Ertl, Barbara Kieslinger, Francisco Sanz Garcia, and Candida G. Silva. 2014. *White Paper on Citizen Science in Europe*. EU: Socientize Consortium. http://www.zsi.at/object/project/2340/attach/White_Paper-Final-Print.pdf.

Sforzi, A. 2017. 'Citizen Science as a Tool for Enhancing the Role of a Museum', *Museologia Scientifica Memorie* 16: 124–8.

Sforzi, A., F. Pezzo, F. Ferretti, and V. Rizzo Pinna. 2013. *Report del primo BioBlitz della Toscana (25-26 Maggio 2013, Oasi San Felice, Grosseto)*. Grosseto, Italy: Museo di Storia Naturale della Maremma.

Shamir, Lior, Derek Diamond, and John Wallin. 2016. 'Leveraging Pattern Recognition Consistency Estimation for Crowdsourcing Data Analysis', *Ieee Transactions on Human-Machine Systems* 46, no. 3: 474–80.

Shanley, Lea, Erin Heaney, Stuart Lynn, Lina Nilsson, Jake Weltzin. 2013. 'New Visions for Citizen Science Roundtable'. Filmed 20 November 2013 at the Woodrow Wilson International Center for Scholars, Washington, DC. Last accessed 14 December 2017. https://www.youtube.com /watch?v=RzIaZS6VOxQ.

Shapin, Steven. 2010. *Never Pure: Historical Studies of Science as if It Was Produced by People with Bodies, situated in Time, Space, Culture, and Society, and Struggling for Credibility and Authority*. Baltimore: Johns Hopkins University Press.

Shapin, Steven, and Simon Schaffer. 1985. *Leviathan and the Air-pump: Hobbes, Boyle, and the Experimental Life: Including a Translation of Thomas Hobbes, Dialogus physicus de natura aeris by Simon Schaffer*. Princeton, NJ: Princeton University Press.

Shaw, Edward, Dan Surry, and Andre Green. 2015. 'The Use of Social Media and Citizen Science to Identify, Track, and Report birds', *Procedia-Social and Behavioral Sciences* 167: 103–8.

Shirk, Jennifer L., Heidi L. Ballard, Candie C. Wilderman, Tina Phillips, Andrea Wiggins, Rebecca Jordan, Ellen McCallie, et al. 2012. 'Public Participation in Scientific Research: A Framework for Deliberate Design', *Ecology and Society* 17, no. 2: 29. http://dx.doi.org/10.5751/ES -04705-170229.

Shirk, Jennifer, and Rick Bonney. 2015. *Citizen Science Framework Review for the US Fish and Wildlife Service*. Ithaca, NY: Cornell Lab of Ornithology.

Shohamy, Daphna, and R. Alison Adcock. 2010. 'Dopamine and Adaptive Memory', *Trends in Cognitive Sciences* 14: 464–72.

Sieber, Renée E., and Muki Haklay. 2015. 'The Epistemology(s) of Volunteered Geographic Information: A Critique: The Epistemology(s) of VGI: A Critique', *Geo: Geography and Environment* 2, no. 2: 122–36. https://doi.org/10.1002/geo2.10.

Siegle, Lucy. 2012. 'Invasive Non-native Species: Attack of the Aliens'. *The Guardian*, January 15, 2012. https://www.theguardian.com/environment/2012/jan/15/invasive-non-native -species-extinction.

Silka, Linda. 2013. '"Silos" in the Democratization of Science', *International Journal of Deliberative Mechanisms in Science* 2, no. 1: 1.

Silva, Cândida G., António Monteiro, Caroline Manahl, Eduardo Lostal, Teresa Holocher-Ertl, Nazareno Andrade, Francisco Brasileiro, et al. 2016. 'Cell Spotting: Educational and Motivational Outcomes of Cell Biology Citizen Science Project in the Classroom', *JCOM* 15, no. 1: A02-02.

Silvertown, Jonathan. 2009. 'A New Dawn for Citizen Science', *Trends in Ecology & Evolution* 24, no. 9: 467–71. https://doi.org/10.1016/j.tree.2009.03.017.

Silvertown, Jonathan, Laurence Cook, Robert Cameron, Mike Dodd, Kevin McConway, Jenny Worthington et al. 2011. 'Citizen Science Reveals Unexpected Continental-Scale Evolutionary Change in a Model Organism', *PLoS ONE* 6: e18927. https://doi.org/10.1371/journal .pone.0018927.

Silvertown, Jonathan, Martin Harvey, Richard Greenwood, Mike Dodd, Jon Rosewell, Tony Rebelo, Janice Ansine. 2015. 'Crowdsourcing the Identification of Organisms: A Case-study of iSpot', *ZooKeys* no. 480: 125–46.

Simpson, Robert, Kevin R. Page, and David De Roure. 2014. 'Zooniverse: Observing the World's Largest Citizen Science Platform.'. In *Proceedings of the 23rd International Conference on World Wide Web*, 1049–54.

Singh, Navinder, Kjell Danell, Lars Edenius, and Göran Ericsson. 2014. 'Tackling the Motivation to Monitor: Success and Sustainability of a Participatory Monitoring Program', *Ecology and Society* 19, no. 4: 7.

Skrip, Megan M. 2015. 'Crafting and Evaluating Broader Impact Activities: A Theory-based Guide for Scientists', *Frontiers in Ecology and the Environment* 13: 273–9. http://dx.doi.org/10.1890 /140209.

Smallman, Melanie Lynne. 2014. 'Public Understanding of Science in Turbulent Times III: Deficit to Dialogue, Champions to Critics', *Public Understanding of Science*, 25, no. 2: 186–97.

Smallman, Melanie Lynne. 2017. 'Science to the Rescue or Contingent Progress? Comparing 10 Years of Public, Expert and Policy Discourses on New and Emerging Science and Technology in the United Kingdom', *Public Understanding of Science* 96366251770645 [Epub ahead of print]. https://doi.org/10.1177/0963662517706452.

Smallman, Melanie Lynne. 2016. 'What Has Been the Impact of Public Dialogue in Science and Technology on UK Policymaking?' UCL. http://discovery.ucl.ac.uk/1473234/.

Smolinski, Mark S., Adam W. Crawley, Kristin Baltrusaitis, Rumi Chunara, Jennifer M. Olsen, Oktawia Wójcik, Mauricio Santillana, Andre Nguyen, and John S. Brownstein. 2015. 'Flu Near You: Crowdsourced Symptom Reporting Spanning 2 Influenza Seasons', *American Journal of Public Health* 105, no. 10: 2124–30.

Snik, Frans, Jeroen H.H. Rietjens, Arnoud Apituley, Hester Volten, Bas Mijling, Antonio Di Noia, Stephanie Heikamp et al. 2014. 'Mapping Atmospheric Aerosols with a Citizen Science Network of Smartphone Spectropolarimeters', *Geophysical Research Letters* 41: 7351–8.

Snow, C.P. 1959. *The Two Cultures and the Scientific Revolution*. New York: Cambridge University Press.

Snow, C.P. 1974. *The Two Cultures and a Second Look*. London and New York: Cambridge University Press.

Socientize. 2013. *Green Paper on Citizen Science. Citizen Science for Europe – Towards a better society of empowered citizens and enhanced research*. EU: report to European Commission. http://socientize.eu/?q=eu/content/green-paper-citizen-science.

Socientize. 2015. *White Paper on Citizen Science for Europe*. EU: report to European Commission. http://www.socientize.eu/?q=eu/content/download-socientize-white-paper.

Spoelstra, Kamiel, Roy H.A. Van Grunsven, Maurice Donners, Phillip Gienapp, Martinus E. Huigens, Roy Slaterus, Frank Berendse, Marcel E. Visser, and Elmar Veenendaal. 2015. 'Experimental Illumination of Natural Habitat – An Experimental Set-up to Assess the Direct and Indirect Ecological Consequences of Artificial Light of Different Spectral Composition', *Philosophical Transactions of the Royal Society of London. Series B, Biological Sciences* 370: 20140129.

Spooner, Fiona, Rebeca K. Smith, and William J. Sutherland. 2015. 'Trends, Biases and Effectiveness in Reported Conservation Interventions', *Conservation Evidence* 12: 2–7.

Stanaway, Jeffrey D., Donald S. Shepard, Eduardo A. Undurraga, Yara A. Halasa, Luc E. Coffeng, Oliver J. Brady, Simon I. Hay et al. 2016. 'The Global Burden of Dengue: An Analysis from the Global Burden of Disease Study 2013', *The Lancet Infectious Diseases* 16, no. 6: 712–23. https://doi.org/10.1016/S1473-3099(16)00026-8.

Steinbach, Rebecca, Chloe Perkins, Lisa Tompson, Shane Johnson, Ben Armstrong, Judith Green, Chris Grundy, Paul Wilkinson, and Phil Edwards. 2015. 'The Effect of Reduced Street Lighting on Road Casualties and Crime in England and Wales: Controlled Interrupted Time Series Analysis', *Journal of Epidemiology and Community Health* 206012. https://doi.org/10.1136/jech-2015-206012

Stepenuck, Kristine, and Linda Green. 2015. 'Individual-and Community-level Impacts of Volunteer Environmental Monitoring: A Synthesis of Peer-reviewed Literature', *Ecology and Society* 20, no. 3: 19. http://dx.doi.org/10.5751/ES-07329-200319.

Stephens, Greg J., Lauren Silbert, and Uri Hasson. 2010. 'Speaker–Listener Neural Coupling Underlies Successful Communication'. In *Proceedings of the National Academy of Sciences* 107: 14425–30.

Stephenson, Janet, and Henrik Moller. 2009. 'Cross-cultural Environmental Research and Management: Challenges and Progress', *Journal of the Royal Society of New Zealand* 39: 139–49.

Stevens, Matthias, Michalis Vitos, Julia Altenbuchner, Gillian Conquest, Jerome Lewis, and Muki Haklay. 2014. 'Taking Participatory Citizen Science to Extremes', *IEEE Pervasive Computing* 13: 20–29

Stilgoe, Jack, Simon J. Lock, and James Wilsdon. 2014. 'Why Should We Promote Public Engagement with Science?' *Public Understanding of Science* 23: 4–15.

Stocks, Anthony, Benjamin McMahan, and Peter Taber. 2007. 'Indigenous, Colonist, and Government Impacts on Nicaragua's Bosawas Reserve', *Conservation Biology* 21, no. 6:1495–505.

Storksdieck, Martin, Jennifer Lynn Shirk, Jessica L. Cappadonna, Meg Domroese, Claudia Göbel, Muki Haklay, Abraham J. Miller-Rushing, et al. 2016. 'Associations for Citizen Science: Regional Knowledge, Global Collaboration', *Citizen Science: Theory and Practice*. 1, no. 2: 10.

Strasser, Bruno J. 2011. 'The Experimenter's Museum: GenBank, Natural History, and the Moral Economies of Biomedicine', *Isis; an International Review Devoted to the History of Science and Its Cultural Influences* 102: 60–96.

Strasser, Bruno J., et al. 'Critical Studies of the Citizen sciences.' Forthcoming.

Sturm, Ulrike, Sven Schade, Luigi Ceccaroni, Margaret Gold, Christopher Kyba, Bernat Claramunt, Muki Haklay et al. 2017. 'Defining principles for mobile apps and platforms development in citizen science.' *Research Ideas and Outcomes* 3: e21283.

Sturn Tobias, Michael Wimmer, Carl Salk, Christoph Perger, Linda See, and Steffen Fritz. 2015. 'Cropland Capture – A Game for Improving Global Cropland Maps'. In *Proceedings of the 10th International Conference of the Foundations of Digital Games*.

Sullivan, Brian L., Tina Phillips, Ashley A. Dayer, Christopher L. Wood, Andrew Farnsworth, Marshall J. Iliff, Ian J. Davies et al. 2017. 'Using Open Access Observational Data for Conservation Action: A Case Study for Birds', *Biological Conservation* 208: 5–14.

Sullivan, Brian L., Christopher L. Wood, Marshall J. Iliff, Rick E. Bonney, Daniel Fink, and Steve Kelling. 2009. 'eBird: A Citizen-based Bird Observation Network in the Biological Sciences', *Biological Conservation* 142, no. 10: 2282–92.

Suomela, Todd. 2014. 'Citizen Science: Framing the Public, Information Exchange, and Communication in Crowdsourced Science'. PhD diss., Univ. of Tenessee, Knoxville, USA.

Surowiecki, James. 2005. *The Wisdom of Crowds*. New York: Doubleday Anchor.

Sutcliffe, Hilary. 2011. *A Report on Responsible Research and Innovation for the European Commission*. EU: DG Research and Innovation, European Commision. http://www.rri-tools.eu/documents /10184/106979/Sutcliffe2011_RRIReport.pdf/2601043b-0b34-4575-8870-1c8d82741d48.

Tantek, Celik. 2006. 'Remembering the idea of BarCamp'. *Tantek's Thoughts* (blog), 10 July 2006. Last accessed 28 August 2017. http://tantek.com/log/2006/07.html#d10t0805.

Tengö, Maria, Eduardo S. Brondizio, Thomas Elmqvist, Pernilla Malmer, and Maria Spierenburg. 2014. 'Connecting Diverse Knowledge Systems for Enhanced Ecosystem Governance – The Multiple Evidence Base Approach', *Ambio* 43, no. 6: 579–91.

Tengö, Maria, Rosemary Hill, Pernilla Malmer, Christopher M. Raymond, Marja Spierenburg, Finn Danielsen, Thomas Elmqvist, and Carl Folke. 2017. 'Weaving Knowledge Systems in IPBES, CBD and beyond – Lessons Learned for Sustainability', *Current Opinion in Environmental Sustainability* 26–27:17–25. http://dx.doi.org/10.1016/j.cosust.2016.12.005.

Tengö, Maria, Pernilla Malmer, Patrica Borraz, Chico Cariño, Joji Cariño, Tirso Gonzales, Jorge Ishizawa et al. 2012. *Dialogue Workshop on Knowledge for the 21st Century: Indigenous Knowledge, Traditional Knowledge, Science and Connecting Diverse Knowledge Systems*. Workshop Report. Stockholm, Sweden: Stockholm Resilience Centre.

Tenopir, Carol, Suzie Allard, Kimberly Douglass, Arsev Umur Aydinoglu, Lei Wu, Eleanor Read, Maribeth Manoff, et al. 2011. 'Data Sharing by Scientists: Practices and Perceptions', *PLoS ONE* 6, no. 6: e21101.

Terras, Melissa. 2016. 'Crowdsourcing in the Digital Humanities'. In *A New Companion to Digital Humanities*, edited by Susan Schreibman, Ray Siemens, and John Unsworth, 420–39. Chichester, UK: Blackwell.

Tewksbury, Joshua J., John G.T. Anderson, Jonathan D. Bakker, Timothy J. Billo, Peter W. Dunwiddie, Martha J. Groom, Stephanie E. Hampton, et al. 2014. 'Natural History's Place in Science and Society', *Bioscience* 64: 300–10.

Themba, Makani N., and Meredith Minkler. 2003. 'Influencing Policy through Community Based Participatory Research'. In *Community-based Participatory Research for Health*, edited by Meredith Minkler and Nina Wallerstein, 349–70. San Francisco: Jossey-Bass.

Theobald, Elli Jenkins, Ailene K. Ettinger, Hillary K. Burgess, L.B. DeBey, N.R. Schmidt, H.E. Froehlich, C. Wagner et al. 2015. 'Global Change and Local Solutions: Tapping the Unrealized Potential of Citizen Science for Biodiversity Research', *Biological Conservation* 181: 236–44.

Thiel, Martin, Sunwook Hong, Jenna Jambeck, Magdalena Gatta, Daniela Honorato, Tim Kiessling, Katrin Knickmeier, et al. 2017. 'Marine Litter – Bringing Together Citizen Scientists from Around the World'. In *Citizen Science for Coastal and Marine Conservation* edited by J.A. Cigliano, and H.L. Ballard, 104–31. New York: Routledge.

Thompson, Paul. 2017. *The Voice of the Past: Oral History*. Oxford and New York: Oxford University Press.

Thorpe, Charles, and Jane Gregory. 2010. 'Producing the Post-Fordist Public: The Political Economy of Public Engagement with Science,' *Science as Culture* 19: 273–301.

Tidball, Keith, and Marianne Krasny. 2007. 'From Risk to Resilience: What Role for Community Greening and Civic Ecology in Cities?' In *Social Learning Towards a more Sustainable World*, edited by A.E.J. Wals, 149–64. Wagengingen, The Netherlands: Wagengingen Academic Press.

Tinati, Ramine, Max Van Kleek, Elena Simperl, Markus Luczak-Rösch, Robert Simpson, and Nigel Shadbolt. 2015. 'Designing for Citizen Data Analysis: A Cross-sectional Case Study of a Multi-domain Citizen Science Platform'. In *Proceedings of the 33rd Annual ACM Conference on Human Factors in Computing Systems*: 4069–78. New York: ACM.

Tollis, Claire. 2012. 'Bien gérer les 'espaces de nature'. Une éthique du faire-avec. Propositions pour une géographie des associations hétérogènes'. PhD diss., Université de Grenoble.

Tolmie, Andrew, Zayba Ghazali, and Suzanne Morris. 2016. 'Children's Science Learning: A Core Skills Approach', *British Journal of Educational Psychology* 86: 481–97.

Tomblin, David, Richard Worthington, Gretchen Gano, Mahmud Farooque, David Sittenfeld, and Jason Lloyd. 2015. *Informing NASA's Asteroid Initiative: A Citizen's Forum*. Arizona State University, Tempe, Arizona. ECAST Network.

Trautmann, Nancy M., Jennifer L. Shirk, Jennifer Fee, and Marianne E. Krasny. 2012. 'Who Poses the Question? Using Citizen Science to Help K-12 Teachers Meet the Mandate for Inquiry'. In *Citizen Science: Public Participation in Environmental Research,* edited by Rick Bonney and Janis L. Dickinson, 179–90. Ithaca and London: Comstock Publishing Associates.

Tredick, Catherine A., Rebecca L. Lewison, Douglas H. Deutschman, Timothy A.N.N. Hunt, Karen L. Gordon, and Phoenix Von Hendy. 2017. 'A Rubric to Evaluate Citizen-Science Programs for Long-Term Ecological Monitoring', *Bioscience* 67: 834–44.

Tregidgo, Daniel J., Sarah E. West, and Mike R. Ashmore. 2013. 'Can Citizen Science Produce Good Science? Testing the OPAL Air Survey Methodology, Using Lichens as Indicators of Nitrogenous Pollution', *Environmental Pollution* 182: 448–51.

Treise, Debbie, and Michael F. Weigold. 2002. 'Advancing Science Communication: A Survey of Science Communicators', *Science Communication* 23: 310–22.

Trench, Brian. 2006. 'Science Communication and Citizen Science: How Dead is the Deficit Model?' Paper presented at the 9th International Conference on Public Communication of Science and Technology (PCST), Seoul, South Korea, 17–19 May 2006.

Trumbull, Deborah J., Rick Bonney, Derek Bascom, and Anna Cabral. 2000. 'Thinking Scientifically During Participation in a Citizen-Science Project', *Science Education* 84: 265–75.

Trumbull, Deborah J., Rick Bonney, and Nancy Grudens-Schuck. 2005. 'Developing Materials to Promote Inquiry: Lessons Learned', *Science Education* 89: 879–900.

Tse, Rita, and Giovanni Pau. 2016. 'Enabling Street-level Pollution and Exposure Measures: A Human-centric Approach'. In *The 6th ACM International Workshop on Pervasive Wireless Healthcare, Paderborn,* 1–4. New York: ACM.

Tsueng, Ginger, Steven Nanis, Jennifer Fouquier, Benjamin Good, and Andrew Su. 2016. 'Citizen Science for Mining the Biomedical Literature', *Citizen Science: Theory and Practice* 1, no. 2: 14.

Tucker, Melanie T., Dwight W. Lewis Jr., Pamela Payne Foster, Feleccia Lucky, Lea G. Yerby, Lisle Hites, and John C. Higginbotham. 2016. 'Community-Based Participatory Research – Speed Dating: An Innovative Model for Fostering Collaborations Between Community Leaders and Academic Researchers', *Health Promotion Practice* 17, no. 6: 775–80.

Tulloch, Ayesha I.T., Hugh P. Possingham, Liana N. Joseph, Judit Szabo, and Tara G. Martin. 2013. 'Realising the Full Potential of Citizen Science Monitoring Programs', *Biological Conservation* 165: 128–38.

Turnhout, Esther, Bob Bloomfield, Mike Hulme, Johannes Vogel, and Brian Wynne. 2012. 'Listen to the voices of experience', *Nature* 488: 454–5.

Tweddle, John C., Lucy D. Robinson, Michael J.O. Pocock, and Helen E. Roy. 2012. *Guide to citizen science: developing, implementing and evaluating citizen science to study biodiversity and the environment in the UK.* Natural History Museum, NERC Centre for Ecology & Hydrology for UK-EOF. http://www.nhm.ac.uk/content/dam/nhmwww/take-part/Citizenscience/citizen -science-guide.pdf.

Tyson, Elizabeth, Anne Bowser, John Palmer, Durrell Kapan, Frederic Bartumeus, and Eleonore Pauwels. 2018. 'Global Mosquito Alert: Building Citizen Science Capacity for Surveillance and Control of Disease-Vector Mosquitoes'. Washington: Woodrow Wilson International Center for Scholars. https://www.wilsoncenter.org/publication/global-mosquito-alert -building-citizen-science-capacity-for-surveillance-and-control.

UNDP. 2009. *Supporting capacity building development. The UNDP approach.* New York: United Nations Development Programme.

UNESCO. n.d. 'Enrolment by Level of Education'. Last accessed 27 December 2016. http://data.uis .unesco.org/.

Unit, S. C. 2013. *Science for Environment Policy Indepth Report: Environmental Citizen Science.* University of the West of England, Bristol: European Commission DG Environment.

United Nations. 1993. *Agenda 21*. Chapter 25, on the role of Children and Youth in Sustainable Development, 25.1 – 25.17. https://sustainabledevelopment.un.org/content/documents /Agenda21.pdf.

United Nations. 2008. *Declaration on the Rights of Indigenous Peoples.* United Nations. http://www .un.org/esa/socdev/unpfii/documents/DRIPS_en.pdf.

United Nations. 2009. *State of the World's Indigenous Peoples.* United Nations Department of Economic and Social Affairs, Division for Social Policy and Development, Secretariat of the Permanent Forum on Indigenous Issues ST/ESA/328. www.un.org/esa/socdev/unpfii /documents/SOWIP_web.pdf.

United Nations Economic Commission for Europe. 1998. 'Convention on Access to Information, Public Participation in Decision-Making and Access to Justice in Environmental Matters'. Adopted 25 June 1998. http://live.unece.org/fileadmin/DAM/env/pp/documents/cep43e.pdf.

United Nations Environment Programme. 2016. *Report of the IPBES Multidisciplinary Expert Panel on its 8th Meeting*. IPBES/MEP-8/15. https://www.ipbes.net/sites/default/files/downloads /pdf/ipbes-mep_8_report_final.pdf.

University of West of England. 2013. *Science for Environment Policy In-depth Report: Environmental Citizen Science*. Bristol: European Commission DG Environment. http://ec.europa.eu/science -environment-policy.

US Fish and Wildlife Service. *Endangered and Threatened Wildlife and Plants; Proposed Endangered Status for 21 Species and Proposed Threatened Status for 2 Species in Guam and the Common- wealth of the Northern Mariana Islands*. 50 CFR § 17, 2014. https://www.fws.gov/policy /library/2015/2015-00259.html.

US GEO. 2014. *National Plan for Civil Earth Observations*. Washington, DC: White House Office of the Science and Technology Policy. https://www.whitehouse.gov/sites/default/files /microsites/ostp/NSTC/2014_national_plan_for_civil_earth_observations.pdf.

US National Oceanic and Atmospheric Administration. *Endangered and Threatened Wildlife and Plants; Final Endangered Listing of Five Species of Sawfish Under the Endangered Species Act*. 50 CFR § 24, 2014. https://www.federalregister.gov/documents/2014/12/12/2014-29201 /endangered-and-threatened-wildlife-and-plants-final-endangered-listing-of-five-species-of -sawfish.

USEPA. n.d. 'Citizen Science Projects Supported by EPA'. Last accessed 13 March 2018. https://www.epa.gov/citizen-science/citizen-science-projects-supported-epa.

USGRCRP. 2014. *Tracking a Changing Climate: Workshop Report*. Washington, DC: US Global Climate Change Program. http://www.globalchange.gov/sites/globalchange/files /Citizen%20Science%20Workshop%20Report%20Final.pdf.

Vallabh, Priya, Heila Lotz-Sisitka, Rob O'Donoghue, and Ingrid Schudel. 2016. 'Mapping Epistemic Cultures and Learning Potential of Participants in Citizen Science Projects', *Conservation Biology* 30: 540–9.

van Mierlo, Trevor. 2014. 'The 1% Rule in Four Digital Health Social Networks: An Observational Study', *Journal of Medical Internet Research* 16, no. 2: e33.

Vann-Sander, Sarah, Julian Clifton, and Euan Harvey. 2016. 'Can Citizen Science Work? Perceptions of the Role and Utility of Citizen Science in a Marine Policy and Management Context', *Marine Policy* 72: 82–93.

Vaughan, Hague, Graham Whitelaw, Brian Craig, and Craig Stewart. 2003. 'Linking Ecological Science to Decision-Making: Delivering Environmental Monitoring Information as Societal Feedback', *Environmental Monitoring and Assessment* 88, no. 1: 399–408.

Vayena, Effy, and John Tasioulas. 2016. 'The Dynamics of Big Data and Human Rights: The Case of Scientific Research', *Philosophical Transactions of the Royal Society A* 374, no. 2083: 20160129.

Verhulst, Stefaan, and Andrew Young. 2016. *Open Data Impact: When Demand and Supply Meet. Key Findings of the Open Data Impact Case Studies*. New York: GovLab.

Vitone, Tyler, Kathryn A. Stofer, M. Sedonia Steininger, Jiri Hulcr, Robert Dunn, and Andrea Lucky. 2016. 'School of Ants Goes to College: Integrating Citizen Science into the General Education Classroom Increases Engagement with Science', *Journal of Scientific Communication* 15: 1–24.

Vogels, Chantal B.F., Lennart J.J. van de Peppel, Arnold J.H. van Vliet, Marcel Westenberg, Adolfo Ibañez-Justicia, Arjan Stroo, Jan A. Buijs, Tessa M. Visser, and Constantianus J.M. Koenraadt. 2015. 'Winter Activity and Aboveground Hybridization Between the Two Biotypes of the West Nile Virus Vector Culex Pipiens', *Vector-Borne and Zoonotic Diseases* 15, no. 10: 619–26. https://doi.org/10.1089/vbz.2015.1820.

von Ahn, Lous. 2006. 'Games with a Purpose', *Computer* 39: 92–4.

von Schomberg, René. 2013. 'A Vision of Responsible Research and Innovation'. In *Responsible Innovation*, edited by Richard Owen, John Bessant, and Maggy Heintz, 51–74. Chichester, UK: John Wiley & Sons, Ltd.

Wadsworth, Barry J. 2004. *Piaget's Theory of Cognitive and Affective Development: Foundations of Constructivism*. White Plains, NY: Longman Publishing.

Wal, René, Nirwan Sharma, Chris Mellish, Annie Robinson, and Advaith Siddharthan. 2016. 'The Role of Automated Feedback in Training and Retaining Biological Recorders for Citizen Science', *Conservation Biology* 30: 550–61.

Wallerstein, Nina B., and Bonnie Duran. 2006. 'Using Community-based Participatory Research to Address Health Disparities', *Health Promotion Practice* 7, no. 3: 312–23.

Warncke-Wang, Morten, Vladislav R. Ayukaev, Brent Hecht, and Loren G. Terveen. 2015. 'The Success and Failure of Quality Improvement Projects in Peer Production Communities. In *Proceedings of the 18th ACM Conference on Computer Supported Cooperative Work & Social Computing, Vancouver, Canada,* 743–56. New York: ACM.

Waterhouse, Robert M., Xiaoguang Chen, Mariangela Bonizzoni, Giuliano Gasperi, and Giuliano Gasperi. 2017. 'The Third International Workshop on Aedes Albopictus: Building Scientific Alliances in the Fight against the Globally Invasive Asian Tiger Mosquito', *Pathogens and Global Health* 111, no. 4: 161–5. https://doi.org/10.1080/20477724.2017.1333560.

Watson, David, and Luciano Floridi. 2016. 'Crowdsourced Science: Sociotechnical Epistemology in the E-research Paradigm', *Synthese* 195, no. 2: 741-64.

Weaver, Scott C., and Marc Lecuit. 2015. 'Chikungunya Virus and the Global Spread of a Mosquito-Borne Disease', *New England Journal of Medicine* 372, no. 13: 1231–9. https://doi .org/10.1056/NEJMra1406035.

Weigold, Michael F. 2001. 'Communicating Science: A Review of the Literature', *Science Communication* 23: 164–93.

Weiss, Carol Hirschon. 1995. 'Nothing as Practical as Good Theory: Exploring Theory-based Evaluation for Comprehensive Community Initiatives for Children and Families'. In *New Approaches to Evaluating Community Initiatives: Concepts, Methods, and Contexts,* edited by James P. Connell et al., 65–92. Washington, DC: Aspen Institute.

Wells, Nancy M., and Kristi S. Lekies. 2006. 'Nature and the Life Course: Pathways from Childhood Nature Experiences to Adult Environmentalism', *Children, Youth and Environments* 16, no. 1: 1–24.

Wells, Nancy M., and Kristi S. Lekies. 2012. 'Children and Nature: Following the Trail to Environmental Attitudes and Behavior'. In *Citizen Science: Public Participation in Environmental Research,* edited by J.L. Dickinson and R. Bonney, 201–14. Ithaca, US: Comstock Pub. Associates.

Wernand, Marcel R. 2010. 'On the History of the Secchi Disc', *Journal of the European Optical Society-Rapid Publications* 5. http://www.jeos.org/index.php/jeos_rp/article/view /10013s.

Wells, Nancy M., Beth M. Myers, Lauren E. Todd, Karen Barale, Brad Gaolach, Gretchen Ferenz, Martha Aitken et al. 2015. 'The Effects of School Gardens on Children's Science Knowledge: A Randomized Controlled Trial of Low-income Elementary schools', *International Journal of Science Education* 37, no. 17: 2858–78.

Wernand, Marcel, Luigi Ceccaroni, Jaume Piera, and Oliver Zielinski. 2012. 'Crowdsourcing Technologies for the Monitoring of the Colour, Transparency and Fluorescence of the Sea', *Proceedings of Ocean Optics* 11: 8–12.

West, Sarah, and Rachel Pateman. 2016. 'Recruiting and Retaining Participants in Citizen Science: What Can Be Learned from the Volunteering Literature?' *Citizen Science: Theory and Practice* 1, no. 2: 15.

West, Sarah, Rachel Pateman, and Alison Dyke. 2016. *Data Submission in Citizen Science Projects.* Report for Defra PH0475. University of York.

West, Sarah E. 2014. 'Evaluation, or Just Data Collection? An Exploration of the Evaluation Practice of Selected UK Environmental Educators', *The Journal of Environmental Education* 46, no. 1: 41–55.

White, Randy, and Vicki L. Stoeck. 2008. 'Nurturing Children's Biophilia: Developmentally Appropriate Environmental Education for Young Children', *Collage: Resources for Early Childhood Educators,* November. http://citeseerx.ist.psu.edu/viewdoc/download?doi=10.1.1 .453.5868&rep=rep1&type=pdf.

Whitelaw, Graham, Hague Vaughan, Brian Craig, and David Atkinson. 2003. 'Establishing the Canadian Community Monitoring Network', *Environmental Monitoring and Assessment* 88, nos. 1–3: 409–18.

Whittle, John, Erinma Ochu, Maria Angela Ferrario, Jen Southern, and Ruth McNally. 2012. 'Beyond Research in the Wild: Citizen-led Research as a Model for Innovation in the Digital Economy'. In *Proceedings of Digital Futures 2012 Conference.*

WHO. 2014. 'A Global Brief on Vector-Borne Diseases', World Health Organization, 9. https://doi .org/WHO/DCO/WHD/2014.1.

WHO. 2016a. 'World Malaria Report 2016'. World Health Organization. http://apps.who.int/iris /bitstream/10665/252038/1/9789241511711-eng.pdf?ua=1.

WHO. 2016b. 'Zika Strategic Response Plan'. World Health Organization. http://apps.who.int/iris
/bitstream/10665/246091/1/WHO-ZIKV-SRF-16.3-eng.pdf.

WHO. 2017. 'World Malaria Report 2017'. World Health Organization. http://www.who.int
/malaria/publications/world-malaria-report-2017/report/en/.

Wickson, Fern, and Anna L. Carew. 2014. 'Quality Criteria and Indicators for Responsible Research
and Innovation: Learning from Transdisciplinarity', *Journal of Responsible Innovation* 1, no. 3:
254–73. https://doi.org/10.1080/23299460.2014.963004.

Wiedemann, Peter M., and Susanne Femers. 1991. 'Public Participation in Waste Management
Decision-making: Analysis and Management of Conflicts', *Journal of Hazardous Materials* 33,
no. 3:355–68.

Wiggins, Andrea, Greg Newman, Robert D. Stevenson, and Kevin Crowston. 2011. 'Mechanisms
for Data Quality and Validation in Citizen Science'. In *e-Science Workshops (eScience W), 2011
IEEE Seventh International Conference*.

Wiggins, Andrea. 2013. 'Free As in Puppies: Compensating for ICT Constraints in Citizen Science'.
In *The 2013 Conference on Computer Supported Cooperative Work (CSCW '13)*, 1469–80.

Wiggins, Andrea, Rick Bonney, Eric Graham, Sandra Henderson, Steve Kelling, Richard Littauer,
Gretchen LeBuhn et al. 2013. *Data Management Guide for Public Participation in Scientific
Research*. Albuquerque, NM: DataONE.

Wiggins, Andrea, and Kevin Crowston. 2011. 'From Conservation to Crowdsourcing: A Typology
of Citizen Science'. In *44th Hawaii International Conference on System Sciences (HICSS) 2011*,
1–10.

Wiggins, Andrea, and Kevin Crowston. 2015. 'Surveying the Citizen Science Landscape', *First
Monday* 20, nos. 1–5. http://dx.doi.org/10.5210/fm.v20i1.5520.

Wiggins, Andrea, Greg Newman, Robert D. Stevenson, and Kevin Crowston. 2011. 'Mechanisms
for Data Quality and Validation in Citizen Science'. In *Proceedings of the 2011 IEEE Seventh
International Conference on e-Science Workshops*, 14–19. Washington, DC: IEEE Computer
Society.

Wikipedia. n.d. 'Citizen Science'. Last modified 10 December 2017. https://en.wikipedia.org/wiki
/Citizen_science.

Wilderman, Candie, Alissa Barron, and Lauren Imgrund. 2004. 'Top Down or Bottom Up?
ALLARM's Experience with Two Operational Models for Community Science'. In *Conference
Proceedings of the 4th National Monitoring Conference*. Chattanooga, TN: National Water
Quality Monitoring Council. http://acwi.gov/monitoring/conference/2004/proceedings
_contents/13_titlepages/posters/poster_235.pdf.

Willis, Charles G., Elizabeth R. Ellwood, Richard B. Primack, Charles C. Davis, Katelin D. Pearson,
Amanda S. Gallinat, Jenn M. Yost, et al. 2017. 'Old Plants, New Tricks: Phenological Research
Using Herbarium Specimens', *Trends in Ecology & Evolution* 32, no. 7: 531–46.

Wing, Kate. n.d. 'A Citizen Science Manifesto'. Last modified 16 February 2014. https://medium
.com/openexplorer-journal/a-citizen-science-manifesto-287f67f007e0#.vho3nkmb0.

Wood, Harry. 2014. 'The Long Tail of OpenStreetMap'. Last accessed July 2015. http://harrywood
.co.uk/blog/2014/11/17/the-long-tail-of-openstreetmap/#slide18.

Woolley, J. Patrick, Michelle L. McGowan, Harriet J.A. Teare, Victoria Coathup, Jennifer R.
Fishman, Richard A. Settersten, Sigrid Sterckx, Jane Kaye, and Eric T. Juengst. 2016. 'Citizen
Science or Scientific Citizenship? Disentangling the Uses of Public Engagement Rhetoric in
National Research Initiatives', *BMC Medical Ethics* 17, no. 1: 33.

Worthington, R., D. Cavalier, M. Farooque, G. Gano, H. Geddes, S. Sander, D. Sittenfeld, and
D. Tomblin. 2012. *Technology Assessment and Public Participation: From TA to pTA*. Expert and
Citizen Assessment of Science and Technology (eCAST) Report. http://www.loka.org
/Documents/ecast-report-ta-to-pta.pdf.

Wright, Dale. 2011. 'Evaluating a Citizen Science Research Programme: Understanding the
People Who Make It Possible'. MSc diss., University of Cape Town. http://www.adu.org.za
/pdf/Wright_D_2011_MSc_thesis.pdf.

Wynne, Brian, 1996. 'May the Sheep Safely Graze? A Reflexive View of the Expert–Lay Knowledge
Divide'. In *Risk, Environment and Modernity: Towards a New Ecology*, edited by Scott Lash,
Bronislaw Szerszynski, and Brian Wynne, 47–83. London: Sage.

Wynne, Brian. 2006. 'Public Engagement as a Means of Restoring Public Trust in Science - Hitting
the Notes, but Missing the Music?' *Community Genetics* 9, no. 3: 211–20.

Yan, Katy. 2012. 'Interview: China's Green Hunan Trains Citizen Scientists to Fight River
Pollution'. Last accessed 6 June 2017. https://www.internationalrivers.org/resources

/interview-china%E2%80%99s-green-hunan-trains-citizen-scientists-to-fight-river-pollution
-7757.

Yong, Ed. 2017. 'Shazam for Mosquitoes.' *The Atlantic*, March 31. https://www.theatlantic.com
/science/archive/2017/03/shazam-mosquitoes-cellphone-citizen-science/521505/.

Zastrow, Mark. 2014. 'Crisis Mappers Turn to Citizen Scientists', *Nature* 515, no. 7527: 321.

Zentrum für Citizen Science. 2016. Top Citizen Science Prgramme. Last accessed 14 July 2016.
http://www.zentrumfuercitizenscience.at/en/citizen-science.

ZeroWaste Scotland. n.d. 'How to Use Citizen Science in Litter Prevention'. Last accessed 13
March 2018. http://www.zerowastescotland.org.uk/litter-flytipping/citizen-science.

Zevin, Michael, Scott Coughlin, Sara Bahaadini, Emre Besler, Neda Rohani, Sarah Allen, Miriam
Cabero, et al. 2017. 'Gravity Spy: Integrating Advanced LIGO Detector Characterization,
Machine Learning, and Citizen Science', *Classical and Quantum Gravity* 34, no. 6: 064003.

Zhang, Jian, Shengbin Chen, Bin Chen, Yanjun Du, Xiaolei Huang, Xubin Pan, and Qiang Zhang.
2013. 'Citizen Science: Integrating Scientific Research, Ecological Conservation and Public
Participation', *Biodiversity Science* 21, no. 6: 738–49.

Zhao, Mingxu, Søren Brofeldt, Qiaohong Li, Jianchu Xu, Finn Danielsen, Simon B. L. Læssøe,
Michael K. Poulsen et al. 2016. 'Can Community Members Identify Tropical Tree Species for
REDD+ Carbon and Biodiversity Measurements?' *PLoS ONE* 11, no. 11: e0152061.

Zhao, Yijiang, Xiaoguang Zhou, Guangqiang Li, and Hanfa Xing. 2016. 'A Spatio–Temporal VGI
Model Considering Trust-related Information', *ISPRS International Journal of Geo-Information* 5, no. 2: 10.

Zhao, Yuanyuan, Duole Feng, Le Yu, Linda See, Steffen Fritz, Christoph Perger, and Peng Gong.
2017. 'Assessing and Improving the Reliability of Volunteered Land Cover Reference Data',
Remote Sensing 9, no. 10: 1034.

Zoellick, B., S.J. Nelson, and M. Schauffler. 2012. 'Participatory Science and education: Bringing
Both Views into Focus', *Frontiers in Ecology and the Environment* 10: 310–13.

Zooniverse. n.d. 'People powered research'. Last accessed October 2017. https://www.zooniverse
.org/.

Index

Note: Numbers in *italics* refer to figures, tables, illustrations.